TIME RESTORED

TIME
RESTORED

The Harrison Timekeepers and R.T. Gould,
the man who knew (almost) everything

Jonathan Betts

Senior Specialist, Horology
Royal Observatory
National Maritime Museum, Greenwich

OXFORD
UNIVERSITY PRESS

AND THE
NATIONAL MARITIME MUSEUM

OXFORD
UNIVERSITY PRESS

Great Clarendon Street, Oxford OX2 6DP

Oxford University Press is a department of the University of Oxford.
It furthers the University's objective of excellence in research, scholarship,
and education by publishing worldwide in

Oxford New York

Auckland Cape Town Dar es Salaam Hong Kong Karachi
Kuala Lumpur Madrid Melbourne Mexico City Nairobi
New Delhi Shanghai Taipei Toronto

With offices in

Argentina Austria Brazil Chile Czech Republic France Greece
Guatemala Hungary Italy Japan Poland Portugal Singapore
South Korea Switzerland Thailand Turkey Ukraine Vietnam

Oxford is a registered trade mark of Oxford University Press
in the UK and in certain other countries

Published in the United States
by Oxford University Press Inc., New York

British Library Cataloguing in Publication Data

Data available

Library of Congress Cataloging in Publication Data

Betts, Jonathan.
Time restored : the Harrison timekeepers and R.T. Gould, the man who
knew (almost) everything / Jonathan Betts.
p. cm.
Includes bibliographical references and index.
ISBN-13: 978–0–19–856802–5 (alk. paper)
ISBN-10: 0–19–856802–9 (alk. paper)
1. Gould, Rupert Thomas, 1890–1948. 2. Chronometers. 3. Nautical
instruments. I. Title.
TS544.8.G68B48 2006
681.1′18092—dc22   2006000426

Typeset by Newgen Imaging Systems (P) Ltd., Chennai, India
Printed in Great Britain
on acid-free paper by
Biddles Ltd., King's Lynn

ISBN 978–0–19–856802–5 (hbk); 978–0–19–960671–9 (pbk)

1 3 5 7 9 10 8 6 4 2

For my mother

Contents

Front end paper: Associated places in Britain
Back end paper: Rupert Gould's family tree

Preface to the Paperback Edition

It is very gratifying that the success of *Time Restored* has meant a paperback edition is viable. In fact little has needed changing in this new edition except the correction of a number of typo's and other small errors.

If I had been starting again, I confess I would probably have put Chapter 4 – the story of Harrison and his timekeepers – in an appendix rather than in the main text of the biography. As it is, some readers have said they feel that placed there, the background to the Harrison timekeepers rather interrupts of the story of Gould's life. Having said that, I suppose it could be argued that that is what actually happened to Gould – his obsession with Harrison effectively broke his life in two, and perhaps having the Chapter there introduces an appropriate symbolism into the storyline.

A welcome result of making Gould's story better known is that attention was drawn to the deteriorating state of his tombstone in Ashtead churchyard, and the Clockmakers' Company, in conjunction with Gould's grandchildren Simon and Sarah Stacey, are now kindly funding the restoration of the marble and the lettering.

This year at the National Maritime Museum sees the completion of detailed research into the design and construction of all Harrison's marine timekeepers. The close study of these incredibly complex machines, in conjunction with Gould's eighteen notebooks, has been a poignant reminder of his trials and tribulations and of what an extraordinary achievement those restorations really were.

Jonathan Betts
Greenwich
March 2011

Preface and Acknowledgements

Rupert Gould was an achiever in many varied fields. My interest in him, naturally enough, has always been principally focused on his work in *Horology*: the study of timekeeping and time telling.

One of the most frequently asked questions we horologists face is 'How did you first get into that'? Looking at my friends and colleagues in the professional horological world, it seems to be the case, at least these days, that most come into the subject from other occupations, more often than not in early middle age. Sadly, there are very few youngsters in the world of horology. I myself however, was one, once.

At the tender age of eleven, in the mid-1960s, I decided to opt out of the family camping holidays in the New Forest and asked my father for a holiday job in the family business. Betts Ltd was a firm of wholesale and retail watchmakers and jewellers, based in Ipswich in Suffolk, but with a number of branches round the country. As an errand boy, I would run between the three retail shops round the town and the company's watch repair workshops carrying bags full of customers' watches for repair (it wouldn't be allowed today!). Having an inclination towards mechanical things, I took an interest in what the men behind the benches were working on, and one of the watchmakers, Philip Hancock, was asked to look after me. Patiently he showed me the ropes, supplying me with large ex-government pocket watches to have a go with and, I believe, I didn't make too bad a job of cleaning them and getting them together again. I do know that I was immediately hooked, and never had any doubts about what I wanted to do when I was older.

It was not long after that I first heard the name of John Harrison, through a family friend, Kit Welford (who was also our GP). Welford was a keen sailor: he vainly tried to interest me and my four siblings in the joys of getting cold and wet (in spite of my employment at the National Maritime Museum, I'm afraid I have never been able to discover a passion for boats). Hearing of my new-found interest in clocks, he urged me to visit Greenwich and pay homage to the great timekeepers by John Harrison 'the most important clocks ever constructed'. He also alluded to the fact that, had it not been for the dedication of a retired Navy Officer, they may not have survived at all, though I was not told this man's name, nor what it was he had done to save them.

At the time the family had a flat in Marylebone in London where we stayed sometimes over holidays, and it was during one of these visits that I made the first pilgrimage to the Royal Observatory, Greenwich (ROG) and the National Maritime Museum (NMM). The historic Royal Observatory, intimately linked with the history of clocks, watches, and chronometers, was my first port of call. For nearly three hundred years it was the home of the Astronomer Royal who, in the seventeenth and eighteenth centuries, was tasked with developing an astronomical solution to the greatest scientific problem of the age: finding the Longitude, ones east-west position, when at sea. The Observatory was also where, in about 1728, John Harrison had first visited when seeking his fortune. He too was developing a solution to the Longitude problem, but by creating marine timekeepers, of which much more will be said in this book.

The professional astronomers had moved out of the Observatory after the Second World War and, when I first visited in 1968, the buildings had just been restored and fully reopened as a museum. In those days the buildings were called the Old Royal Observatory and it had recently become an outstation of the NMM. Standing for the first time at the centre of these buildings in Christopher Wren's Flamsteed House, I found the sense of history quite overwhelming. There are few buildings standing today with such intimate and direct links to the history of precision clock making.

In 1968 however, the Harrison timekeepers were displayed 'down the hill', in the Navigation Gallery in the South West wing of the National Maritime Museum. So, following a visit to the horological 'Mecca' of Greenwich Observatory, a stroll down through Greenwich Park was necessary to take in the four 'Holy Grails' of our subject. Needless to say, I soon understood Dr Welford's insistence that I see the Harrison timekeepers. Others have described first setting eyes on them as akin to a religious experience. They are truly astonishing machines and, nearly forty years on, they still have the ability to amaze and thrill me. What they led to, the developed marine chronometer, is also a pretty amazing instrument. Made to the highest standards of horological craftsmanship, the chronometer is the most carefully made, the most beautifully finished and yet the most delicately vulnerable of horological instruments; it is the 'aristocrat' of the clock-making world.

I would visit Greenwich many times more in my youth, never imagining for a moment that one day I might be lucky enough to work at the museum and actually have the Harrison timekeepers themselves, among many other chronometers, in my care. From 1972, after ducking out of taking Pure and Applied Mathematics and Physics 'A' Levels, and refusing to turn up for an interview my father had arranged for me to join the Royal Air Force, I decided, for better or worse, that my future lay in the world of Horology.

Attending the British Horological Institute's two-year course in Technical Horology at Hackney College, in the East End of London, I knew immediately I'd made the right decision, though commuting the eighty miles from Ipswich to Hackney and back was not fun. This meant a 5.30 a.m. start, five days a week, something this lazy 17 year old could only ever have managed thanks to his mother, who got me up and cooked me breakfast everyday! Until college I had never been a fast learner and was beginning to wonder if I was in fact, like Winnie the Pooh, just 'a bear with a small brain'. But at Hackney, for the first time in my life, things really worked and I couldn't get enough of the subject. Naturally enough, horological history was covered in the course and John Harrison's pioneering work was studied and the timekeepers themselves visited at Greenwich on one of the students' trips out.

Of the various books which were either referred to or recommended by the teachers, one was universally revered, *The Marine Chronometer*, and it was then that I began, little by little, to learn more about its author, Lt Cdr Rupert T. Gould. Not only was he evidently a fine illustrator, as well as an engaging and amusing writer on a potentially dry subject, but also he was apparently a first rate practical restorer. This then was the man that Dr Welford had alluded to, the man who virtually single-handedly rescued Harrison's masterpieces.

As students of horology, it was necessary for us to learn a little of the immense complexity of those machines and we heard of all manner of elaborate devices such as 'remontoirs' and 'temperature compensation'. It thus gradually dawned on me just what an extraordinary achievement their restoration had been. It was said that Cdr Gould had dedicated much of his life to this task, yet I was also told that, in the course of that life, he had followed many other deep interests and was well known in circles completely removed from horology. I was intrigued by this unusual and rather unlikely figure in a trade in which achievers would typically be serious and dry, one-subject specialists, and hoped one day to find out more about him.

Following a short period with a London firm of clock restorers, Keith Harding's, I decided to specialize in the restoration of watches and chronometers, and the next 5 years were spent in self-employment, in my home town of Ipswich. The supply of chronometers for overhaul was initially sparse and I was at times obliged to also take on such treasures as bedside alarm clocks and chiming mantel clocks from the 1930s. But it has to be said they made a wonderful apprenticeship in horology's real world. Only with clocks such as these does the practical clockmaker learn to trouble-shoot. The craftsman who has only ever worked on top-grade watches and clocks has probably learnt very little as, once cleaned and given fresh oil, they almost always go well, *by design*.

During the period with Keith Harding in London, I first met clock-maker Martin Burgess and it was Burgess and Harding who first introduced to me the idea that antique clocks might be conserved (i.e. maintained and made presentable, but with minimal change to the object's historic integrity) rather than restored (i.e. rebuilt to a conjectural 'as-new' condition). Of the two philosophically very different approaches, conservation was the concept I instinctively found more satisfactory, and since the late 1970s have always followed that path. A steady supply of chronometers for conservation did eventually begin in the Ipswich workshop, but then the opportunity of a lifetime occurred. In 1979 the post of Senior Conservation Officer (Horology) at the NMM became vacant, an opportunity to care for the world's largest collection of chronometers. Perhaps owing to my firm convictions on the question of conservation, I was offered the job and have been at the Observatory ever since.

Although the Curator of Horology at the museum, my colleague Beresford Hutchinson, had curatorial responsibility for the Harrison timekeepers at the time, they were not maintained by museum staff: up until the early 1980s they were always looked after by the Ministry of Defence Chronometer Section (the successors to the Royal Observatory's chronometer staff). Then, soon after my arrival in the new post, the care of the Harrisons was transferred 'in house' and I was given responsibility for their conservation.

Just about the only source of technical information then available on the timekeepers was Cdr Gould's notebooks, which had been deposited in the NMM's library some years before, and reading them was an absolute inspiration. Not only are they packed with technical data, but include a great deal of personal asides and notes; a rich source of interest for any potential biographer. In fact it was almost as if Rupert had been subconsciously writing with just such a work in mind, and I took the hint, though it would be several years before these first inklings crystallized into definite intentions.

The first suggestion of a biography occurred as far back as 1948, in one of Gould's obituaries in the Horological Journal in fact. In his appreciation of Gould's life, the electric clock pioneer, Frank Hope Jones, remarked that he hoped someone would do a fuller biographical note 'if a biography is not contemplated'. Soon after I had spotted this remark by Hope Jones, I had a letter from the celebrated watchmaker George Daniels, in which he commented on what a fine researcher Gould had been, describing him as 'a Prince among men' (coincidentally appropriate for a man who had been a navigating officer—see page 44). I think it was probably at that point that the scales of decision tipped in my mind, and I began to collect information more seriously. But it was the publication of Dava Sobel's book *Longitude*, which really focused my attention on getting the Gould biography completed.

Following Harvard's *Longitude Symposium*, which Will Andrewes had organized in 1993, Dava had visited Greenwich, gathering information for a new book on the story of Harrison. Showing her the Harrisons, I naturally had to include a mention of my other 'hero' in that story, but little expected her to include mention of Rupert in the new work. But there Gould was, at the end of the book, appropriately reflecting Harrison's dramatic story with a brief mention of his own trials and tribulations in restoring the timekeepers to their former glory.

With all due modesty Dava told me subsequently that the wonderful success of *Longitude* came as a surprise to her. It was of course a very pleasant surprise to us all, and has been the most extraordinarily good ambassador for us at Greenwich too. Day after day, over ten years after the publication of *Longitude*, the Harrison Gallery at the Observatory is still usually peopled by Sobel fans, paying homage to the timekeepers and recognizing Gould's work. Dava's portrayal of the Astronomer Royal Nevil Maskelyne as 'the villain of the piece' has, as Dava knows, always been something we have been keen to redress, but that's a small matter in such a very positive achievement.

The film of the book, made in 1999 by Charles Sturridge and Granada Film, was of course a further tremendous boost for the story of John Harrison. I simply cannot imagine a more convincing and realistic portrayal of Harrison than Michael Gambon's amazing performance.

But the film had a dramatic element not present in the book, in that it ran Gould's story, much expanded by biographical material provided from my files, alongside that of Harrison's. Although the choice of Jeremy Irons to play Gould was controversial—he neither looked nor sounded at all like Gould—Irons did a magnificent job and conveyed much of the emotional drama of Rupert's life, echoing the difficulties Harrison had experienced two centuries before, in the most moving and uncanny way. As well as raising Gould's profile, the film increased demands for the biography to be completed, and it was particularly gratifying to find support from the National Maritime Museum when I sought permission to finish the writing in 'museum time'.

It was only once the writing was official, of course, that I stopped to think what it was I had taken on, and slight panic set in. Here was one of the twentieth century's great polymaths, and I had just agreed to write about his whole life! But then I reasoned that there were probably very few who could have written authoritatively on *all* the subjects Gould covered anyway, and if the writing had to be done by a specialist in any *one* of those subjects, it ought perhaps to have been an horologist. I do not pretend to be an authority on, nor, in some cases, even to understand, some of the topics Gould studied; for those, I have relied on the help of others. But if errors or glaring omissions remain in this biography, then they are certainly mine. A full study of the life and interests of a man such as Gould is in itself almost a lifetime's work and this biography would not have been produced at all if the timescale was any longer.

During the research, I was extremely fortunate in being able to interview Rupert's children, Cecil and Jocelyne, who were both as helpful as could be. It was a great privilege to be presented with a copy of Cecil's unpublished autobiography, and a copy of his edited diaries, from which much of the family side of the story has been culled. Then, both Cecil and Jocelyne allowed me to interview them with a tape recorder running, and several hours of their reminiscences of family life have proved invaluable. Jocelyne's children, Simon and Sarah Stacey, have both been enthusiastic supporters. Thanks to them I have had access to many of the family photograph albums, images from which have helped illustrate and enhance the book, for which I am exceedingly grateful.

As has been said, although I felt confident in writing about Gould's work in the world of horology, and particularly on the Harrison timekeepers, dealing with the many other areas of his interest was another matter altogether. When considering his contribution to crypto-zoology and some of the other scientific mysteries, I was on particularly shaky ground, and Mike Dash and Peter Costello very kindly read the draft text, providing many useful comments and corrections. In fact Mr Costello had intended to do something on Gould himself, but, on hearing of my plans, he very generously handed over all his notes and encouraged me to press on with the writing. Others who have been particularly helpful with the text are Ian and Betty Bartky, and David Grace who between them have provided many very useful steers on the presentation of the facts, and Dava Sobel has made a number of helpful comments.

The staff at many institutions have given generous help, including Bob Headland at the Scott Polar Research Institute; Julia Poole at the Fitzwilliam Museum; Adrian Webb at the UK Hydrographic Office; Peter Meadows and Adam Perkins (RGO Archives) at the Cambridge University Library; Arthur McDonald and Alan Midleton at the British Horological Institute, Peter Hingley at the Royal Astronomical Society, Bert West and Robin Thatcher of the old M.O.D. Chronometer Section, and staff at the Royal Geographical Society (RGS) and the Wimbledon Lawn Tennis Museum.

David Penney very generously allowed free use of his fine illustrations of parts of the Harrison timekeepers

Philip Whyte and Richard Stenning of Charles Frodsham Ltd kindly gave support in funding colour illustrations.

Many National Maritime Museum staff, past and present, have also given assistance, including Robin Catchpole, Geraldine Charles, Roy Clare, Gloria Clifton, Richard Dunn, the late Cdr Derek Howse, Beresford Hutchinson, Peter Linstead-Smith, Nigel Rigby, the late Michael Robinson, David Rooney, Ann Shirley, Janet Small, Brian Thynne, Bob Todd, Liza Verity. I am also grateful to the staff of the Museum's Publications Department and the editing staff at Oxford University Press.

I have also received much generous help from a wide variety of other people, all of whom I would like to thank: Charles Allix, Jim Arnfield, Andrew Barsby, Doug Bateman, Chris Beetles, John Black, Neil Brown, Mike Budd, John Burgess, Roger Carrington, Margaret Clark, Dr Ros Cleal, Patrick Cole, Loren Coleman, Andrew Cook, S.E.T Cusdin, Richard Dalby, Cdr Andrew David, Aubrey de Bordenave, Peter Dineley, Hilary Evans, Jeremy Evans, Maurice Fagg, George Feinstein, Tim Flower, Jeff Formby, Errol Fuller, Jock Gardner, Michael Gibbon,

Gary Glynn, Mary Godwin, Herb Gold, Peter Gosnell, Jeremy Gray, Wallace Grevatt, the late Professor Teddy Hall, The Earl of Halsbury, David Harries, Stephan Harris, Peter Hastings, Derek Higginson, Christal Hippisley, Elizabeth Holbrook, Merlin Holland, Bob Holmstrom, Professor Vinesh Y Hookoomsing, Professor Edgar Jones, John Keegan, Gavin Kennedy, Andrew King, Pauline Kingswood, Jill Lang, Rodney Law, John Leopold, Raymond Lister, J MacBeath, Gary Mangiacopra, John Marten, Arthur Meaton, Tony Mercer, Dr Patrick Moore, Bob Moran, Willem Morzer Bruyns, Fortunat Mueller Maerki, Linda Murray, Dr Joan Navarre, Gordon Norwood, Albert Odmark, Hugh Owen, Alan Partridge, Reg Pennells, Ed Powers, Derek Pratt, Anthony and Ann Marie Randall, Frederick Ratcliffe, Mike Rowe, Anthony Salmon, Rex Sawyer, Catherine Scantlebury, Charles Schwartz, David Sealy, Rita Shenton, Andrew Simpson, David Singmaster, Jonathan Snellenburg, Hans Staeger, Irene Stewart, Charles Stickland, Stuart Tucker, Professor Gerard L'E Turner, Trevor Waddington, Francis Wadsworth, Cdr David Waters, Cliff Watkins, Eric Whatley, Sir George White, Eric Whittle, Malcolm Wild, Bernard Williams, Louise and Pat Wilmot, Ian Wilson, David Winfield, Don Wing and Philip Woodward. Last, and by no means least, are thanks to my partner Steve, who has stoically tolerated Rupert's seemingly constant presence in the household for almost quarter of a century. If there are others who have given help and whom I have forgotten to acknowledge then please accept my sincerest apologies.

As for the Harrison timekeepers, thanks to Rupert Gould and the attention of the staff at the Chronometer Section in the post-war period, they continue in fine working condition and have been running virtually constantly. Study and interpretation of their construction and history continues to reveal new and fascinating insights, of course, but as the man also responsible for their conservation I have—and have had in my 25 years here—very little to do to them. Apart from H4, which has needed several overhauls, all that has been necessary for the larger timekeepers has been their careful winding every day. John Harrison's incomparable designs have been vindicated by the severest test of all, that of time itself.

Jonathan Betts
Greenwich
April 2006

Introduction: Rupert T. Gould

On a summer morning in 1949, Muriel Gould invited her children for a picnic in the village of Ashtead in Surrey. Cecil, now 31, and Jocelyne a year younger, had started lives of their own well before the war, but both son and daughter kept in regular touch with their mother and there was nothing unusual about the invitation. The picnic itself, however, was unusual because of the venue Muriel had chosen for the lunch. The family met that day in Ashtead churchyard, the blanket laid out by the grave of Muriel's late husband, Rupert. Lead capital letters, fixed onto the top of the large, horizontal white marble slab read: 'RUPERT THOMAS GOULD, LIEUTENANT COMMANDER, ROYAL NAVY, HOROLOGIST, AUTHOR AND BROADCASTER, BORN SOUTHSEA 1890, DIED CANTERBURY 1948'. The funeral had taken place 10 months before, but of the family, only Rupert's son Cecil had been there among a small handful of mourners. It was also Cecil who had chosen the epitaph on the tombstone, and the emphasis on Gould's professional achievement and the absence of any memorial to a 'father' or 'husband' was not an oversight. 22 years before, Rupert and Muriel had undergone a bitter judicial separation and had hardly met thereafter.

So the picnic, which finally brought together these two diverse and conflicting aspects of their father's life, his career and his family, was always a bit of a mystery for the children. Their mother never did explain what it had been about; in later years Cecil supposed it might have been a case of his mother, in her own particular way, saying farewell to her husband, reconciling and resolving a 10 year-partnership which had, after all, originally been loving. But the separation had been disastrous for Rupert, who lost his wife, his home, his closest friends, his job and custody of his children.

By the time of his death at just 57 years old, Rupert Gould was enjoying celebrity status; he was nationally famous for a multitude of achievements and Cecil's epitaph characterized his father's public image well. But there was another very different side to this curious life: severe depressions, crippling bouts of overwork and the emotional struggles he faced during his short span, witnessed by his family but unknown to his public audience, directed the course of his life at every turn.

Ironically, Lieutenant Commander Gould's role as a Navy Officer, the title of which he carried with him all his adult life, was the least successful of his contributions. Rupert Gould was however possessed of an exceptional mind; he was a veritable polymath, and his extraordinary knowledge and his dissemination of it in his books and radio broadcasts was what made him a household name in his day. Of all the subjects he became expert in, it was Horology, the study of clock making and timekeeping, where he made his greatest contribution: he had an all-consuming obsession with the subject and was one of the twentieth century's finest antiquarian horologists.

Gould's fascination for clocks and watches began in childhood and, with an interest already present, it is no surprise that when the young naval officer, studying navigation at Greenwich, encountered that indispensable horological instrument, the marine chronometer, he should do so much more than simply learn how to use it. In just 4 years, beginning at the age of 29, Gould prepared and wrote the definitive history of the instrument. *The Marine Chronometer, Its History and Development*, first published in 1923 and reprinted many times since, was so beautifully written and so thoroughly researched it still has no equal on the subject in the twenty-first century. We tend to associate accessible books on technical subjects as very much a recent phenomenon and in this sense Gould was a long way ahead of his time.

In studying the subject it was Gould who rediscovered the great marine timekeepers of John Harrison, corroding in store at the Royal Observatory at Greenwich. These incredibly complex and intricate machines were the prototypes which led to the successful marine chronometer and are of monumental importance in the history of science. In fact, as Dava Sobel reminds us in her best-seller *Longitude* (Walker/4th Estate, 1995), Harrison's fourth timekeeper 'H4' proved to be the first of all precision watches, winning for him the famed £20,000 Longitude prize money, and marking him out as one of the eighteenth century's greatest scientific achievers. Although famous in his day, Harrison's life had virtually been consigned to history until Rupert Gould retold his extraordinary story in 1923. Not only did he see Harrison's greatness recognized once again, Gould then dedicated over fifteen years of his life to restoring the timekeepers to their former glorious working condition. The first three large machines were the greatest challenge, with H3, the most complex of all, having over 700 parts needing restoration. Every stage of the work was recorded by him in eighteen meticulously detailed notebook/diaries and his work has not only preserved these timekeepers for posterity, but has enabled us to understand them in every aspect. What is even more remarkable is that these Herculean tasks were all done, to use Gould's own words, as a labour of love, and carried out in his own time; he would only accept repayment of expenses. For his contribution to horology alone the name of Rupert Gould should be remembered in perpetuity.

But Gould was much more than simply an horologist. Blessed with an almost photographic memory and an insatiable appetite for knowledge, in the course of his life he became an expert in many other fields, accumulating a truly astonishing breadth of learning. He was the epitome of the popular philosophers and 'men of knowledge' who were a feature of the mid-twentieth century, revered by the public for their knowledge and encouraged by establishment bodies such as the BBC as examples of intellects to be admired and emulated. And the fact that Gould was essentially an amateur, albeit a consummate one, made his broad knowledge all the more remarkable. Always the strictest of academics, he nevertheless appeared to be the typical gentleman dilettante in the many things he did. He had no formal academic degrees and on the only occasion when he entered a profession, his first career as a Navy officer, he was obliged to bow out, being quite unsuitable emotionally and psychologically to cope with the rigours of wartime.

He may have appeared to play the role of dilettante, but anything he took an interest in was studied in extraordinary detail. Had he not been well recognized as a polymath, he could easily have been remembered independently by small, specialist-interest groups. He was a pioneer in, for example, the study of tennis, of clocks, typewriters and on unsolved scientific mysteries such as the Loch Ness Monster phenomenon, of which, as will be seen, he wrote the first systematic study. These same wide interests led to Gould being asked to give a weekly wireless broadcast on *The Children's Hour*, the BBC's programme for younger 'listeners in'. Gould was billed as 'The Stargazer', though astronomy was only one of many subjects he chose to talk about. Few whose childhoods included listening to The Children's Hour forgot the talks given by 'the man who knew everything'. And in the 1940s it was inevitable he would be invited to join that select coterie who broadcast on the BBC's Home Service as *The Brains Trust*, the celebrated panel of experts answering questions of all kinds from listeners. Gould was notable as the welcome foil to the famous Professor of Philosophy, Cyril ('it all depends what you mean by . . .') Joad. Brains Trust producer Howard Thomas remarked that Gould was the only member of the panel never, in the history of the programme, to have been contradicted.

Only in exceptional men and women does one find a brain that combines a profound scientific understanding with the vivid imagination and artistic talent necessary to write and illustrate with the skill demonstrated by Rupert Gould. His published works are notable for their engaging style and characterful clear illustrations. There is no doubt then that he was exceptional, yet in many ways he was Everyman, subject to the times in which he lived. It is a cliché to say that the personalities of talented people often contain a mass of contradictions, but it is certainly true in this case. Gould was handsome and very large in stature, standing 6 ft 4½ in. tall without his size 13½ shoes. Being intellectually strong and socially confident, with something of a commanding presence, an acquaintance might have expected him to possess a tough and rugged mentality. But his depressions and overwork led to a complex and emotionally chaotic life, punctuated by four severe nervous breakdowns. And it was not only overwork which kept Rupert from the marital home. A mind like his needed regular intellectual stimulation, something which, in spite of many fine qualities, his wife was unable to provide in large measure. *The Sette of Odd Volumes*, a literary dining club founded by the noted antiquarian bookseller, Bernard Quaritch in the 1870s, was ideally suited to Gould's needs and regular meetings of the Sette provided another forum for his intellectual pursuits.

Had fate served him different opportunities who knows what one might have been able to say about R.T. Gould. Had his academic schooling continued on conventional lines it seems certain he would have excelled, probably resulting in a University professorship, a perfect role for a personality like his. In fact, if the Harrison timekeepers hadn't stolen his heart in 1920, Gould was considering studying for the Bar which, with his great memory and pungent wit, would have been another magnificently appropriate career for him. The reality however, though important and productive in its own way, was not to be so straightforward.

In the early twenty-first century, over fifty years after Gould's death, studies in the history of Science and Technology are increasingly teaching us to appreciate and commemorate the vital role played by the backroom boys of technological history: the ones who preserve and record the evidence. In conserving, illustrating and interpreting the scientific and technical evidence left behind, men such as Gould are now beginning to take their proper place alongside the great inventors and scientists; they are just as important in our understanding of the advances that were made.

Gould's was certainly not an ordinary life, but from whatever aspect one considers it, his was an interesting and important contribution. He knew the value of his achievement but he also knew he'd made many mistakes along the way; Cecil recalled that at the end of his father's short span, when dying in Canterbury Hospital Rupert talked about the many triumphs and disasters that had brought him to that end. One wonders whether, that day in Ashtead churchyard one year after his death, his family were also reflecting on his extraordinary life. That however, almost sixty years later, is the simple aim of this book.

1

Childhood 1890–1905

Rupert Thomas Gould was born, on 16 November 1890, into an upper middle class English family that was about as typical of its kind as could be imagined. It was a period in late Victorian society when there was an unprecedented move towards social mobility and Rupert Gould's forebears, who range from an artisan gardener to an upper-middle class surgeon, provide a perfect example of the various ways families from different social strata might aspire to, and sometimes achieve, a higher standing in the community. In the case of the Gould side of Rupert's ancestors, this was achieved through education, hard work, and by marrying well. Over the previous two generations a wide variety of professions, vocations, and trades had contributed to putting bread on the various family tables, including music, medicine, teaching, auctioneering and the textiles industry.

By the time of Rupert's birth, the Gould family had settled down into the upper-middle classes, wealthy in family connections but not itself particularly well off. Typically for such Victorian families, bringing up the children was the single focus. Rupert's parents' priority was ensuring a sound education and a firm but caring childhood, as was imbuing a sense of family continuity in the off-spring by emphasising the importance of knowing their ancestry. This was evidently a tradition passed on to Rupert's son Cecil who was equally well informed and aware of his family's notable forebears.[1,2]

Rupert's father was William Monk Gould (1858–1923) an organist, teacher and composer, and his mother Agnes Hilton Gould, née Skinner (1860–1937), always known affectionately within the family simply as 'Dodo'.[3] Both parent's names included that of their respective mother's family—in both Agnes's and William's cases the mother's was the wealthier side of the partnership and in these typically heritage-conscious families both sides of the conjunction were included in Christian names for the sake of commemoration.

So it was the Monks and Goulds, the somewhat 'upwardly mobile' of Rupert's ancestors, and the Skinners and Hiltons, the professional and monied part of his family tree, who combined to produce R. T. Gould. Although it was the female line, Agnes's family, who were much the wealthier of the two, in fact little of it came to the Goulds until after William's death.

William's maternal family, the Monks, had been a well established, middle class Devon family[4] and it was in Abbey Place in Tavistock that William Monk, shown in the 1851 census as a 48-year-old parish clerk and auctioneer, and his wife Mary Ann Monk (52 years old that year) brought up their family of five children.[5] Parish clerk was a responsible and honoured position in the community and the profession of auctioneer would have provided a good living, so the family's position seems to have been one of comfortable and sober, if unexciting, respectability in the relatively small community of Tavistock. One of the four girls was Eliza who, as a 'Pupil Teacher at the Church National School' at just 16 years of age, stands out among the siblings as bright and the most likely to be looking for advancement. Good teaching skills have always been in demand, in pastures old or new. In 8 years time, Eliza would marry Thomas John Gould and give birth to William, Rupert's father.

Looking at William's paternal family, the Goulds, the 1841 census finds William's grandfather, Thomas Gould, living in Town Hall Street, Dover, with his wife Ruth, 2 years younger, and five children, Thomas John (William's father), then 4 years old, being one of three boys.[6] Thomas Gould was then a 35-year-old gardener, shown as originally from Canterbury. Although the title gardener could at the time include those with managerial responsibilities, it seems likely from the status of the Goulds' neighbours that in this context it indicated a relatively lowly occupation. It would appear then that the Goulds were almost certainly of the respectable but staid working, or lower middle classes. Although Cecil Gould claimed well-connected forebears for them,[7] it seems most likely that this branch of the family came from relative obscurity.

At some stage soon after 1851[8] the young Thomas John Gould, who would become Rupert Gould's grandfather, flew the family nest in Dover and in 1858, by which time the 21-year-old was earning his living as a teacher, he married Eliza Monk at Tavistock. Perhaps the teacher Eliza and he met as colleagues at the same school, but this is not known for certain. Their son William Monk Gould was born in Tavistock on 5 October that year, the first of five children, all of whom (given their parents' profession) would have been provided with a careful and full education.[9]

In 1861 the family had moved to Lidford (still in Devon) where Thomas was teaching. Then, after a period spent in Malvern in Worcestershire, doubtless to take up new teaching opportunities, the Goulds moved to Rye in Sussex, on the South Coast of England. The move probably occurred in 1867 as in that year a new mixed school was built in Mermaid Street in Rye, and both Thomas and Eliza would be teachers there. In 1874 a new girls' boarding school was built in Lion Street and in 1876 Eliza became headmistress, with Thomas becoming Headmaster of the school in Mermaid Street which was now exclusively a boys' boarding school. In spite of Thomas and Eliza's senior positions, the family at this stage were still by no means wealthy. For Thomas John the marriage would have been regarded as a social step forward, but if Eliza had been seeking a chance to elevate her new family in prosperity, it does not seem to have succeeded.

The family may have been relatively poor, but there were at least compensating talents revealing themselves in the children. As early as 12 years old William Monk Gould played the organ at Rye church[10] and was organist there from sixteen. The church of St Mary the Virgin, Rye has an exceptionally fine and early turret clock in the tower, famous for its automata 'jacks' on the outside, which strike bells at the quarters, and the clock is notable for having one of the longest pendulums in the country. The pendulum is in fact so long it swings well within the nave of the church and was, according to Rupert, a constant nuisance for his father as its regular beat was extremely distracting for the organist, trying to maintain a very different tempo!

In 1885 William's father Thomas John Gould died, at the early age of 48, leaving just £53 (equivalent today to about £2,500 in total)[11] in his estate to his widow Eliza who outlived him by a further 22 years. By the time of his father's death William had already left home. The 1881 census shows that the 22-year-old bachelor had moved west, along the coast to the prosperous and rapidly expanding town of Portsea, adjacent to Portsmouth, to make his way in the world. Following in his parents' footsteps, William's primary income was from teaching and that year we find him earning a living at 21 Kings Terrace as a Professor of Music. But William Monk Gould was also a composer: the printed Music Catalogue[12] records some 56 items he published between 1883 and 1920, including many popular songs, mostly of the very old fashioned, romantic kind. His most notable work, in its day, was the Victorian parlour song *The Curfew*, though many of his other songs were popular too and were regularly performed at the time.

Unlike his parents, William's teaching was therefore just a part of a specialist professional vocation in one of the liberal arts. By his choice of an academic career he had placed himself among, and to some extent part of, the Victorian middle classes. He was also better placed to advance further, both socially and financially, if he happened to make the right match and, when he married Agnes Skinner on 8 August 1888, he gained a considerable improvement in the fortunes of this branch of the Monk-Goulds.

Agnes was the daughter of the celebrated Scottish doctor, Thomas Skinner and wife Hannah, née Hilton (d.1897), the daughter of the successful Lancashire silk merchant Henry Hilton (d.1878) of Harpurhey in Manchester. So the young family William created was now connected to a well-established upper middle class family, The Skinners, and had the financial security of the wealthy (albeit probably a touch *nouveau riche*) Hiltons. At the time of Henry Hilton's death in 1878, he had amassed

1. The Hiltons, c.1885. Hannah Skinner (nee Hilton), in the dark silk dress, is seated, centre stage, with her sister, Jane Hilton, on the right holding the parasol. The young Dodo (Rupert's mother) stands at the left. (© Sarah Stacey and Simon Stacey, 2005)

an estate valued at the considerable sum of £160,000—he was a multi-millionaire by modern standards. A share of this fortune became the source of the Gould family money, but access to the funds was severely restricted. Henry's will very carefully tied up the capital in trust funds allowing only the income to be distributed to his three daughters, Hannah, Jane (who would never marry), and Mary, who would eventually marry Edward Baynes Badcock, Canon of Ripon Cathedral.

As an aside, it's worth noting that the Badcock's daughter, also Mary, has a curious claim to fame. When 9 years old, her photograph, seen in a Ripon shop window in 1864, inspired Lewis Carroll to propose her image to his illustrator, Sir John Tenniel, as the ideal model for Alice in his newly written children's book.[13] One could therefore say, whimsically, that Rupert T Gould's great aunt was Alice in Wonderland (at least in image; the character herself was of course based on the real-life Alice Liddel, later Mrs Hargreaves).

Thomas Skinner

Rupert Gould's maternal grandfather, Thomas Skinner MD (1825–1906) (see colour plate 1), was born and educated in Edinburgh, the second son of Edinburgh solicitor John Robert Skinner. Thomas Skinner became one of the country's foremost gynaecologists and obstetricians, as well as a celebrated specialist in anaesthetics (the 'Skinner mask' for administering chloroform was invented by him).

Latterly, Skinner was also an important figure in homeopathic medicine. Until the 1870s he had been a firm traditionalist on that currently controversial practice. Anaesthetics and obstetrics were at the cutting edge of the official medical profession and in spite of support from one or two influential places (e.g. the Royal Family), the practice of homeopathy was looked upon by most professionals as little more than charlatanism. In fact Skinner himself saw through a regulation prohibiting the practice of homeopathy by the Liverpool Medical Institute's membership. However, in 1875, after a successful homeopathic treatment for his own, apparently incurable health problems, Skinner was won over and, obliged to resign under his own regulation, he became one of homeopathy's leading and most passionate advocates.[14]

Rupert was very proud of this distinguished member of the family, with his pragmatic and open-minded approach to his practice. Skinner was evidently 'his own man', and no doubt took this professional course in the face of much criticism from his peers. It was an approach that Rupert admired and learnt from and, emulating his grandfather's admirable academic tradition, it was one he adopted in his own independent and open-minded research into scientific mysteries in later years. Cecil recalled that Rupert was nevertheless always amused by the irony of his grandfather's self-dismissal from the Liverpool Institute.

From Edinburgh, Thomas Skinner had moved to Liverpool[15] in 1859 and, on 12 April the following year, he safely delivered Hannah of their daughter, Agnes Hilton Skinner. In a letter to his brother-in-law that day he noted that his wife 'was nine hours under chloroform and of course she knows precious little of the stormy sea she

2. Agnes Hilton Skinner, 'Dodo', Rupert's mother, c.1890. Described by Cecil Gould as 'by far the best member of the family', she was the emotional mainstay of the Gould household. (© Sarah Stacey and Simon Stacey, 2005)

3. William Monk Gould, c.1890. Rupert's father, the somewhat eccentric and 'happy-go-lucky' composer and music teacher, enjoyed the shooting and fishing traditions of his in-laws. (© Sarah Stacey & Simon Stacey, 2005)

has passed through . . .'.[16] One year later the couple had a son, Hilton Skinner (1861–1928), their only other child. In 1881 the Skinners moved to the outskirts of London settling in Beckenham in Kent, at *Waylands*, a large house in 'The Knoll', a very select private road in the town. Skinner now established a large homeopathic consulting practice in the West End of London, and carried on, at various addresses, for a quarter of a century. Very fond of fishing and shooting, for many years he leased a moor in Scotland, a practice that his daughter Agnes and son-in-law William Monk Gould continued into the twentieth century with their sons Rupert and Harry.

These then were the rather mixed group of people who made up Rupert's relatives, all of whom he would either have met or would have learnt about from his parents; quite a varied assortment from which to consider his roots and derive his identity as he grew up in the Gould family home in Southsea at the end of the nineteenth century.

When Agnes Hilton Skinner (Dodo) and William Monk Gould married, in the summer of 1888, Agnes was already 28 years old. Rupert's son Cecil described his grandmother as 'absolutely admirable' in nature, and 'by far the best member of the family'. Utterly down to earth and dependable, Agnes was however rather plain in appearance and she had not inherited her father's intellectual brilliance. In fact the family were beginning to think of her as probably 'on the shelf', though they considered that a connection with the Goulds represented something of a social 'compromise'. There was little socialising between the families after the marriage. In addition, although Dodo's family were well to do, all the money was tied up in trust, so there was very little in the way of a dowry for the wedding.

As it turned out though, the marriage was an excellent match. Dodo's eminently sensible, no-nonsense attitude balanced William's rather eccentric, extroverted, and rather 'happy-go-lucky' temperament, and she would prove to be the emotional mainstay of what was a wholly matriarchal family.

The couple decided to settle where William was already living, in Southsea, and 5 November 1889 saw the birth of their first child, Henry Hilton Monk Gould, known in the family as Harry, with Rupert being born just over a year later.

Why the Goulds should have chosen that first name is unknown. Rupert is the English version of Rupprecht, the Low German form of the ancient English name Robert, literally meaning 'Fame Bright'. The surname Gould incidentally is a version of Gold, originating either from those who worked in gold, or from those of flaxen hair. If the origins of Rupert's names suggest one who would be famous and of yellow hair, it would seem fate served him rather appropriately.[17]

Rupert's birth certificate shows the Goulds to be living in St Edwards Road in Southsea and, a few months later, the 1891 census reveals the house to be that known as Dunedin, probably quite a recent (possibly new) house, built on the south side, and towards the western end, of the street. The day of that census (5 April) Agnes and Harry were away and the record shows only William and young son Rupert, with two young

4. Harry and Rupert Gould, taken in Bromley in August 1892. The Goulds would have been staying with Dodo's parents, the Skinners, in Beckenham. (© Sarah Stacey and Simon Stacey, 2005)

domestic servants.[18] Later that year, with an expanding family, the Goulds moved round the corner to a larger, 10-year-old four-bedroom detached house, with garden, named 'Bedworth House'. It stands at No.11 Yarborough Road (it too survives today), and it was here in Southsea that the two boys were to spend the whole of their childhoods.

Southsea

William and Agnes had chosen the town wisely: the place they had decided to bring up the children was a rapidly developing and prosperous town. A suburb of Portsmouth, the principal naval base of the world's largest colonial power at the time, the Goulds were part of a vibrant and affluent community, and one absolutely dominated by the Royal Navy. At the time, Southsea itself was a relatively young, residential development. Much of it built on reclaimed marshlands, and growing south and east from the old town of Portsea during the early nineteenth century, the population of the area doubled between 1841 and 1881, the Goulds being just a part of this rapid influx. This early part of the new town had always been an elegant, middle class area inhabited chiefly by naval and military officers and civilian staff of the dockyards and armed services, many of the houses being owned by investors, leasing for profit. There were literary residents too; for example, Arthur Conan-Doyle lived just round the corner from the Goulds, in Elm Grove, though it would be well into the next century before Rupert would begin correspondence with him on supernatural matters.

By the second half of the century however, Southsea was not just a dormitory town. In 1847 the railway had reached Portsmouth and Southsea quickly became one of Victorian England's most popular, south-coast seaside resorts with all the social and economic advantages such an attraction brings. It was a thriving, 'happening' place the Goulds had moved to; with a burgeoning economy and well-to-do residents, Southsea was the fifth wealthiest town in England in the mid-nineteenth century.[19]

It is not known for certain why William first came to Southsea, nor why the young newlyweds decided to settle there, but it is much more likely to have been the general prosperity and the atmosphere of 'opportunity', rather than the Navy or sailing connection. Neither William nor Dodo had any interest in recreational sailing, as far as is known, and it would not appear to have been with careers for the boys in mind as Harry and Rupert were born well after the move and only one would go on to have a Navy career anyway.

Owing to qualities in both parents (and especially to Dodo's influence), the Gould boys' upbringing fits the idyllic, Victorian stereotype: completely stable and loving while being very disciplined and highly organized. Strict periods of reading were always demanded by Dodo (Rupert's daughter Jocelyne recalled her grandmother even insisting on this for her grand daughter when she visited during the 1930s) a discipline which goes a long way to explain how Rupert managed to pick up so much knowledge in so few years.

But theirs was certainly a happy childhood; Harry and Rupert wanted for nothing and were particularly close as brothers. Years later Rupert's son Cecil noted that Dodo had secretly always been more attached to Harry, the less brilliant but more gentle, 'needy' and emotional of the two.[20] Similarly, Rupert, the more competitive and outgoing of the brothers was closer to his father, though always more self-sufficient, cerebral, and a little more 'aloof', as Cecil put it. Whatever their private feelings, the parents naturally never allowed such preferences to be obvious at the time and both children were doted on. As a tiny child

Rupert had a thicker head of blond locks and it was he who was thereafter known affectionately within the family as 'Bunny'.

Following the tradition of Dodo's father, the family holidays were usually spent in Scotland, the Goulds holding a long lease on a house on the Isle of Mull (following the lease of one on Skye some years before). Each summer the whole family (William's role as teacher allowing an extended break at this time) spent several weeks here at the house, called *Druimgigha*, a smallholding in Dervaig, five miles west of Tobermory.

For the average family looking for somewhere to spend the summer, there were of course plenty of attractive and beautiful resorts much closer to home in England. But, ever since Queen Victoria and Prince Albert had conceived a love for North Britain (as Scotland was often called at the time), and had built the residence at Balmoral in Aberdeenshire (in 1856), Scotland was the destination of choice for upper-middle class Victorians. The Skinners, firmly traditionalist by inclination, were no exception to this, and the next generation, the Goulds, were proud to continue that tradition.

In the years to come Rupert cherished many fond memories of happy days spent fishing and shooting there with his father, and in his twenties and early thirties spent countless hours in the grounds creating a large and elaborate water garden, with fountains and waterfalls fed by a burn above the house. Druimgigha is still there today, with the remains of Gould's water gardens. The current owners of the house even have Rupert's carefully written instructions on maintaining and operating the hydraulics.

Music

While the holidays provided an excellent means of nurturing and encouraging the boys' outdoor interests, William's own profession enabled him to imbue in his children another of the pastimes traditionally associated with the well educated and civilized Victorian family: a love of music. Accompanying and composition work provided a part of William's income, but the principal family money came from teaching. He had a large number of pupils and Bedworth House must have resounded with music during the boy's childhoods. According to his obituary, William 'took a keen interest in the musical life of the town, and also won a considerable reputation for his songs . . .'.[21] Strangely though, given the artistic abilities that Rupert revealed later, his talents did not extend to practical music making. Neither, Rupert once remarked[22] was he ever able to whistle, though in his adult life he would often be heard humming to himself. It was Harry who, doubtless taught by his father, took up the clarinet and became a proficient amateur musician as a boy.

Although not a practitioner, Rupert certainly developed a great love for music. His parents had bought an early gramophone and a good collection of records of classical music were regularly played at the house. In spite of his father trying to steer him towards J.S. Bach, the classical and romantic composers were Rupert's particular passion, though he followed contemporary works too and was very fond of the music of the Russian composer Alexander Glazunov (1865–1936). He did however have one particular aversion: Cecil remembered that his father 'neither liked nor approved of Wagner'.

The Goulds were nominally of the Church of England faith, though it is doubtful if anyone in the family was particularly devout, and church services on Sunday were in all probability, as for many parishioners, simply part of their weekly routine in the community. Certainly, in later years Rupert was vehemently anti-clerical and, while he maintained a clearly agnostic stance (he was never atheist) he was very critical of any tendency to substitute religious faith for sound scientific learning. Particularly scathing of Christian fundamentalism, he once commented on *The War on Modern Science* by the American writer, Maynard Shipley, that it 'shows what a welter of bigotry and stupidity the whole business is'.[23]

The Church did play an important role in the life of the family though because, once settled in Southsea, William picked up his voluntary duties again as a church organist, for over twenty years fulfilling that role at St Michael's in Portsea (destroyed in the heavy bombing of Portsmouth in 1941). St Michael's stood as one of the town's more 'ritualistic' Church of England places of worship. Although far from Roman Catholic, the ritualistic form of churchmanship laid strong emphasis on ceremony, music and choir and William's role would have been important in the services. It is therefore most likely at St Michael's that the rest of the Bedworth household worshipped on Sundays too. As the family of the organist, Dodo and the boys would have been well known and would have been expected to appear at Church on Sundays. In later years when living in Surrey, Dodo was involved in voluntary charity work, and may well have contributed to Church activities in Southsea too, though keeping the family going at home would have taken much of her time.

Family Money

William's music teaching provided enough to pay the mortgage and put bread on the table but, throughout the boys' childhoods, the family was, by the upper-middle class standards of the day (but only by those standards), relatively short of money. The Hilton family money may have provided security, but the capital was always just out of reach. Three months after Dodo and William had married, Dodo's mother Hannah wrote her will. She evidently had her daughter's new family in mind, but the terms of the will would not prove particularly beneficial to Dodo's family in the short term. Following family tradition, the estate remained in trust, with the beneficiaries only receiving the income. Hannah directed half of this income to her husband, Thomas Skinner, during his lifetime, with the other half to be divided between Dodo and her brother Hilton.

In May 1897, probably when it became clear that Hannah was not long for this world (she died on 14 July that year), Dodo managed to persuade her mother to add a codicil to her will. This granted her daughter £100 a year until the death of Dodo's father (another 10 years in fact) allowed the interest on the remainder to transfer to her and Hilton. So, thanks to Dodo's intervention on her family's behalf at this critical time in the boys' upbringing, she enabled the family to maintain something of the lifestyle she wanted. Costs of keeping a large household were still high though, and Dodo had to scrimp and save, giving the boys extensive and demanding pre-school teaching herself and making and mending their clothes. Naturally, they still had servants to look after them: staff in service were generally very poorly paid and even impecunious middle class families could manage a maid and a cook. The 1901 census for their house records the parents and boys accompanied by three female servants,[24] all unmarried and living in (tax had to be paid on male servants' salaries, and footmen and butlers were generally only employed by households far better off than the Gould's).

Schooling

Many of Southsea's residents were young families and there were of course a large number of schools. No less than 87 private (fee paying, 'public') schools in 1857, 22 of them boarding and many of them 'naval crammers' giving concentrated 'crash courses' specifically designed to get pupils through the Royal Navy's entrance examinations.

Just as with the social structure of English society as a whole at this period, the social background of a typical Royal Navy officer was much more flexible than had ever been the case before. Gone were the automatic commissions for aristocratic young gentlemen (whatever their capabilities) and in were many more opportunities for talented and motivated young men from other backgrounds to advance into a prestigious career. But there was, of course, still plenty of scope for prejudice and the pulling of social strings to influence acceptance. In fact, being accepted was not straightforward for any applicant, whatever their background, as the number of young 'hopefuls' (in many cases the 'hopefuls' were in fact the parents of the boys) far outstripped the limited number of cadetships available.

Whether or not William and Dodo had originally intended either of their boys to enter the Navy is not known, but it seems likely, by their choice of school, that they at least had in mind such careers as a possibility for one or both of their sons. Once of school age, Harry and Rupert

5. Eastman's Royal Naval Academy, Southsea, in 1904. Rupert attended school here for the whole of his childhood, leaving in 1906 when he gained a place at Brittania, Dartmouth. (Courtesy of Central Library, Portsmouth)

were sent to Eastman's, one of the private preparatory schools, the full title of which was Eastman's Royal Naval Academy, or 'ERNA'. Though it had no formal links with the Royal Navy, and some pupils were not destined for Navy service of any kind, many boys attended Eastman's before going on to Dartmouth as cadets. Perhaps not as celebrated as Dr Burney's Royal Academy, or Stubbington House School, Eastman's was nevertheless considered one of the top schools for boys intended for the Navy and a considerable number of its alumni distinguished themselves in Naval service.[25]

While at Eastman's however, the elder son Harry managed to win a scholarship to go to Charterhouse (the famous public school in Godalming, Surrey) but, much as they wanted a public school education for Rupert too, this was beyond the family's means at the time. A career in the Royal Navy was thus a way of giving Rupert an equivalent, prestigious and thorough education, at lower cost. There had been no naval service in any wing of the family within memory, so this was a new departure, based on expedience rather than tradition.

Eastman's

Primarily for boarding pupils, the ERNA had been founded in 1851 by the retired naval instructor Thomas Eastman. The main school was built in 1854 and situated right down on the front at South Parade. There were a few day boys, the Goulds among them, and South Parade was only a 10-min cycle ride from Yarborough Rd for Harry and Rupert. Whether boarder or not, by all accounts the schooling was tough. Later alumni described it as 'a decent place to learn the elements of an English education . . .', though subjects were 'well taught within certain narrow, useful limits'. The curriculum naturally included the classics. Rupert later deplored the time he was obliged to spend on Latin and Greek (though he made full use of Latin quotations in his later writings—some may say excessively so), but remarked how much he had appreciated having studied the masters of English literature.[26]

Reading of all sorts, in fact, became something of a permanent occupation for Rupert. When coupled with what turned out to be a 'photographic' memory, the extent of his literary and general knowledge was soon quite remarkable. Rupert was also possessed of the gift of a vivid imagination (typically, one of his favourite authors at this time was Jules Verne), but he was just as happy reading the cheap boys' magazines known as 'Penny Dreadfuls'. So avid was his desire to learn, that as a young man he systematically read through the whole of *Chambers Encyclopaedia*, followed by *Encyclopaedia Britannica*, noting blithely in later years 'Candidly, I greatly prefer the former—not merely because it is shorter, but because, in my opinion, it is planned on far sounder lines and with a much better sense of proportion'.[27]

Swimming, gymnastics, and rowing were naturally on the curriculum at Eastman's too and Rupert enjoyed, and excelled at, all types of sports, though he always preferred solo games, such as tennis and billiards, to those of the team variety. Subjects with a more naval accent were also taught including the rudiments of theoretical navigation— how to work out the course and distance between two given places, how to calculate the ship's position from celestial observations and how to plot the ship's course on a chart.[28] Practical skills were obligatory too, such as knotting, splicing and carpentry, the latter something Rupert remained highly proficient in for the rest of his life.

In fact from the first, Rupert showed extraordinary promise and interest in all subjects at school, usually coming top of the class. One has the distinct impression that, at least at this stage in this highly competitive young man's education, winning was the primary aim and motivation. Consideration of what he would actually make of a Navy career was probably not uppermost in his mind at this stage beyond the idea of being a good officer. Wary of his becoming complacent, the schoolmasters never allowed him to relax though; on one occasion he slipped to second from top and was simply told: 'Gould, not quite up to your usual standard this week!'.[29]

Having established that Rupert was the brighter of the two brothers, one may reasonably ask why it was that Harry, not Rupert, won a scholarship to Charterhouse? There is insufficient evidence now to answer this with certainty, but we know that the two brothers were very different in their approach to learning, with Harry the methodical and careful reader, and Rupert the more unpredictable student, but capable of intensely hard, concentrated work when he put his mind to it. Assuming Rupert was put forward for a scholarship too, one possible scenario is that on the day he simply didn't put his all into the exams, while the much more predictable Harry had revised fully beforehand and got through on his preparation.

The failure to get a scholarship may even have spurred Rupert on in his determination to succeed in the later years at Eastman's, and may even, ironically, have contributed to his later successes. This phenomenon does seem to crop up a number of times later in Rupert's life: the failure to achieve something is followed by a hugely determined effort to compensate, which in turn, is crowned with a remarkable success. Another, more practical result of Rupert not getting a scholarship was that his education at Eastman's would now be geared towards entry into the Navy. A career as an officer in the Royal Navy was by no means a 'second best' alternative for a young man at the time. The British fleet was the largest and strongest in the world, and the schooling at Eastman's would have left the boys in no doubt that, in going for officer training, the opportunities for distinguished careers of all kinds were unprecedented.

With Eastman's right by the water and Spithead, the scene of Royal Reviews of the Fleet, just a couple of miles distant across the Solent, the pupils were encouraged to take an interest in the day-to-day naval activity around Portsmouth harbour.[30] And Rupert took to the culture of the sea and the practical detail of boats and ships like the proverbial duck to water, though, as has been noted, this interest was not passed down from any member of his family.

An interest in mechanism

Similarly, the origins of his interest in all things technical, apparent from a very early age, are uncertain. Even as a boy he was fascinated by technology, especially mechanisms such as clocks and watches, and he read his first horological book, Britten's *Old Clocks and Watches and their Makers*, when he was twelve, in 1902. Considering how important these technical interests were in his later life, it's interesting to speculate where such inclinations came from.

There was certainly a wide variety of professional experience in the family's background, but very little which could be described as specifically 'technical'. The only exception was Rupert's Uncle Hilton who was, in his short career (he retired at just 39, owing to ill-health), a civil engineer and who apparently encouraged his nephew. It was his uncle's copy of Britten's book that Rupert first read and from which he first heard of the chronometer-making pioneer, John Harrison.

One can only assume Rupert's main inspiration came from a general awareness of, and interest in, the many new technologies appearing at the time. The Navy would have been at the forefront of many of these developments and the boys at Eastman's would have been well placed to hear of the latest ideas and improvements. For a boy with an enquiring mind this period in late Victorian England must have seemed an incredibly modern age with an almost endless supply of new and fascinating devices to look at and understand.[31] Naturally, there were great scientific advances occurring at this time too, but it was the technical advances and the associated hardware that were of particular interest to the young Rupert Gould.

In spite of the Gould family's financial difficulties, once some of the Hilton inheritance was available, the family managed to afford some of these new devices themselves. By 1914, for example, the Goulds were using a telephone, a phonograph and a photographic camera at Bedworth House and it would not be long before Dodo acquired a motor car.

Motor cars and motor cycles, with the internal combustion engine, all of which were being developed during Rupert's childhood in the 1890s, were bound to interest him too, not just as an end-user, but in studying their construction and repairing them. He later (1934) noted proudly: 'For some twelve years before the War (i.e.1902, as a 12 year old) I was a keen motor-cyclist, amassing much hard-bought experience upon mounts now seen only in the Pioneer Runs. While I have never actually ridden a Holden (the first motor-cycle ever manufactured in England, and the second anywhere in the world . . .),[32] I have held its handlebars while the owner filled up; undeterred by chronic backfiring, I have wound a tube-ignited Bollee for a solid twenty minutes; I have been cast with great suddenness from a Singer, whose tiny engine was housed inside the back wheel, into the path of an oncoming tram-car; and I know, as well as most people, what the euphemism "light pedal assistance" used to connote'.[33] Later, he joined that select group known as the Association of Pioneer Motor Cyclists.[34]

About the time of the Gould's first largish family inheritance, resulting from the death of Dodo's aunt Jane in 1918, Dodo bought a Model T Ford (which, typically, she drove herself), quickly followed by Rupert buying his own, very smart American sports car, the Scripps-Booth which he enjoyed dissecting and repairing when necessary. According to his son Cecil, Rupert would like nothing more than to discover some hapless motorist on the roadside, needing assistance, so he could 'get out and get under' as the old music-hall song has it.

Technology at sea and in the air

The year 1897, when Gould was 7 years old, saw the invention by
Charles Parsons' of that remarkable advance in marine propulsion,
the steam turbine, the great power of which was famously demonstrated by Parsons at the hugely successful Diamond Jubilee Review
of the Fleet that year.[35] That particular Review must have been
something every resident of Southsea, even the seven-year-old
Rupert, would never forget, as by all accounts it was very extravagant.
W. G. Gates, a local News Editor described it as an event 'the like
of which neither this nor any other port in the world had ever before
witnessed'.[36]

The birth of mechanical flight occurred at this time too. Looking
back over 50 years, in July 1942, The Stargazer told the 'listeners in' to *The
Children's Hour*, the BBC's daily programme for its younger audience,
about his interest in early mechanical flight.[37] He described how in 1899
he had bought a book entitled *Progress in Flying Machines*, from which it
was made clear that, with the new light and powerful petrol engines
then becoming available, mechanical flight would very soon become
possible.[38] He also recalled, when 'a very small boy . . .', being taken to
see the experimental steam-driven flying machine designed and built by
Hiram Maxim (of machine gun fame) at his house near Bexley in Kent
(needless to say, at 4 *tons*, the machine was quite uncontrollable and when
it did once 'accidentally' get briefly off the ground, it did a large amount
of damage on landing!).

Just 14 years later in the summer of 1913 Gould would
experience flying himself, taking a trip up in an aeroplane when
visiting the aerodrome at Hendon. Few young men could have
more closely experienced the excitement of reading about a new
technology, having the patience and ability to understand its
intricacies, witnessing its realisation and then participating in it
first hand.

So, with a combination of a thirst for knowledge, his ability to understand potentially abstruse subjects, a vivid and active imagination and
the technological age he was born into, it's perhaps not surprising
Rupert developed such a fascination for, and knowledge of all things
technical.

Artistic ability

Not everyone with a talent for technical subjects has an equal, artistic ability to illustrate. However this Rupert also managed exceptionally well, not just in the form of technical illustration, but also with freehand drawing from his own imagination. In the succeeding years he proved himself capable of impressive artistic expression, evidently inspired by up-to-the-minute trends in illustration such as the works of Aubrey Beardsley and his followers, in ground breaking publications such as John Lane's *Yellow Book*.

Working very much in this highly fashionable and sometimes 'decadent' style, Gould even had drawings commissioned, and ultimately did virtually all the illustration for his own books. His ability to use a very fine mapping pen (soon to be put to good professional use by Navigating Officer Gould in the chartroom) enabled drawings with incredibly fine detail, perhaps a pointer to his tendency for deep introspection at certain times. Reflecting another of his favoured pastimes, his inspiration was often the music he listened to at concerts and on the gramophone, many of the drawings being imaginative interpretations of pieces he loved.

A parallel skill he learned with his mapping pen was the novelty of writing in extremely small script. He enjoyed astonishing his friends by writing the Lord's Prayer in capitals *three times* on a piece of paper smaller than a sixpence (19 mmØ), a lens then being provided for their confirmation that it was fully legible. Early drawings by him, up to the age of about 19, were signed with a peculiar snail-like figure, said by his son Cecil to have been in some way a monogram for RTG. In adult life he signed drawings, where there was room, with a depiction of a man standing with two poles, in imitation of the prehistoric chalk cutting on the South Downs known as the 'Long Man of Wilmington', a reference to his own considerable height of 6 feet $4\frac{1}{2}$ ins. Where no 'trademark' was possible, his initials or name can usually be found somewhere—in tiny script, naturally.

The Royal Navy

If the turn of the century was an inspiring time for a boy following technical advance, it was an even more exciting period to be starting a career in the Royal Navy. By the time Rupert was old enough to enter

Dartmouth, in 1906, Britain's new, improved (and still expanding) Navy, overseen by its inspirational and flamboyant First Sea Lord (The First Lord of the Admiralty), Sir John Fisher, was the largest and strongest in the world. It must have been many a schoolboy's dream to be a part of it.

But there were sinister overtones to this rapid expansion and everyone, young and old, in the town of Southsea, would have been increasingly conscious of them. Above all, the principal impetus for the unprecedented British expansion was the unwelcome emergence and rapid growth of the German Imperial Navy. The German High Seas Fleet was still not the equal of Britain's, and it is still argued today among military historians whether Britain's domination and its own expansion was in fact the main cause of the arms race at the time.

Britain's traditional military rivals in the nineteenth century had in fact been the French and the Russians, but as the end of the century approached, the balance of power was dramatically shifting. The 1889 Naval Act had seen the active rebuilding and strengthening of the Grand Fleet, and during the last years of the century continuing German arms growth, in spite of the Kaiser's close blood connection with Britain's own Royal family, exacerbated the British Government's fears.[39]

Nowhere could this arms race have been more obvious than in Portsmouth and Southsea. During Rupert's childhood in the last decade of the century, the demographics of the town changed significantly, the area to the north of Southsea quickly filling with relatively humble, terraced housing specifically to provide homes for a great influx of dockyard workers, shipbuilders (Portsmouth dockyards built many of the Royal Navy's ships), mechanics, clerks, administrators and sailors and their families, all in response to this perceived menace across the North Sea.

Should any threat prove real, Britain had to be ready and the Royal Navy's role as the first line of defence was crucial. The activities of the majority of the community of which young Rupert was a part, were defined by one clear subtext: the growth of the Royal Navy in the face of a potential military threat.[40] Rightly or wrongly, it would have been taken for granted by Rupert and other potential cadets, that a career in the Royal Navy at this time meant a contribution to protection from Germany and doubtless coloured his view of that nation in his childhood.

With experimental new 'all big gun' ships, including the famous Dreadnought class of battleship, first launched in 1906 (the year Rupert entered the Navy), the First Sea Lord Jackie Fisher demanded, amongst other qualities, to be able to 'hit a thimble on the horizon'. One imagines Rupert and the other young 'hopefuls' attempting to get places as cadets, were excited at the prospect of serving on these wonders of naval technology.

2

Navy Training 1906–1913

Once Rupert was 15, William and Dodo began the enrolment for his formal Navy schooling. The first step involved obtaining a nomination from the Admiralty, by writing to the Assistant Private Secretary to the First Lord of the Admiralty. In deciding on nominations, a boy's academic performance at school was considered and his general fitness and sporting prowess would have been a significant factor too. Personal, anecdotal evidence from staff and relatives would have contributed to the decision to nominate, and here families with established navy backgrounds would often have an advantage, though Rupert managed well without such connections.

Once nominated, the boy faced a very tough, and competitive, written examination divided into two parts or 'classes'.

Class I included exams on mathematics (arithmetic, algebra and geometry with a pass in all three parts obligatory); English; English history; geography and two languages (choice from Latin, French or German). The aggregate of the marks from each of these subjects then needed to add up to a certain minimum figure. However, all was not lost if the figure was not quite reached, as Class II provided 'top-up' marks if required, from exams in a choice of drawing, another language and 'natural science' or mechanics, combined with either physics or chemistry.

Thus, simply being accepted as a cadet in the Royal Navy, *The Senior Service* as it was known, put the successful candidates in an exclusive intellectual bracket of their own. Even before any training took place, here was an exclusivity the boys and their families had every reason to be proud of. Needless to say, given his academic performance so far, Rupert passed these examinations with flying colours and easily gained his place at the cadet school, HMS *Britannia*, at the Royal Naval College, Dartmouth, in Devon.

As in more conventional schooling, the academic year at Dartmouth consisted of three terms, but unlike the conventional system, a new

6. Rupert T. Gould, in April 1906. The 15-year-old cadet was already over 6 feet tall. (© Sarah Stacey and Simon Stacey, 2005)

group of cadets was introduced every term, rather than just once a year. Each group of cadets was itself therefore called a 'term' and to distinguish one from another, these terms were each named after a famous British Admiral. Rupert Gould joined Greynville term, one of 34 cadets, on Monday, 15 January 1906. This then was the first day of his Navy career.

Up to now Rupert had been a day boy, so attending this boarding school at Dartmouth was the first stage of his 'flying the nest'. But at 15 years old this competitive and self-contained young man would have had no difficulties in settling down and was perfectly capable of standing on his own two feet at the new school.

As originally created, the Royal Navy's cadet school, HMS *Britannia*, was a venerable, three-deck wooden battleship (in company with another wooden two-decker, the HMS *Hindostan*, also part of the cadet school accommodation) moored at Mill Creek, on the mouth of the River Dart, and the cadets began their naval careers living together on board.

The Selbourne Scheme

In December 1902 however, arrangements for cadets entering schooling for the Navy underwent fundamental reforms, the First Lord of the Admiralty at the time, The Earl of Selbourne introducing the 'Selbourne Scheme'.

Unfortunately for Rupert however, his age meant that he fell exactly between the old and the new schemes and he missed most of the benefits of the new scheme. As well as concentrating on a 'broad and liberal' education for cadets who may not have had the benefit of a public (fee-paying) preparatory school, one of the aims of the 'new scheme' was to give all cadets a better training in the basics of marine engineering. New technologies abounded in the Navy and more qualified engineers were needed, but there was a reluctance for new officers to go into that branch of the service. The reason was that there were professional and social distinctions, and prejudices, over Engineer Officers, who had up to then been trained at a separate establishment at Keyham. Although they could rise in rank, even to Admiral, Engineer Officers were never allowed to command a ship.

The new scheme trained all the cadets together, giving everyone a basic education in these technical subjects. It also left the choice of branch, made by individuals, until they reached the rank of Lieutenant, approximately 6 years after joining as a cadet. This, they hoped, might contribute to a levelling of the perceived status among officers, increasing the number of Engineer Officers and ensuring a modicum of practical, technical skill in those choosing the other branches of Gunnery, Torpedo, Navigation (the 'Executive' branches) or the Royal Marines.[41]

Including this broader education and basic technical training for all however meant two additional years of training at a newly established cadet school at Osborne on the Isle of Wight, the young cadets then going on to *Britannia* at the usual age. The result was that the age for starting a naval education was reduced to between the ages of 12 and 13, and the new scheme, introduced late in 1903, occurred just a few months too late for Rupert to be included. This must have been a great disappointment for him; one can imagine how much he would have enjoyed the lessons in the engineering workshops and all the technical training at Osborne, rather than continuing at Eastman's a further 2 years.[42]

As part of the new scheme, in September 1905 the cadet school known as *Britannia* was transferred from the two old wooden battleships to newly constructed buildings on the headland. *Britannia* became one of the 'stone frigates' as these 'dry land' establishments were known, and Rupert's term was only the second to occupy the new buildings. Although they missed out on the new Osborne curriculum, the Greynville cadets would certainly have benefited from the move towards social levelling and the new emphasis on technical expertise. Rupert's achievements and approach to work would soon reveal that he particularly appreciated this new culture at the Navy's cadet school.

His first year at *Britannia* saw him doing well and his personal file remarks that at the time his performance was: 'Good, promising, keen on all games'.[43] He was not the only star that term though: fellow cadets—also friends of Gould's—included Denis W Boyd (1891–1965), later a full Admiral; Henry E Horan (1890–1961), later a Rear Admiral; and John Hely Owen (1890–1970), who would be a well respected naval historian.

Passing out in December 1906, the term had their photograph taken, sitting on the steps in front of *Britannia*, all obediently smiling for the camera except Gould who, apparently inattentive, looks at his feet. By coincidence, another of Gould's contemporaries at *Britannia*, though from a different term, was his close friend H.C. Arnold-Forster, a direct descendant of John Harrison's *bête noire*, the Astronomer Royal Nevil Maskelyne.

Two deaths in the family occurred during Gould's first year away from home. In September 1906 Gould's maternal grandfather, Dr Thomas Skinner, died at his home in West London and Rupert's parents finally received an income from the interest from Hannah Skinner's trust fund (though nothing directly from Thomas's estate). In February 1907, Rupert's paternal grandmother, Eliza Worth Gould, who by then had been living at Deal in Kent, died leaving £461 between her three children, so the Gould family came into a couple of legacies at this time. It's not known how the deaths of these grandparents of Rupert's affected him, he certainly admired his paternal grandfather academically, but he left no record as to how close he was emotionally to any of his grandparents. His paternal grandfather had died before he was born anyway, but Eliza, his paternal grandmother did not mention him specifically in her will. What evidence there is suggests that Dodo and William's family kept in closer touch with Dodo's side of the family, so Rupert was probably not particularly emotionally disturbed by her passing.

After Christmas leave in December 1906, Rupert and the other cadets in Greynville term were sent for training in the cruiser, HMS *Isis*. Given the time of year, from January to April 1907, *Isis* would have gone to sea, rather than staying in home waters, the West Indies being a likely destination. The focus at this early stage in the boys training was seamanship, learning the core practice of what would be their profession, the day-to-day practical running of a ship and working as part of a team, understanding the many different roles of the officers and crew and ensuring one is part of an effective management structure.

Following this period on the training cruiser, the cadets faced their first academic examinations to qualify for promotion to midshipmen. The results of these exams were then combined with an element of continuous assessment, taken from the whole of the cadets' schooling so far, to determine the boys' seniority (their competitive ranking) in the term.

The results of the cadets' examinations (routinely published in *The Times*)[44] for Easter 1907, thus show the boys ranked in order of their seniority, and R.T. Gould is at the top of the page, in first place. But even this exemplary mark gives little clue to Rupert's outstanding performance during his cadetship.

The Prize List, shown after the results, reveals that Gould qualified for 1st Prize in Seamanship; 1st Prize in Charts and Instruments; 1st Prize in Drawing; 1st Prize in Gunnery & Applied Mechanics; 1st Prize in French; 1st Prize in Scripture; 2nd Prize in Mathematics; 2nd Prize in Navigation and 3rd Prize in Steam. An asterisk printed next to all but three of these results indicates that those 1st prizes were not awarded to Gould. The reason was that no cadet was allowed to take more than three prizes. Apparently, for one boy to take more than this would have risked excessive self-esteem and would not have been good for the overall morale of the class. Gould was probably irritated by failing to get firsts in Navigation and Mathematics as he always prided himself in his ability in these two subjects.

Rupert's friends did well from his losses however: Henry Horan, who took the 1st in Navigation and in Physics, picked up Gould's 1st in Gunnery and Applied Mechanics while only getting a 2nd in the exam; John Owen took the 1st in Scripture after a 2nd in the exam and, as a knock-on effect, Denis Boyd was enabled to take two 2nd prizes, in Gunnery and Applied Mechanics, and Scripture, though only passing 3rd and 4th in the exams.

The Navy Career

The 15 May 1907 was a proud and exciting day for Rupert and his family as he officially became a midshipman:[45] It was his first day as a Navy Officer and he would soon be serving at sea in a Royal Navy ship. With such a tremendous start to his career, the sixteen year old must have felt immensely empowered. He surely had no doubts that the coming years would be similarly crowned with professional achievement. His family, and *Britannia* staff too, believed he was looking forward to many years of successful advancement through the ranks: how long before he would have his own command, and how soon after that might he reach the rank of Admiral? He was an exceptional pupil and everything in his academic achievements so far suggested it would happen, perhaps sooner rather than later.

Midshipman Gould's first posting was to the battleship HMS *Formidable*[46] under Captain Herbert Lyon, serving in the Mediterranean and based on the island of Malta. Continuing his fine service, Gould is described on his personal file as 'very good, promises well'. He enjoyed his time on shore leave too and it was at this time that he became a fan of movies, recalling 'there were some six cinemas in Valetta (the capital of Malta) . . . not so many then in the British Isles' a somewhat strange place to find such a concentration of cutting edge recreational technology! He would be a regular cinema-goer for the rest of his life.

HMS *Queen*

In April 1908, midshipman Gould was transferred from *Formidable* to HMS *Queen*,[47] also serving in the Mediterranean at the time. Unfortunately for him, in July 1908 Rupert was admitted to military hospital on Corfu, diagnosed with mumps. In the early twentieth century this was a serious illness, the treatment for which usually entailed extended isolation and rest in a darkened room, something Rupert would have found extremely frustrating. Convalescence, which took much of the rest of the year, was in England, Rupert taking passage on *Queen* when she returned to Portsmouth that summer. According to his son Cecil, part of Rupert's recovery was spent in hospital at Greenwich during the autumn of 1908.

In December HMS *Queen* was being 're-commissioned' (being given a new ship's company and a refit) and, coincidentally, midshipman Gould was due to return to the same ship. His posting was announced in the Times as 15 December,[48] but in fact this did not properly take place until the beginning of 1909, as Rupert had another period of illness.

Gould's personal file reveals that on Boxing Day 1908, the day he was supposed to have rejoined *Queen* after the Christmas break, he was in fact admitted to Haslar hospital, just across the harbour from Portsmouth, but with no specific reason given. On 16 January 1909 he was discharged and sent home to recuperate, but on 28 January he was 'surveyed and found unfit, to be unemployed for a time'. On 6 February he was re-surveyed at Haslar, presumably still unemployable, as he was surveyed again on 27 February 1909, this time being 'Found fit'.

Depression

It is uncertain what this illness was. It is possible it was connected with his recovery from the Mumps in the previous year. However, as will become evident later in the story, there was a darker side to Rupert's psyche that may be connected with this period of sickness. At regular intervals throughout his life Rupert suffered from bouts of nervous depression and, in spite of his triumphant academic successes at this period, the illness may have been the beginnings of these psychological problems. Gould's son Cecil recorded that a family friend[49] told him Rupert had always had a morbid fear of three things: 'lightning (I remember the sensation that was caused when he insisted on having lightning conductors installed on our small Epsom House), revolution and Hell'. Rupert had evidently examined his feelings carefully on the matter and was quite prepared to defend his stance, as Cecil recalled the family friend telling him that Rupert 'argued perfectly logically (up to a point) on the legitimacy of his fears on all three heads'.

Of his three fears, the dread of 'revolution' seems to have given him the greatest trouble. With a world becoming dominated by increasing nationalism and political posturing, it was not a good period for those worried by conflict and civil unrest. Some form of national or international crisis was to dog him throughout his life and, even as early as 1908, a general awareness in his home town of the slowly accelerating arms race may have contributed to depression.[50]

Some of the rather hellish drawings done by the 19-year-old Midshipman during this period also suggest a macabre state of mind and a preoccupation with Hell, one of his stated fears. It is not unusual of course for teenagers, in the course of growing up, to display an interest, an obsession even, in death and the 'black arts'. The late twentieth century phenomenon of 'Goths', and several generations of 'Hell's Angels', are just two very familiar examples of the many ways people have of presenting the issue of death, in these cases using it to shock. Gould's teenage obsession may have had a similar root, but the result of his obsession was much greater than these 'lifestyle expressions'; he was actually debilitated by the anxiety he felt.

With such proven ability, one might have expected a young artistically inclined midshipman in these circumstances to be taking inspiration from the many interesting scenes and images he was witnessing during his busy life at sea, or doing more in the way of caricatures and portraits of his colleagues (something Gould later proved very good at). Resorting instead to such depressing imagery, albeit inspired by a love of music, would seem to confirm an introspective motive and more than just a teenage intention to shock. Perhaps the process of creating the drawings may itself have provided some form of relief. Expressing his inner imaginings explicitly in such a creative way may have been a means of reducing or removing the fear he experienced.

A pen and ink sketch he drew that year and titled *'Ronde du Sabbat'* after the penultimate scene from Berlioz's *Symphonie Fantastique* is an example. This programmatic work, a favourite of Rupert's, tells of a young lovesick man who attempts suicide with an overdose of opium but fails to die. Instead he suffers a nightmarish dream in which he is ultimately executed and buried as the Witches' Sabbath, surrounded by a ghastly host of monstrous creatures, celebrates. Gould signed this sketch with his monogram superimposed over the profile of a skull. And the scene from Berlioz was not an isolated choice. For example, another drawing done at this time, titled *The King Orgulous* and obviously influenced by the work

7. *'Ronde du Sabbat'*, 1909. Rupert's macabre interpretation of the scene of that name from the *Symphonie Fantastique* by Hector Berlioz. (© Sarah Stacey and Simon Stacey, 2005)

of Aubrey Beardsley, also has a nightmarish quality about it and suggests a not-altogether-happy state of mind.

Nevertheless, if his illness had been a kind of nervous depression, by the end of February 1909 he was sufficiently recovered to join HMS *Queen* and the following couple of years' service would be the most rewarding of his career at sea.

David Beatty

From December 1908, *Queen* was part of the Atlantic Fleet under Prince Louis of Battenburg (1854–1921), and was captained by a man who would make a profound impression on the young midshipman Gould. This was Captain David Beatty (1871–1936), later *Earl Beatty, Admiral of the Fleet, Viscount Borodale of Wexford, Baron Beatty of the North Sea and of Brooksby, P.C. G.C.B. O.M. G.C.V.O. D.S.O. D.C.L. LL.D.*, to give him his full style!

When making one's way in the world, it is natural as a young adult to look for role models. For a midshipman on one of his first postings, an important part of aspiring to effective leadership is observing and assessing the varied styles of his commanding officers. Gould considered himself exceedingly lucky to have found, right at the beginning of his career, the best possible inspiration, in the form of Captain Beatty (see colour plate 2). For Rupert, no one would ever command such respect as this revered leader of men, and he positively idolized him. In a few years' time, Gould's first, and best, piece of research and writing, his 'magnum opus' *The Marine Chronometer*, was dedicated, not to his wife, mother, nor any of his close friends, but to Beatty.[51] In the 1930s Gould would even begin writing a biography of the great man, though other work prevented it from seeing the light of day. Beatty evidently brought out the best in his men too, Gould's personal file during his service on *Queen* describing him as: 'Very zealous, loyal'.

The Captain's letters to his wife at this period survive[52] and reveal what was uppermost in Beatty's mind, and what life was like for Gould and his shipmates in *Queen*. On 13 February 1909, soon after taking up his new command, Beatty wrote: 'There is good material, and if I can't make a smart ship out of it I'll give up the sea and take to growing cabbages . . .'. Needless to say he did make a smart ship of *Queen*, writing on 20 February: 'We had our first competition with the Fleet yesterday in mooring the ship and we beat them all to blazes'. Prince Louis, writing to Beatty's wife

that July, remarked: 'It will probably give you pleasure to know that
your husband's handling of *Queen* is the best in the squadron'.

Beatty's approach was intelligent and highly successful. He understood the need to act as a role model and taught by example, recognizing
that his junior officers were not simply there to obey orders but in
turn were learning how best to command men. Beatty believed that
particularly the Midshipmen under his command should be given
special treatment. Speaking to them personally on a regular basis in the
Gun Room of the ship, he ensured they developed the way he felt they
should, and that they understood clearly what was required of them.
It has to be said though that Beatty's own rather well known and
pessimistic view of the troubled international situation at the time, a
view all his officers would have been aware of, would not have helped
Rupert during his periods of introspection and depression.

Nevertheless, Gould had the fondest of memories of his 20 months
serving under Beatty at this time, and had a fund of stories about life
aboard *Queen*. On one occasion soon after he joined the ship, Gould was
patrolling one of the upper decks of the ship in the early hours.
Indulging in a snack (a favourite pastime of his) to while away the time,
Gould was eating an orange and, fancying a little sport, decided to aim
the large seeds, of which there were many, at a ventilator nearby. A
minute or so after Gould's very successful game had seen every one hit
the inside of the ventilator and disappear somewhere into the ship, a
very flustered Midshipman appeared on deck with the message that
'Captain Beatty wishes to know why the vent *above his pillow* appears to be
raining orange pips'. Typical of Beatty, such a restrained and humorous
rebuke 'by proxy' was guaranteed to correct the problem while ensuring
the respect of the offender.

There were lighter moments on shore leave too. On 2 March Beatty
regaled his wife with the story of a dance at Queenstown in Ireland,
attended by most of the officers in the Fleet: 'Such a debauch . . . Admirals and Midshipmen, with intermediate ranks thrown in, vied with
each other as to who should last the longest. Needless to say the
Midshipmen won in a canter, in fact the girls nearly stampeded the lot
in their endeavour to secure a Mid each time'. Beatty invariably took a
careful and human approach to his junior officers, ensuring loyalty
by treating them as people first. This was something Rupert Gould
certainly emulated in later life, never displaying social snobbery and
always treating others, of whatever background, with respect unless

they gave him reason not to. However, Rupert was never able to 'suffer fools gladly', especially not ones 'who should have known better' and, unlike Beatty, he often managed to rub people up the wrong way.

In April, this time from the town of Kirkwall in the Orkneys, a letter from Beatty to his wife tells of soothsayers in the town, saying that 'All my midshipmen except four have gone . . .' One wonders if Gould had a go, he was not at all of a superstitious nature, but it's likely his curiosity got the better of him. By the end of June, *Queen* had circled Britain clockwise and was on summer manoeuvres off Scotland, the ship then anchoring at Oban for a week. They must have sailed right past Mull, with Rupert literally within sight of his family's summer home, Druimgigha. In July *Queen* had sailed south down the eastern side of Britain and was anchored, with the whole Fleet, off the east coast, by Southend, giving the public an opportunity to view the Royal Navy's protective strength.

Gould's time in *Queen* was certainly rewarding, but his concerns about international crises and the likelihood of war with Germany, and perhaps Russia too, must have remained ever present. As has been observed already, Beatty was absolutely convinced something of the kind was inevitable, and his officers, with whom he communicated quite freely, would have known of his pessimistic view. While the Fleet was anchored off the East coast that July 1909, Beatty and a party of 1,200 officers, Gould among them, were invited to lunch at the Guildhall in London. The whole company marched ceremonially through the City and past the Lord Mayor at the Mansion House. In his speech in reply to the toast to the Navy, Beatty stated: 'Never in the History of the world had there been so overpowering a preparation for war. The time must be drawing very close when the efficiency of the Navy may be put to the test . . . . We are watching and praying . . . that should that time ever come we shall be found to conform to the words of the immortal Nelson that we have done our duty'.

The end of the year saw *Queen* sail to its base at Gibraltar. With war now a probability for Beatty, who was equally convinced the British Government was not taking the threat seriously enough, he wrote prophetically to his wife on 10 December 1909: 'When the blow falls we shall be unprepared, suffer many losses, and lose many lives and valuable assets . . . and out of the debris we shall dig our way to a successful issue'. With such a view held by his commanding officer, perhaps it's only surprising that Midshipman Gould wasn't more depressed at the time.

While most of the fleet returned to England for Christmas, *Queen* remained at Gibraltar and Gould spent his first Christmas away from home. The beginning of the new-year saw Beatty promoted to Rear Admiral (at the extraordinarily early age of 39 years old, the youngest flag officer in over 100 years) and Gould's new boss, Captain Ernest F.A. Gaunt, was appointed on 4 January.

Drawing

Not all the drawings Gould did during this year were introspective, though all were marked by intense detail and must have taken huge patience and many hours of intense, close work. If his activities in later life are any indication, much of this work would have been done late at night. There were some periods of leisure time during the day and evening too and these were not required for academic studies as his training was, for the time being, largely practical. But Rupert was no introvert and he would have taken part in social life on board, so his drawing was more than likely done while others slept.

A highly detailed study titled *Valses des Fleurs*, inspired by the scene of that name from Tchaikovsky's ballet *The Nutcracker*, was published the following year (16 March 1910) in *The Sketch*, the English weekly art magazine.[53] To Gould's great annoyance, this drawing was then lifted, without his permission or payment, coloured up, and used by a commercial company to advertise a gadget known as 'The Colour-phone' (see colour plate 3).[54] It is clear from his reaction to this breach of copyright, and his subsequent observance and interest in such matters that, even at this early stage, Gould considered his artistic skill as a valuable asset and possibly a potential source of income in due course. As will be seen, in a couple of years time he did sell his first drawings, more imaginative sketches of his based on a musical theme, this time inspired by the great violinist Paganini, a subject he would return to many times in later life.

On 15 July 1910, having passed the next part of the seamanship examinations (with marks of 866/1000) in *Queen*, Gould was promoted to Sub Lieutenant (acting) and, discharged from HMS *Queen* in the summer, he went forward for his one-year Sub Lieutenant's courses. These courses were divided into two parts. Part I involved studies at Portsmouth, followed by Part II which took place at the Royal Naval College at Greenwich.

These courses were not to be taken for granted however. It was necessary to take examinations to qualify before one could even start Part I. These entrance tests involved two groups of exams. The Group A examinations included separate tests covering the topics of Navigation and Nautical Astronomy, Practical Navigation, Trigonometry and Observations. The second battery of tests, the Group B examinations, included tests on Mechanics, Heat and Steam (virtually all the Navy's ships were steam driven and there were many new technologies developing in this area), Electricity and Magnetism, and French. Each group totalled 500 marks, and a minimum of 250 marks in each group was required to go forward, 'subject to reasonable discretion on the part of the examining authority'.

Gould took these entrance examinations from 13 September, doubtless needing no 'discretion' by the authorities, and his first term on Part I, from 21 September to 10 December 1910, was spent at HMS *Excellent*, the Gunnery School at Portsmouth, coming out with creditable marks of 859/1000. After Christmas leave, from early January to 11 February 1911 he continued his Part I courses, covering Pilotage on the Navigation Schoolship, the torpedo gunboat HMS *Dryad*, reaching first rate marks of 926/1000 and evidently impressing the Captain, Edward Booty, who would later recommend Gould for advanced navigation training. From there he studied at HMS *Vernon*, the Torpedo School (also covering studies in electricity and radio) where his marks, after examination on 11 March, were 176/200 (equivalent to 880/1000). The results for that whole term showed that, out of a class of 25 Acting Sub Lieutenants, Gould was third from the top. With such outstanding results he soon received confirmation of his promotion to Sub Lieutenant, along with Horan, Boyd, and R.B. Wilson.[55]

As usual, promotion in the Navy was dependent on previous achievement. Being allowed to take the Part II courses at Greenwich depended very much on the results of Part I, and similarly, the results from all of these Sub Lieutenants courses determined how soon the officer would then be promoted to Lieutenant (retrospectively decided the following year). This date of seniority, which also determined the level of the officer's pay, then decided the time of promotion to Lieutenant Commander, automatic after 8 years as a Lieutenant.

Between April and June 1911 Gould attended the second part of the Sub Lieutenants' courses, at RNC Greenwich. These classes covered two theoretical subjects, navigation and mathematics. This time, for some reason Gould's performance does not seem to have been quite so outstanding. His marks in the three mathematics examinations only averaged a rather poor 52 per cent. The 'term marks' (out of 40), given for each of the two subjects, represent a continuous assessment for the whole term. In these, Gould managed 35/40 for Navigation but a surprisingly lacklustre 20/40 for Maths. Given his established abilities in this subject, one wonders if he simply got bored and failed to take the exams seriously enough.

Following Annual Manoeuvres at sea during the summer of 1911, Gould took his final term of the Sub-Lieutenants courses at Greenwich, consisting mostly of academic subjects including naval history and languages, from September to December 1911. The top that year was Henry Horan, but Gould, Denis Boyd and eight others were not far behind, Gould and Boyd being promoted to Lieutenant with a seniority of 15 October 1911, just 3 months after Horan.

Harry Gould

During this time, Rupert's brother Harry had continued his education by going up from Charterhouse to Clare College, Cambridge. Here he studied to practise as a surgeon and, with B.A. Cantab, went on to St Thomas's Hospital in London where doubtless he would have qualified, had war not intervened. Harry, like his brother, was a sociable young man (he joined the Masons a couple of years later) and took an active role in the University's, sports, social, musical and literary life, continuing to play the clarinet, and fencing for Cambridge.

While he would have been proud of his brother's academic achievement, Rupert would have envied Harry his University education. Rupert's literary and artistic interests contrast vividly with his Navy training and suggest that he saw his Navy career as just one part of a much more rounded, 'three dimensional' potential in himself that he wanted to develop and expose. While he worked hard to ensure his achievement in officer training was fully up to his own high standards, he evidently felt there was more to life than the Navy and, whenever he could, he searched elsewhere for experience and learning. Thus he valued his visits to see his brother, both socially and academically. He recalled in later years[56] that on one occasion in 1911 he and Harry dined at King's College with the scholar M.R. James, the Provost of

Eton and famous writer of ghost stories. After dinner, James read to them the reports of the ghostly *Berbalangs of Cagayan Sulu*, and Rupert included a chapter on this sinister 'phenomenon' in his first book on unsolved scientific mysteries, *Oddities*, published in 1928 (see Appendix 4). Rupert was very taken with James' ghost stories, doing a preliminary drawing (whether for James or for his own amusement is not known) to illustrate the ghost in James' short story 'Whistle and I'll come to you'.[57]

The China Station

Rupert may well have envied his brother's life in academia, but the next stage of his own career soon redressed the balance and gave him the interest and stimulation he craved. The one thing a Navy officer was guaranteed was travel to exotic places, and it didn't get much more culturally and geographically exotic than China, where Rupert was now posted. He must have been thrilled at this new opportunity to broaden his knowledge and experience.

As he was now a professional part of the British presence in Asia, Gould would also have discovered that the Royal Navy's policy on diplomacy in the far reaches of the empire was considerably more bullish than in home waters, though he does not appear to have been uncomfortable with this. In as much as he had political views, they were generally conservative and traditionalist in tendency.

On 25 January 1912 Gould joined cruiser HMS *Hawke*,[58] for passage to China,[59] the intention being that he should initially serve on the Yangtze River. He was first posted to the river gunboat HMS *Kinsha*,[60] on the Yangtze, joining the ship on 28 March. *Kinsha* was in fact a paddle steamer, and had a shallow draft, very useful for navigating the upper reaches of inland waterways. Being armed she could have more of a sting than her appearance might have suggested, though the duty was only river patrol and must have been uneventful most of the time. Gould served in *Kinsha* less than 3 months and on 16 June 1912, he was given a much more challenging and responsible role as an officer on a larger vessel on the China coast. That day he was transferred to the gunboat HMS *Bramble*[61] under Lt. Cdr Bernard E. Pritchard, serving as one of just six officers. Uncertain he was able to meet this new challenge, Gould was nevertheless being given an opportunity to show what he was capable of.

Navigation

By this time Gould had decided which discipline within the Executive
Officer branch he would eventually specialize in: it had to be Navigating
Officer, the man responsible for directing the ship safely and effectively
at sea. The reason for choosing this rather cerebral of the branches would
have been quite simple: his interests and skills fitted perfectly and he
knew he would be good at it and enjoy it. He already had a great
fascination for the instrumentation and now he could combine the
practical use of these devices with mathematical and geometrical calcu-
lation, something else he enjoyed. Then again, as an accomplished
draftsman, Gould knew that another of the essentials of a good Navigating Officer was well within his capabilities.

He also would have been aware that, of all the officer branches, that of
the Navigator is inclined to be the most singular activity.

In his history of the Navigation School ship, HMS *Dryad*, B.B.Schofield
cites C.L. Lowis, in his book *Fabulous Admirals* (to which Gould's friend
Denis Boyd contributed), where Lowis describes the role of the Navigator: 'In harbour they led a detached life in some quiet nook, carefully
correcting charts with the Admiralty Notices to Mariners . . . they even
escaped the daily Divisions or Church Parade by arranging to wind their
chronometers at that time . . . Tankie, the Midshipman attached as
Assistant Navigator, having collected the key from the keyboard sentry,
used to report to the Pilot, generally in the middle of his breakfast, and
the two would descend to the gloomy depths where the chronometers
lay in state. Most Pilots wound their clocks in profound silence, but one
of the old school made quite a ritual of the daily occasion . . . every day
he would turn to his Tankie and say 'Who are the salt of the earth?' and
the Tankie was obliged to respond 'the Navigating Branch, sir', the short
ceremony being brought to a close by the Pilot's response 'And the
princes among men'. In fact the Sergeant Major of Marines being the
least concerned with the ritual surrounding the chronometers, was
made responsible for checking with the Navigating Officer at 0900 each
day that his officer had done his duty and of subsequently reporting to
the Captain 'Chronometers wound, Sir'.

Naturally, in an efficiently run ship, all the officers would generally
work as a team, but once the ship is under way the Navigating Officer has
his mind constantly on his role and, except when training others, his

8. *HMS Bramble*, c.1913. A river gunboat, *Bramble* was 180 feet by 33 feet. Lt Gould was one of six officers in charge of eighty ratings on board. (© National Maritime Museum, London)

tasks tend to keep him alone with his thoughts and calculations. His one important interrelation would be a close liaison with his Captain, but his role required less team effort than the other branches. Gould was undoubtedly the kind of man who preferred to work alone, wanting all the credit if a task was well achieved, but equally being prepared to take criticism if not (something he rarely had to face). One important aspect of the Navigating Officer's role which Gould may not have fully appreciated was that, at times, it involved a great deal of stress, with a heavy burden of responsibility if things went wrong, but at the time everything was going well for Lt Gould and there was no need to go looking for problems.

Unlike *Kinsha* on the Yangtse, *Bramble* was on the open seas, up and down the coast of China and would provide excellent experience for a budding navigator. Lieutenant Edward Wise was *Bramble's* Navigating Officer and at various times he and Gould worked closely together during Gould's time in the ship which was for a period of a full year, his longest posting in his naval career.

HMS *Bramble*

What would life have been like for Rupert while in *Bramble*? Built in 1898, HMS *Bramble* was one of the very last of the gunboats to be supplied to the Navy before policy changed from gunboat diplomacy towards greater use of larger cruisers and capital ships. At 180 ft in length and just 33 ft wide, she was not a big ship and with a complement of eighty ratings (ordinary seamen) as well as the six officers (Cdr. Pritchard had Wise, Gould, a Surgeon, a Gunnery Officer and an Artificer Engineer under him), it must have been pretty cramped for the crew. Although quite modern in appearance, the ships equipment was very simple; all ground tackle was worked by hand and all lighting in the ship was by candle lamp.[62]

Gould's diary for 1913 has survived, and provides a fascinating and highly detailed insight into this particular period of his life.[63] The entries cover his many day-to-day activities and thoughts during this year, the first half of which sees the crew of *Bramble* on patrol on the east coast of China. In summary, the diary tells of a rather privileged, relatively comfortable existence, utterly redolent, in fact, of the image one has of outstations of the late Victorian British Empire. Gould's days were taken up with the usual naval duties such as Officer of the Watch and managing the crew, interspersed with occasional exercises and inspections on board, but rarely anything that one might describe as really stressful.

For a man with a lively mind, life on the China Station, even when on coastal patrols and not on the Yangtse, would have had great potential for boredom if he did not, as Gould evidently did, made careful arrangements to occupy his mind. Having said that, the very reason Gould was there, as part of the British Navy's military presence in the very sensitive political arena that Asia was, meant he was always aware of the potential for a crisis. At every stage Gould's ship encountered vessels from rival navies including the Russians, the Japanese and the Germans. So while it may have been a relatively comfortable and possibly hum drum routine, his new responsibilities and the presence of military tensions meant it would certainly not have been a relaxed time for Rupert.

The Sunday service when at sea was usually held on the mess deck (sometimes standing, in which case Gould's diary simply noted 'Perpendicular Church'!), but when in port a church party went ashore, usually consisting of most of the officers, and normally including Gould. Although hardly religious by inclination, he invariably conformed to the traditional routines such as the Sunday service for the sake of appearances.

The rest of the time on board was filled with reading (much of the time), drawing, and playing cards with fellow officers, though card games never interested him much and he was never, in his whole life, a gambling man. A favourite pastime was listening to music on his gramophone, the use of which he gladly shared, even if his choice of music was not to everyone's taste (e.g. 'Henderson pining for ragtime').

Reading between the lines, it is clear his fellow officers did not really provide much in the way of intellectual stimulation for him and he often resorted to the solitary pursuits of reading and listening to music alone. Most of the major ports had Officers' Clubs with facilities for overnight accommodation, bathing, dining, sports and parties, and it was when visiting these, sometimes meeting up with old friends from *Britannia*, that he found the social life more rewarding. Visits ashore (sampan and rickshaw were the usual transport), which were usually only for a day or two at a time, also included social events (afternoon 'tiffin', the light afternoon snack originating from India, being a regular), dinner parties, tennis and billiards ('pills'). From the 'Siberian' and 'paper' mails, there were fairly regular letters from family (Dodo and Harry usually) back home, sometimes with small sums of money enclosed.

Being of an adventurous nature, on occasions Rupert would take time to go further a field, exploring on foot around the city or port, sometimes taking in the occasional bout of shopping for books and curios. Years later, he compared notes with his friend Professor Stewart on the relative effects of taking various drugs, and stated that when younger he had experimented with smoking Hashish. It was probably during this period on the China Station that the opportunity arose.[64]

1913, an important year

A few quotes from the diary will give a flavour of what life was like, and how officer Gould was performing, during this critical period in his training.

The year starts with *Bramble* in Hankow (Hangzhou), and due to remain there until 8 February. That first day of the New Year the ship was being painted 'weather being providentially fine'; it seems small vessels like *Bramble* were almost constantly being repainted to keep the required pristine appearance up to scratch. It was payday: 'My pay $152, clearing $80', Gould also remarking: 'Gramophone was polished'. Gould's gramophone was a machine he was re-engineering at the time in *Bramble's* tool-room, to be driven, not by a spring but by a heavy weight on a multiple gut line, giving it a much longer duration between windings. The weight descended down vertical guide-rods; Gould described it as being reminiscent of a weight-driven chronometer by Ferdinand Berthoud.

He was constantly increasing his collection of records, having records sent by mail and buying them when he could find suppliers in ports. His love of music (which evidently did not stretch to ragtime!) was also satisfied by concerts put on at various times in the principal ports they visited. For example, on 17 January he attended a concert of the Hankow Musical Society at the Victoria Hall listening to chamber music by Beethoven and Saint-Saens, some piano solos and two songs. But he missed home and craved advancement. Almost from the day he arrived in the Far East, Gould began petitioning strongly to be sent home to go on the qualifying course at Portsmouth to be a Navigating Officer, writing to Beatty, Selater, Booty and others for recommendations.

Germany was not forgotten by Gould during this time of course. Among several other navies, the German fleet had ships on the China coast too (they also had possessions here, further up the coast at Kiao Chow). Whether or not there were tensions, official diplomacy dictated that the two Navies must at least put up a pretence of friendship. Gould noted ruefully on 26 January: 'Shall have to dress ship to-morrow, Kaiser's birthday', but a few days later, on 6 February, he expressed himself more clearly, reporting that among the many ships they encountered in port: 'A damned German destroyer blew in about 10.30 . . .' Although his diary usually suggests a perfectly well balanced demeanour from day to day, he suffered from regular headaches, and introspection occasionally makes itself known. Thus, with his love of great literature, it is not at all surprising to find him buying Robert Burton's classic three-volume work *The Anatomy of Melancholy* (1621) to read at this time.

On 8 February 1913 Bramble made her way down to Kiukiang and over the next few days on to Shanghai where they moored for a week. At the end of the month they were under way again, this time out to sea: 'First day at sea since June last year'. Over six months without practice at sea was a long time and it seems they had become somewhat rusty in their anchoring procedures.

Reaching Pagoda Anchorage on 1 March, Gould records they had 'the most edgeways [nerve-racking] day we have had this commission', arriving at 3 p.m. with the wind blowing a gale and accidentally anchoring on a sandbank, everyone getting irritable as they weighed the anchors again and got the cables fouled, the ship nearly ramming a

Japanese neighbour, 'then the Pilot blamed everybody & everybody blamed the pilot . . .'. Eventually Gould noted, 'So, having placed the sheet anchor, I went rounds, gave the hands a lie in till 5.30, & turned in, thanking Providence it was no worse . . .'. He was right to be relieved; had *Bramble* rammed the Japanese neighbour, who knows what kind of an incident it might have sparked, and severe damage to a navy ship would have very serious consequences for the officers responsible. A pilot working in a Navy ship was only ever acting as an 'adviser' on the ships navigation, the commanding officer always 'carried the can'.

Two days later finds *Bramble* arriving several hundred miles down the coast at Swatow (Shantou), Gould noting 'Pretty Place, mountainous and rugged'. The following day, 4 March, saw the ship's paintwork all cleaned down, after which Gould enjoyed 'The best afternoon's tennis I've ever had' at a party put on by the Commissioner (probably the Commissioner of Police) in Swatow, playing no less than three 3-set matches, and noting 'I used my reverse overhead service for the first time. It puzzled people' (no doubt it did; Gould's son Cecil maintained that although his father's knowledge of the history and rules of the game were unequalled, his skill at play was 'more theoretical than practical' and he was actually unable to serve conventionally). That evening *Bramble's* officers: 'Dined with Consul. B.P. [Cdr Bernard Pritchard] very cheery and bright'. Friday 7 March finds them 300 miles further down the coast at Hong Kong, where they tied up for 12 days in the port, followed by 2 days at sea heading west along the coast, via the Hainan strait, to the port of Pakhoi (Beihai), arriving just after midday on Good Friday, 21 March 1913.

Here *Bramble* stayed for 5 days, one of the first things Gould needing to do was alter the clocks on board: '1.15 put clocks back to 12.15. They keep Singapore time here (nonsensical)'. On 25th Gould visited a leper colony at Pakhoi 'not so terrible a sight as I expected' and later that day 'Skipper gave a tiffin party aboard' for the Consul. 'We were all snapshotted on the fcsle by Mrs Metzethin'. A print (now very faded) of the snapshot was later sent to Gould by Mrs Metzethin and is pasted in the diary entry for that day.

The next few days saw *Bramble* heading back to Hong Kong. On 29th Gould 'spent the dog watches in doing a rough sketch of my new notion for Rachmaninoff'. This would have been another of his drawings characterising music he was fond of, though which piece by Rachmaninoff in this instance is unknown. Probably in his 'Beardsleyesque' style (he was reading the early volumes of *The Yellow Book* at the time), the image did not impress his commanding officer: 'B.P. dropped in and was pleased to say he considered my Rachmaninoff "disgusting" & that he would not care to show it to his wife. And all this because one (adolescent) figure is undressed!' Reading between the lines, it's unlikely Rupert would have worried too much about the artistic opinions of his boss!

After a short trip up river to Canton, the ship returned to Hong Kong on 24th and underwent an official inspection on 28 April, including exercises involving battle situations: 'control officer killed, fires, 1st aid, etc . . . a sorry sight! He [Inspecting Officer] was quite calm about it, though and left complimenting the doctor. B.P. very fed up'.

Gould fell ill on 3 May 1913. Neither his diary nor his service records specify the cause, but it was almost certainly Malaria: the mosquitoes were often very bad, even within the Club buildings. Furthermore, the disease had stricken Cdr. Pritchard for several days at the beginning of April, and Gould had the occasional recurrence of Malaria in later life. The following day Rupert was taken to Hong Kong Hospital in a rickshaw. His diary entry reads: 'Bed very short. Milk diet. Headache'. Gould usually found English beds too short; one can only imagine how uncomfortable a Chinese one must have been for him! Having suffered this for 10 days, he was discharged for 3 days recuperation at the Hong Kong Sanatorium where he resumed his drawing and reading, keeping up with the news, as usual. After reading the papers on 15 May, he noted 'I see Lord Alfred Douglas is suing Ransome & T.B.C. for libel, but I think Ransome will justify his words'.[65] Three months later he annotated the entry: 'Result was a verdict for defendants'.

Coincidentally, during the weeks Gould had been ill, Navy engineers from Hong Kong dockyard had been refitting (overhauling) *Bramble's* engines and gear. At the end of May this was finally complete in preparation for trials on the following Monday, but by this time Rupert was beginning to have had enough of his posting to the Far East and was dreaming of home. The diary entry for 6 June 1913 reads: 'I was thinking all this forenoon what a flap there would be if I were suddenly discharged to the "*Tamar*" for passage home in the "*Kent*" ' And it seems his thoughts were prophetic: that afternoon the signal was received: 'Commodore to *Bramble*. Lieutenant Garlick of *Kent* has been appointed to *Bramble*. Lieutenant Gould returns home in *Kent*. *Kent* has been ordered to meet *Bramble* at sea & exchange officers', Gould entered joyously in his diary. 'I dashed up on deck & met B.P. coming down, very sick at getting Garlick'. A.M. Garlick had been in Greynville term with Gould at *Britannia*, but had put in a poor performance. Garlick's personal file reveals that from quite an early stage in his career he had been having problems with reliability and 'intemperance'. Pritchard would have known what a contrast this new officer would be to his high performing predecessor, Lt Gould.

The next day Gould writes, looking back: 'so my year's work in the "*Bramble*" is ended . . . to sum up my year—I was pitch-forked into a job much more strenuous than any I had ever taken on. I had no previous experience and I was both unfitted and disinclined for it. Naturally I have made many mistakes through carelessness and laziness. It has however done me a lot of good and I dare to hope I have done as well in the job as some would have and better than others.' Gould is being hard on himself here as it is evident he coped magnificently: On 9 June he notes, just before leaving *Bramble*: 'B.P. gave me a most flamboyant flimsy last night and also showed me my confidential report, which was good'.[66]

Another purchase he made, just before he left the China station, was the ordering of a smart new motor bicycle, to be ready for his return to England. A week or two before, he had received literature by mail on the new range of motorcycles by Triumph as well as the BAT Motor Company's advance list. While waiting to transfer to *Kent* he was 'still swithering whether to get a "Bat" or a 3 speed Triumph & save money. Triumph has it at present'. But the more glamorous machine won the day and the last letter he sent from *Bramble* was to order a Bat 'No.2' 8hp model from Wauchopes, the agents in London.

HMS *Kent*

Within hours he was on board the cruiser *Kent*,[67] writing: 'The "*Bramble*" has vanished like a dream—for some time I could not convince myself that the whole thing was not a dream—and already my memories of her are losing their edge. Here in this big ship, steady as a rock, one finds a new standard, less strenuous and more companioned . . .'. A little hint here of some of the things Gould liked less about navy life—in truth he was never going to make the archetypal sea dog.

A few days more were spent in Hong Kong, meeting his old friend Boyd and going to the Victoria Cinematograph on the Wednesday night, *Kent* sailing on the Saturday.

Arriving in Singapore early on 19 June, Gould had time to go ashore and buy some Malacca canes and tiffin at the Hotel de l'Europe, before departing at 5.00 p.m. for another British crown colony, Colombo, Ceylon (as the island was known until 1948, when it became Sri Lanka). His was the first dog watch (4.00 p.m. to 6.00 p.m.) and Gould records 'swung ship'. This is the process whereby, once under way, but before leaving sight of the coast, the ship's compasses are calibrated to ensure they are as accurate as possible. It is done by turning, or 'swinging' the whole ship about its centre, pointing towards specific coastal landmarks of known orientation, and adjusting the compass accordingly.

The 24 June 1913 saw *Kent* arrive at Colombo, to stay for 6 days. It says something of the extent of the British Empire and its commercial connections at that time that while visiting this city on an island in South Asia, Gould was able to visit an agent for Triumph motorcycles! Looking at the model he had been considering, he decided it would definitely have been too small for him.

Seven days sailing took the ship around the Indian subcontinent to the British protectorate of Aden, the chief port on the southern coast of the Arabian Peninsula, near the entrance to the Red Sea. *Kent* stayed in port for just one day, 7 July, and, reading fresh newspapers that day, Gould's mind was already back in Europe. He was worrying about 'fresh trouble in the Balkans', and was carefully noting the results of the World's Lawn Tennis Championships.

Kent entered the Suez Canal on 12 July ('Pilot *very* slow and not at all Chatty'), arriving in Port Said the following day, staying for 2 days, during which time Gould and fellow officers managed several sets of tennis doubles on shore. The crew appeared not to have been quite as disciplined as in smaller ships; the next day Gould records 'two cases of "drunk aboard," one the butcher, after dinner and the other an AB (Able Bodied Seaman, the lowest rank) about 6 p.m. . . . had to send a patrol for a stoker at the police station . . . caught a LS (Leading Seaman) relieving himself on the Mess deck. He bunked but I had him alright. An AB was run in for the same thing on the upper deck about half an hour later'. Kent sailed from Port Said at 3.45 p.m. on 15 July. At 7 a.m. that morning eighteen sailors were still absent, though only one eventually had to be left behind.

On course to Malta, there were gunnery exercises, 'I only had 6 rounds to fire from my turret & we got them off without the slightest trouble'. *Kent* remained 6 days at Malta, where some of the crew were behaving no better: 'two men returned aboard tight, one of them naked'. Letters that had been missing the ship finally caught up and Gould learned that the family were to leave for the summer holidays on Mull on 28th, too soon for him to join them, eliciting the comment 'DAMN'.

Another letter brought better news, as it announced his new motorcycle, the BAT No.2, was finished and waiting collection. The voyage from Malta saw further exercises, including several simulated 'submarine attacks' and, after an afternoon's halt at Gibraltar, Kent made her way to England arriving early on the 31 July 1913. The excitement was not over however, as the ship experienced a dramatic approach past the Eddystone lighthouse in a massively violent thunderstorm, a very uncomfortable experience for Rupert with his phobia about lightning. '5.50am Fore t'gallant mast, main ditto and aerial came down about 5.00am, main hanging by Jacobs ladder from main top mast'. Later that day, masts still hanging, Kent dropped anchor at Portsmouth.

Gould's 18 months posting in China had been an important proving ground. Having detected potential in this new Lieutenant, Gould's commanding officers had given him an opportunity to demonstrate his abilities and show his mettle. He had not disappointed them and there was now every prospect of rapid promotion. He had also decided that his future lay in one of the more cerebral specialisations within the service. As a Navigating Officer his drawing and mathematical skills would be put to good use, and his growing interest in precision instrumentation would find rewards.

He was always proud of his time in China and on his return home his cabin in Kent was laden with souvenirs, curios of all kinds, and a new folder of drawings, many inspired by the places and people he had seen. If there was anything negative he took from this period of achievement, it was perhaps that he was now more aware of the physical and intellectual privations required of a Navy Officer serving on a smaller ship on a distant station; exotic memories and souvenirs were fine, but the 23 year old was glad to be home and looked forward to concerts, motoring and a fuller social life during the next two full months of shore leave.

Holidays

Determined to get straight back into life at home, Rupert planned to head directly off to join his family. With the parents and servants in Scotland, and the Southsea home closed for the summer, Rupert checked into a local inn to stay overnight. The next day, 1 August 1913, having collected his bicycle from Yarborough Rd, and telegrammed his parents on Mull that he was on his way, he took the train to London, bicycle in one hand, bag in the other. While in the city he visited Wauchopes (the motorcycle agents) to inspect his new purchase, the Bat No.2, and paid a further £12/10s off the price with an extra £8 spent on accessories, to be fitted by the time he returned from Scotland. Staying overnight in London, (at the Langham Hotel in Portland Place) he dined at Paganis and went to the ballet at the Palace Theatre, on Cambridge Circus. Evidently he was happy to be back and thoroughly enjoyed the performance, his diary noting 'Very good, Pavlova better than ever'.[68]

The next day, Gould's train north did not leave until evening, so that afternoon he made his way up to Hendon to spend the afternoon at the aerodrome. From an evening of culinary and artistic pleasures, to a day satisfying his technical interests: Rupert prided himself on his broad knowledge and appreciation of these very different spheres of learning. 'Booked a passenger flight S.W. biplane, 50 hp Gnome, Manton pilot. Enjoyed it. Watched flying for about twenty minutes afterwards then left'. Typically interested in the technicalities, Gould notes the manufacturer of the rotary engine used, its power output, and the manufacturer and type of the aeroplane.[69] It was not the route, nor the views on his flight that interested him, but technical information gleaned from chatting with the pilot that he chose to record in his diary. It is also interesting to note that, for someone of an occasionally nervous disposition, this potentially dangerous activity seems to have held no fears for him.

Aunt Jane

En route north, he visited his great aunt Jane in Manchester.[70] There's no reason to suppose that he didn't look forward to seeing his aunt, though she may well have been rather strict and abstemious in her lifestyle: unmarried, she dedicated much of her life to the church and left large sums of money to various Christian causes. Whether or not Rupert looked forward to seeing her, his parents probably reminded him anyway that this particular aged, and exceptionally wealthy, maiden aunt would expect visits from her grandnephews if she was also to remember them in her will. Much to their mutual surprise and delight,

Rupert was met by his brother Harry at the door, who was also visiting great aunt Jane that day. Harry was staying with friends, the Veys, at Blundellsands, near Crosby in North Liverpool, and invited Rupert to join them for a trip out to North Wales in the Veys' motorcar the next day. The trip would end at the Veys' house afterwards for tennis, then dinner followed by billiards and listening to their gramophone. Motoring, socialising, tennis, dinner, billiards and music—it was the perfect agenda for Rupert.

They covered an amazing amount of ground in the car, driving from north of Liverpool, down through to Ruthin for lunch, on to Betws-y-Coed and getting as far as Beddgelert, on the way to the Lleyn peninsula to have tea, before turning back to Blundellsands. Rupert stayed overnight at the Veys and, after a short game of golf on a nearby links the next morning, he took a train from Liverpool to Glasgow. Staying at the Central Station Hotel overnight, he planned to get an early train to Oban from Glasgow the following day. Unfortunately, the hall porter forgot to wake him in the morning, in spite of a specific request, and Gould 'missed train and only steamer to Tobermory that day. Relieved my mind to hotel management'. As in later life, if he felt someone had caused problems through lazy or thoughtless behaviour, he would really let off steam at them.

Mull

The next day the early boat, the *Gael*, ferried Rupert to the island, bags, bicycle and all. The family seem to have had (or more likely, had the use of) two motor cars on the island, and one of these, with Dodo and 'Hugh' (probably one of the local gillies they hired), met Rupert as he cycled on his way to Druimgigha.

Rupert had joined his family just in time to celebrate an important event. Friday 8 August 1913 was Dodo and William's silver wedding anniversary and that day, taking the 'large car', the family celebrated with a visit to Iona. This stunningly beautiful, atmospheric and historic island just off Mull appealed immensely to Rupert's vivid imagination and his sense of history, and a visit to the Abbey (the burial place of early Scottish Kings), combined with lunch at the St Columba hotel, made it a very special day for all concerned. Gould did however note rather haughtily in his diary that the ferry across to Iona had been delayed 'by arrival of the steamer—"the" event of the day—and *droves* of tourists'.

The Gould's were friends with a number of local families on the island, including the Mackenzies and the Forsyths, and the summers on Mull were spent in a relaxed mixture of social calls, shooting and fishing, reading (especially when it rained which, of course, was often) and carrying out work on the house and grounds. Rupert was particularly proud of his elaborate water gardens, only finished a few years later.

3

The War, a Breakdown, and Marriage 1914–1920

On Tuesday 23 September 1913, Rupert left holidays on the island of Mull to return to his Navy commission, arriving in London the next morning. Sending his luggage and bike on to Portsmouth, he returned via his Uncle Hilton's (Dodo's brother), now settled at Hayes in Kent with his wife, Rupert's Aunt Emily and their sons, Rupert's cousins Douglas, now 21, and Charles 14. The main reason for this brief visit was to collect Rupert's new motorcycle (accessories all now fitted) from The Bat motorcycle works, which were just up the road in Penge, in South London.[71] Rupert noted in his diary that after lunch that day, with the help of cousin Douglas, he 'generally keyed her up. Did a six mile run round common' (Hayes Common), and Friday the 26th saw him making an early start off south on the Bat. His plan was to stay the weekend with his close friend Val Hirst at Woodlands, the Hirst family home close to Eastbourne, fifty miles or so east along the coast from Portsmouth.

Valentine D Hirst, the son of Colonel Thomas E. Hirst, had been one term ahead of Rupert on *Britannia*, where they first met, and the two Midshipmen went on together to serve under Beatty on HMS *Queen*, becoming the closest of friends.

Rupert and Val had many interests in common, and their time together was occupied that weekend with tennis, billiards and, on the Friday afternoon, a walk down to the Promenade at Eastbourne to see the new Hydroplane (an early sea-plane) in use. They stayed up late talking after dinner that night, Rupert then reading in his room and finally getting to sleep at 2.00 a.m.

The diary entry for Saturday, 27 September begins: 'Refreshed with my morning cup of tea, I went up and called Val. *Voila une lutte* [wrestling match] *herculanienne*! Subsequently I bathed and dressed in flannels'. The following morning the diary reads: 'Val came down to call me this morning, and some quick scrum work in and around the bath room culminated in his scratching me heavily amidships with the window hook'. Evidently the old friends still enjoyed the occasional schoolboy-style scrapping; though Rupert's coy use of French (*à la* Pepys) might just suggest an element of physical attraction too.

After a day of more tennis and trying each other's motorcycles, Rupert filled up the Bat and left for Portsmouth, paying 3d (three pence) toll to cross the Old Shoreham Bridge and stopping in Chichester to refill the tank. As dusk came on, he stopped to light up his acetylene lamp[72] though on approaching Portsmouth he encountered some fellow motorcyclists; he notes: 'picked up a club run and let them light me in'. Motorcycling was still very much an exclusive yet social activity, both aspects that Rupert relished.

The next day, Monday 29 September 1913 was the day Gould started his qualifying course in Navigation on the navigation school-ship *Dryad*. On arrival in Portsmouth, after garaging his bike, he went directly to the Dockyard, supped at Nab House (the Officers' Mess) and turned in early to his new posting in *Dryad*, noting 'Cabin rather small'.

Monday 13 October saw the beginning of the practical Navigation training at sea. On this date, Gould's diary records 'Sailed in *Dryad*. Navigated from spit onwards'. *Dryad* passed up the Irish sea, arriving in Bangor on the Wednesday where Gould found spare time to do a 'rough sketch for Aria on G String'. While there is no evidence that Rupert was not enjoying his Navy training, it was clearly never enough to keep his mind fully occupied and stimulated: in his spare time he was invariably engaged in non-maritime activities and interests.

After a gap in the diary entries, a little more appears at the end of the year when, on Tuesday 16 December he dined at the Queen's Hotel in Portsmouth with friends, including Val Hirst. They were also joined by a friend named Barwell who invited Gould to an Arnold Dolmetsch concert in London the following evening, including the offer of a lift back by motor car the same night to Portsmouth. The Dolmetsch family were quite a musical phenomenon at the time, perhaps the closest in their day to exponents of the concept of authentic performances on original instruments, and Gould gladly accepted the offer for the following evening.[73] On returning to *Dryad* that night Gould continued his studies, working the way he liked best, by burning the midnight oil: 'cleared decks for action & set about my fair chart. Worked till 6.00am, when it was all but finished. Had an hour's sleep'. In spite of his great intelligence, Rupert was a terrible procrastinator. His son Cecil recalled that, throughout his life, his father would always leave things until the very last moment, usually working under great pressure to get tasks finished, in spite of having had plenty of time initially.

The next day Gould duly took the train to London after school, dined with Barwell at the St James's Hotel and went on to Clifford's Inn in a taxi. He reviewed the concert in his diary: 'Concert splendid. All Dolmetsch family in costume & Neville Lytton *ébouriffé*, with large ivory flute. I liked the tone of the harpsichord immensely, but the clavichord was almost too faint to be heard . . . Drove back to West Kensington Mansions with Barwell, Neville Lytton & a French Artist. Had supper with Barwell . . . *inter alia* received Barwell's cheque for £2.2s for Nicole Paganini, the first drawing I've ever sold'. His interpretations of Paganini were very much in the Beardsley style; the surviving drawings on this subject are imbued with satanic undertones, echoing the suggestion that to achieve his impossibly difficult technique, the great violinist had been in league with the devil.[74] Rupert was evidently now beginning to establish social connections (in this instance a rather wealthy one) outside his Navy circle. He would also have been reassured that his artistic skills had begun to provide a potential supplement to his Navy pay and perhaps one day an alternative living.

They returned to Portsmouth overnight in the height of Edwardian-style luxury: 'Left W. Kensington Mansions at 1.45am in Barwell's limousine, equipped with rugs, sandwiches, flask etc, & a spare chauffeur. No trouble till Godalming, when it was suspected petrol tank was leaking. Twenty minutes delay, then on . . . after Butser I must have slept for I woke just in time to see the Portsmouth lights from the crest of Portsdown Hill. The chauffeur took the long way round to the Dockyard where we arrived at 5am. Turned in at 5.45am'. Even with two chauffeurs, it is unlikely Gould would have resisted the temptation to help sort out the problem with the petrol tank. Apart from the delays, they seem to have managed an average of about 25–30 miles an hour for the 80 miles or so of the journey, not bad going for an overnight journey by motor in 1913. The same morning, he notes that, with less than 2 hours sleep, he was 'Up at 7.30 & to Whaley on "Bat." Had breakfast with Val & left Xmas presents with him to take home. He goes on leave today, I tomorrow'.

This is the last entry in his 1913 diary; a critically important year in his training was over and he had acquitted himself splendidly. In spite of apparently 'burning the candle at both ends', Gould passed his

9. Lt. Gould in full dress uniform, probably taken early in 1914 when he was Assistant Navigator in HMS *King George* V. (© Sarah Stacey and Simon Stacey, 2005)

navigation exams with excellent marks, being awarded 2017/2500, and in January 1914 Lieutenant Gould was officially a fully qualified Navigating Officer. Dodo and William must have been extremely proud; their son had excelled at every stage and clearly had a distinguished future in the Royal Navy. It was at this time that Rupert was photographed, in his Lieutenant's full dress uniform. In all his finery, Lt Gould looks every bit the dashing and handsome young officer.

In February he was posted for a couple of months as Assistant Navigator to HMS *King George V*,[75] the Flagship of Vice Admiral Sir G Warrender, of the 2nd battle squadron, in the Mediterranean at the time. The ship's log contains several entries in Gould's hand as officer of the watch and, from 27 February, his signature as Assistant Navigating Officer countersigning the entries.[76] Having proved himself very capable as a practical navigator, on 4 April 1914 he was posted to the Torpedo Boat Destroyer HMS *Achates*,[77] his first posting as a full Navigating Officer under Commander Walter Allen.

Navigating Officer

The role of Navigating Officer was vital to the ships safety. Working intimately with the ship's Captain, the Navigating Officer gave critical advice on the safe and efficient 'driving' of the ship; the role was a hugely responsible one and could often be very intense and stressful. If, up to now, Rupert's responsibilities had only occasionally been challenging, from now onwards they would totally occupy his mind. At this time *Achates'* duty was patrolling the Irish Sea. With growing international tensions, she was just one of many Royal Navy ships playing an important deterrent and observational role around the coasts of the British Isles.

It was at this point that things suddenly began to go terribly wrong for Lieutenant Gould. What actually happened to Rupert in the first few weeks of his posting in *Achates* is not wholly clear. Unfortunately the ships log of the *Achates* for the period April to June 1914 is missing, but in all probability, during the first part of this period things were going well enough for the new Navigating Officer.

A later annotation in Gould's 1913 diary, dated 23 May 1914, records that the ship was at Ballycastle, off the coast of Northern Ireland.[78] We also know, from a remark he made in one of his later books[79] that on 20 June he had 'the doubtful pleasure', as he described it, of navigating *Achates* at 30 knots (full speed for the ship) through the north end of Jura Sound off the south west coast of Scotland. Those waters are littered with small islands and rocks and this was an extremely risky and stressful exercise at such speed. But it was a mercy mission, as the Hospital Ship *Maine* had run ashore on Lockwood's Island, Mull, and desperately needed relief.[80] The coincidence of such an emergency in his professional life with the Isle of Mull, a place he had always regarded as a personal haven of peace and quiet, was a strange association for him.

When in the Irish Sea the previous evening, the crew of *Achates* sighted a huge basking shark, over thirty feet long, right alongside the ship. Gould commented: 'If our ship's company regarded this encounter as a bad omen, they were not so far wrong'. Ostensibly he was of course alluding to the loss of the *Maine* the following day, but he may well also have been obliquely referring to other, more personal, problems which were soon to reveal themselves.

At that particular time then, it seems *Achates'* Navigating Officer, Lt. Gould, was fit and serving well, even under the considerable stress of an emergency. However, a few days later, towards the end of June, Rupert began to show the signs of mental health problems, the final result of which was to permanently change the course of his life.

World War

Given the worsening international situation at that time, no one in the British forces could have been in any doubt that a crisis of some kind would develop soon. So, when, on Sunday 28 June 1914, Archduke Franz-Ferdinand of Austria-Hungary was assassinated, it was very clear that armed conflict in Europe was escalating and that this time Britain could not stand on the sidelines. Whether or not this single act and its implications were what precipitated the events that followed in Rupert's life, we may never know.

What we do know is that by the start of *Achates'* new ship's log, on 1 July, when Gould was apparently still on board and theoretically still Navigating Officer, he does not appear to be acting in that role. For the whole of July the ship was in and out of the port of Bangor in North Wales, but Gould is making no entries in the log and the Engineer Officer, Leonard Walker, is signing off the notes of the officer of the watch, where one would usually expect to see the Navigating Officer's signature.

It is probable that since the beginning of that month he had been increasingly unwell and was perhaps put on lighter, less stressful, duties. It is clear that Cdr. Allen was generally very impressed by Lt. Gould: if Gould was indeed on lighter duties then Allen was giving him considerable leeway, as the loss of an effective Navigating Officer would have been a very serious matter on a Navy ship on active service.

His personal file indicates that he was then formally relieved of his duty and put on the sick list on 19 July. It is not stated what his ailment was, nor where he was, but it seems almost certain he had remained on the ship as, on 31 July 1914, his personal file records for the first time that he was reported sick in Bangor hospital and that from that day he was officially discharged from his post as Navigating Officer in *Achates*. The Navy lists confirm that on the same day Lieutenant Charles Willis was posted to Achates as the new Navigating Officer in his place.[81]

From Bangor, it would seem that Rupert was sent home to Southsea, remaining on the sick list for the time being. A note on the file then states that on 12 August 1914 his father reported that his son was seriously ill at home with a nervous breakdown. Gould was instructed to attend for survey when fit to travel, and this he evidently did, being placed on the books of Haslar Naval Hospital, at HMS *Victory*.

Haslar, which contained the psychiatric wing for navy staff suffering from mental health problems, was just one part of the 'stone frigate' that was HMS *Victory*. As things turned out, Rupert's illness was so severe that he would remain in hospital right through until the following May 1915.

Sure enough, on 5 August 1914 Britain was at war, the war that it was said 'would be over by Christmas', but which proved so very terrible it would (equally foolishly) be named 'the war to end all wars'. Consideration of this very sad part of Gould's life is potentially fraught with preconceived notions and far-too-rapid judgements of weakness, desertion and cowardice, and one is aware of the vital importance of taking a wholly objective view of the evidence.

In the twenty-first century the terrible psychological effects of battle on the minds of those who fight are becoming much better understood. The history of such illnesses can be traced back through the twentieth century to its beginning where for the first time doctors began to identify specific mental problems associated with military actions in the First World War. 'Shell shock' was the term used to cover a multitude of such psychological damage among Army personnel, and the First World War saw a virtual epidemic of it. That is not to say that such illness has not always been with us as a result of war, but this was the first time it was being given a name and medical treatment, though the psychiatric help servicemen received was basic in the extreme.

The phenomenon of shell shock is primarily associated with psychiatric casualties in the Army and received wisdom has, until recently, been that there was virtually no equivalent phenomenon at sea. Recent research however has revealed that there were significant numbers of these cases, especially later in the First World War, which were diagnosed under the term 'neurasthenia' in an attempt to disassociate them from shell shock.[82] In 1910, in a move to improve the healthcare of its personnel (and ensure servicemen returned to the field of battle sooner, if possible) the Navy added a purpose-built psychiatric unit to Haslar hospital, named 'G Block', comprising two wards of 12 beds and a padded cell. At least 20,000 cases of neurasthenia were recorded during the whole period of the First World War Royal Navy service, the large majority however occurring during the second half of the conflict.

Then there were those for whom the very prospect of battle, the fear itself, became intolerable and as the war progressed a new phenomenon appeared. As the horrific implications of trench warfare became generally understood, an increasing number of men, facing (as they saw it) almost certain death, refused, or were simply too afraid to fight, and chose to flee the battlefield. Those actually fleeing the battlefield were usually soldiers from the lower ranks, because in cases where an officer showed fear at the front line he was usually transferred.[83] Those hapless soldiers who did run, were then labelled 'deserters' and their fate, if caught by their commanding officers, is well known: they were court marshalled and sometimes executed by firing squad.[84]

Thankfully, in more recent times society is making efforts to understand this phenomenon (in 1998 the government agreed to acknowledge soldiers executed for desertion in the First World War as victims, though it stopped short of agreeing to pardon), and people are far less ready to judge without looking closely at the facts of each case.

At sea however (most naval battles began when the ship was not in harbour), desertion during battle appears to have been minimal. Jones and Greenberg cite Meagher: 'It is not improbable that the sailor had a psychological advantage compared with the soldier. The life of the sea habituates him to dangers. He fights on board ship, on his own ground as it were. He cannot fail to recognize that the promptings of the instinct to flee from danger are impossible of fulfilment, and is therefore freed from a mental conflict . . . and, in addition, he probably has the advantage of having experienced many battle practices, and is not the least gun-shy . . .'.[85]

The Case of Lt Gould

The truth is that the case of Lieutenant Gould was unusual and would have been seen at the time as a curious one, possibly even with a primary cause separate from the impending conflict altogether. When, two weeks after Gould's breakdown, war actually broke out, most young men were extremely keen to sign up and, during those first months, the phenomenon of soldiers and sailors reluctant or unable to face battle was not really an issue. Gould's own brother Harry, who was by this time close to taking his finals and qualifying as a surgeon at St Thomas's Hospital, was a good example. So eager was he to 'do his bit', and fearing that the war would indeed be over by Christmas, he joined up and went to France as a dresser, or surgeons assistant.

So Lieutenant Gould's commanding officers were more surprised and perplexed than critical, and as he was a professional (uniformed) sailor, any public opprobrium was avoided too. What was it then that afflicted Gould? In later life he stated that he believed he was slightly schizophrenic, and that once or twice in his life he felt certain he had actually been insane.[86] However, Professor Edgar Jones, Professor of the History of Medicine and Psychiatry, Institute of Psychiatry, Kings College, London, who has kindly considered the known symptoms in this case, has expressed the view that Gould appears not to have been schizophrenic but suffered from mood swings as a result of severe repeating depression. There is also a possibility of mania suggested (in later reports) by delusions and hyperactive behaviour. As for the onset of world war, this would seem to have served as a trigger to his depression, but it would not have been solely responsible for it and may have been partly co-incidental.[87]

Depression

In the case of Gould's first nervous breakdown at this time, the symptoms of depression were apparently severe. This was by far the worst of the four breakdowns he would experience during his lifetime and, from July 1914, he was said to have had a severe memory loss and to have been totally unable to speak for 6 months. His file records the observations of his condition while under treatment in G Block at Haslar: '29 August, Placed on books of *Victory* for (another) 50 days (91 in all from 19 July)'. Then, in October his personal file records two brief reports, both simply dated 10/14, each apparently made by officers who knew Gould and had been asked for a general assessment of him prior to the illness. The comments fall under two nominal headings: 'General Conduct' and 'Ability'.

The first officer, Captain Cameron, noted under 'General Conduct' that Gould was a 'Very good, effective officer' and under 'Ability': 'Very capable navigator. Zeal and aptitude for study. Capable pilot with more experience'.

The second report was added by his old boss, Commander Walter Allen, who had been with him on *Achates* when he first fell ill. Allen's notes were more telling. Under 'General Conduct' he simply noted: 'Satisfactory', which in this context was a decidedly lukewarm rating: any self-respecting officer would be horrified to be given such a comment. But given Gould's failing health on *Achates*, Allen could hardly have been more enthusiastic. However, under 'Ability' Allen enthused: 'Exceptional. Black and White artist of considerable ability. Unusually capable & talented. At outbreak of war was left behind in hospital which has preyed on his mind & seriously impaired his health for the time being'. Commander Allen's remarks are compassionate and understanding, perhaps suggesting that Gould's talents as an individual were his strength while, naturally enough, reserving judgement on his general conduct.

The personal file notes succinctly that on: '18 October 1914 continued on books of *Victory* (221 days)'. And on: '25 February 1915 Resurvey, unfit; Continued on *Victory* (310 days in all)'. Sadly, no further details are given, but Cecil recalled that his father was being treated by a specialist doctor, Sir Thomas Morley, who was eventually responsible for Rupert's recovery.[88] On occasions, patients in the ward were taken for trips into the country as part of the cure, and Cecil remembered in later years being told that the first signs of improvement in his father's mental health had come when, on seeing a pair of rabbits tumbling down a bank, he laughed—the first time he had uttered a sound in 6 months. Over the following few weeks Rupert gradually came out of his shell and by the spring showed signs of definite recovery.

Sadly, any sense of relief the Gould family might have felt on Rupert's improvement, was short-lived, as in April 1915 they received some terrible news. Rupert's brother Harry, who had only been in France for a few months, had developed Tuberculosis and died soon after returning to England (Dodo herself had gone over to Rouen to fetch him) on 5 April 1915. Harry had been Rupert's dearest friend and 'soul mate' during his childhood and this was a devastating blow. The first of all Gould's books, a little work on the seventeenth-century astronomer Jeremiah Horrox, published in 1923, was dedicated 'To the Memory of H.H.M.G. who, like Horrox, was beloved by the Gods'. One can only try to imagine what Dodo and William must have been feeling at this time.

Rupert's file continued: '25 May 1915 Unfit . . . Continued on *Victory* (making 365 days in all) the maximum allowed by regulations'. At this point he was sent home to Southsea, on half pay, in the hopes of his recuperating. The writer and journalist Ralph Straus (1882–1950), who became a very close friend of Rupert's in later years (and had a significant part to play in his changing fortunes), noted that during the war he had looked after men suffering from shell shock.[89] It's tempting to wonder if he met Rupert this way, during this period. Gould's personal file continues: '25 August 1915 Resurvey at Admiralty . . . Unfit, resurvey 26 October'.

By the summer of 1915 Gould was sufficiently recovered to be taken by Dodo to the spa at Harrogate to continue his convalescence. A photograph taken at the time shows Gould in the company of Basil Hallam, then a famous music hall star. Captain Basil Hallam Radford (1889–1916) had schooled at Charterhouse and, being the same age, would have been in the same year as Harry Gould. It's probable Basil had joined Rupert and Dodo in connection with Harry's death a few months before. Meeting a friend of Harry's may well have helped the mother and brother come to terms with their loss and would have been good 'social therapy' to help Rupert re-engage with society. Sadly, a few months later Basil Hallam Radford was another of the unfortunates to be killed in action in France.[90]

Engagement

That summer in Harrogate a rather beautiful young lady by the name of Muriel Estall was also visiting the Spa town with her mother. Every summer it was the custom for Emily Estall to holiday at a succession of English spas, taking her daughter with her as companion. Muriel, always the most 'modern' and fashionable of young women, enjoyed tennis, and it was in Harrogate that summer that she met, and fell in love with Rupert.

Emily Tilley and Thomas Estall had married in 1877 and Muriel, the last of four children, was born in 1894, when her mother was 40 years old. Thomas Estall, from a family of Huguenot descent (originally De Stael), had by this time attained the post of Senior General Manager at the National Provincial Bank. The Estalls, who lived in Evelyn Gardens in Chelsea, were thus very comfortably off and Muriel had had a privileged childhood, including a period at finishing school in Hanover (where she naturally also learned to speak German).

me,   L' Gould.   —   Basil Hallam.
Harrogate -1915.

10. Rupert and Muriel's first meeting in the summer of 1915. Tennis was an absolute passion of Rupert's. (© Sarah Stacey and Simon Stacey, 2005)

Neither family, however, was particularly keen on this match. Both sides regarded the couple as wholly unsuited, with Muriel charming and intuitive but quixotic, and Rupert, while more intelligent, also pedantic and tending towards thoughtlessness, and of course, at that period, psychologically disturbed. According to Cecil, the only things his parents really had in common (apart from enjoying tennis) were their good looks and a good sense of humour.

At the time Muriel's mother was trying hard to push her daughter, unwillingly, into the arms of another suitor, G.C.H. Matthey, later a noted clock collector and heir to the Johnson-Matthey bullion dealing fortune.[91] But Rupert and Muriel were genuinely attracted to one another and Rupert thus appeared at just the right time to prevent a proposal from Matthey. Following the holiday in Harrogate, Muriel was invited to spend some time with the Goulds in Scotland later that Summer, and Rupert appears again in Muriel's photograph album, looking decidedly pensive, sitting in his water garden at Druimgigha.

The courtship was a long one however, probably because Rupert was still unwell, and the official engagement and marriage did not actually take place until 2 years later, when Rupert's mental health was more stable.

Rupert. 1915

11. Rupert at Druimgigha, the Goulds' summer home on Mull, in 1915. At the time, he was building the water gardens in which he sits. (© Sarah Stacey and Simon Stacey, 2005)

In the autumn of 1915, Gould's file records finally: '26 October . . . Found Unfit for further service . . . Placed on retired list (Physically unfit)'. He was thus officially retired from active service and remained on half pay. On 28 April 1916 Gould wrote to the Admiralty: his file notes: 'letter received, states this officer is now [fit] & requests survey with a view to further employment' which resulted, on 3 May 1916, in the entry 'Ordered survey MDG [Medical Director General] in the morning any weekday'. On 7 May he was 'Found fit by MDG for light duty on shore in England', but 4 days later 'Officer reports that he is suffering from recurrence of nerve trouble'.

As if things weren't bad enough, a few weeks later there was further devastating news for the family. On Sunday 16 July 1916, Rupert's cousin Douglas Skinner, 24 years old and a Lieutenant in the West Kent Regiment fighting in France, was killed in action. First Rupert's own brother, his brother's friend and now his cousin; as with just about every family in Britain at the time, the Goulds must have wondered when the carnage would stop. And Rupert's own feelings of guilt, which had been preying so much on his mind since his hospitalization at Bangor, almost exactly a year before, must now have been acute.

There was however a little good news that week too. Rupert was finally found a job, and one that would suit him very well indeed. On 20 July 1916 Lieutenant Gould was posted as one of six Naval Assistants in Hydrographer's Department at the Admiralty in Whitehall. The employment was for a period of 3 years, not to be exceeded without reappointment and he returned to full pay as long as he was in the post. In this new job Rupert was working for the Hydrographer, Admiral Sir J.F. Parry, whom he later described as 'a good friend to me', and who was evidently very understanding at this difficult time.[92]

A New Start

Rupert Gould's successful convalescence was far from meaning a return to life before the events on *Achates*; it was a whole new, and unexpectedly optimistic, beginning for him. Considering the serious consequences of his illness, things had turned out remarkably well. He had emerged from a devastating breakdown with real prospects, a fiancée and an interesting and reasonably fulfilling job. The breakdown, and the recovery from it, was thus a key turning point in his life: perhaps his real transition into adulthood.

As for his new professional role, very sensibly, the Hydrographer decided to employ Gould in areas where his interests and expertise were greatest. Historical research had always fascinated Rupert, and he became the specialist in the Historical Section. A couple of the early tasks he took on were extensive research on the charting of islands and coastlines in the Antarctic regions, and a history of the Hydrographic Department itself. The Antarctic research continued for some years and was a considerable contribution to the Admiralty's work. The departmental history unfortunately proceeded only as far as a 'Sketch Synopsis' and two draft chapters before Gould realized that the task was simply too huge.[93] In spite of being urged to finish it by his ex-boss, Admiral Parry in 1922 and 1923, with other work always taking up more of his time, the project was put on hold, and never advanced further.[94]

The other, more pressing work which intervened included tasks such as the revision of old charts and sailing directions, studying historical papers relating to British and foreign claims to sovereignty, and acting as specialist secretary and representative at Hydrographic meetings and conferences.

Since childhood Gould had been taking an increasing interest in the history of horology, and one project overseen by his department in 1917 would have particularly interested him. This was the subject of Timekeeping at Sea, that is, the actual 'time of day' recognized on board ship and shown on the ships clocks—as distinct from the marine chronometers, which naturally were always kept at GMT.

Before 1916, ocean-going ships the world over did not generally recognize the 24 hour time zone system, which was by that time gradually being adopted on land by most countries (following the 1884 International Meridian Conference in Washington). Both Navy and Merchant service vessels would set their on-board ships-clocks to the exact local time, in hours and minutes, at their position at that moment, the clocks not then being set again for another day at least. This resulted in a great deal of confusion about exact timings of messages, occurrences or accidents at sea observed by different ships, especially those approaching each other from different east/west directions. The loss of the *Titanic* in 1912 was a good example; even today there is some uncertainty as to the exact timing of some of the stages in that terrible disaster, as all the vessels concerned had their clocks set to slightly different times.

In 1916, the French Navy had very sensibly reformed its practice to ensure that its ships would only change on-board clocks when crossing one of the 24-hr time zones, always using same minutes (based on GMT) and simply changing the hour hand. This prompted the British Admiralty to follow suit, and the 'Conference on Time-Keeping at Sea' was held on 21 June 1917, to discuss and coordinate this adoption among those with British interests.[95] The Hydrographer of the Navy, Admiral Parry, chaired the meeting, and Gould's colleague and friend Lt. Cdr Archibald C. Bell represented the Hydrographic Department. Although he would have been too junior to actually take part, Lieutenant Gould would have been active behind the scenes, preparing the papers and perhaps taking minutes.

All of this work would have required a careful, even perhaps a pedantic, emphasis on getting the information right, both in the broader concepts and in the detail, and Lt. Gould was exactly the man they needed. The work was varied and increasingly relevant to Rupert's growing interest in marine chronometry; no wonder then, that in the first few years he was much appreciated for his service in the Department.

Polar Research

Gould's work in the Department soon established him as the unrivalled expert in polar cartography. In November 1919 he discovered a hitherto unknown manuscript by the English navigator, William Smith, master of the brig *Williams*, which established Smith as the first man to discover dry land in the Antarctic (in 1819). The results of this research were summarised by Gould in his Hydrographic Office file titled *The Discovery of the South Shetlands Islands*.[96] The continuation of this work resulted in Gould's 'Report on The Antarctic Charts Published by the Hydrographic Department, February, 1920'.[97] all carefully typed and finely illustrated to incorporate the known charts from other nations, and noting the accuracy and errors in all, including some charted islands which ultimately proved never to have existed as terra firma at all. This work was to be put to great practical use the following year, when it was summarized as 'Information . . . Compiled in the Hydrographic Department . . . for the use of the Shackleton-Rowett Expedition in the forthcoming Antarctic explorations in the S.S. *Quest* (Sept. 1921)'.[98] This publication also served as the basis of an even more politically important conference that Gould would be present at in 1926 (see p. 173).

The cartography of the northern Polar Regions also interested Gould and he was given responsibility for the revision of charts in the Canadian arctic region as well. Here, in the mid-1840s, Sir John Franklin's ill-fated expedition in the ships *Erebus* and *Terror*, sought to discover the fabled Northwest Passage to Asia, but never returned. The tragic end of this voyage of exploration, with all lives lost, remained a great mystery and drew considerable attention in the popular press over the following decades.

Admiralty Chart No. 5101, beautifully drawn by Gould and published in 1927 (and only withdrawn about 1970), incorporated his extensive research on the subject, including data on the several places where the relics of Franklin's expedition were found. In fact, Gould drew this chart very hastily as a result of a last minute visit by Major Lauchie T. Burwash of the Northwest Territories and Yukon Branch of Canada's Department of the Interior. Burwash was planning an aerial survey of the Canadian Arctic coastline and King William Island, partly to search for Franklin's ships. As well as being a very rushed job Gould also faced problems with the Hydrographic Office's own professional cartographers, whose toes he had firmly trodden on by doing all the drawing of this chart himself. Feeling they had been sidelined, they insisted he not be credited on the chart with 'drawn by', but instead demanded that his name accompany the credit 'compiled by', a demand he was obliged to accept.[99] Later Gould would lecture and write on many occasions on polar exploration, and the subject remained a passion all his life.

In his work in the department's historical branch, Gould was often required to consider masses of detailed information, much of it confusing, and some of it occasionally contradictory. He was then expected to present a summary in an easily digestible form to enable important judgements to be made fairly and safely. One such project that was right up Gould's street was researching the history of that most inscrutable of Hydrographic datums, the International Dateline. In 1921 it was proposed that the Pacific island of Samoa should transfer to Eastern Time, involving a change to the position of the dateline. This required extensive briefing notes by Gould for the Hydrographer to use in discussions with the Governor of New Zealand. The resulting work was entitled 'Notes on the History of the Date or Calendar Line in the Pacific Ocean'.[100]

By drawing on his strengths, using his fiercely logical approach and his careful, practical abilities in the specialized craft of redrawing charts, Lt Gould was beginning to reclaim and reform a successful Navy career. Although retired from active service, he still had his rank and title and his work was much appreciated, at least in the first few years of his service, in the Hydrographic Department.

The Wedding

At the same time as his professional life was flourishing, so things continued to bode well personally and, over a year after their first meeting, he and Muriel were still very much in love. With both his work and Muriel in London, Rupert had taken lodgings in town and he moved to 37 Bury St in the heart of fashionable St James's, an area in the centre of the West End particularly favoured by young bachelors.

As has been noted, both the Estalls and the Goulds were uncertain about the suitability of the marriage, but in the autumn of 1916, with his

12. The newlyweds in 1917. The first few years of marriage were to be very happy. (© Sarah Stacey and Simon Stacey, 2005)

13. The young couple in Dodo's Model T Ford. Muriel enjoyed driving as much as Rupert. (© Sarah Stacey and Simon Stacey, 2005)

health much improved, Rupert decided the time had come to propose to Muriel. He was extremely nervous about getting it right however. Later he recalled that the night before he intended proposing to Muriel, he was dining with his old school friend H.C. Arnold-Forster, who on seeing his condition exclaimed: 'Good heavens old man, you look like nothing on earth—have a brandy & soda!'.[101]

Rupert and Muriel's engagement and marriage were duly announced in *The Times*[102] and the marriage took place on 9 June 1917[103] at the Estall's parish church, St Peter's in Cranley Gardens, Chelsea.[104] One of the larger wedding presents was from Rupert's father who bought the newly-weds a gramophone. According to Cecil the model was 'the best that was then available'.

The couple had wanted to honeymoon in Paris, but decided that, until peace was declared, they would defer a proper honeymoon and settle for a short holiday instead. So, immediately following the wedding, they headed off to Newquay, on the north coast of Cornwall. This compromise allowed them to spend the summer in Scotland with Rupert's family as usual.

The first few years of married life were in fact very happy, but things did not get off to an entirely joyous start. Although Rupert's mental health was much improved by this time, he never did conquer all of his fears. Once they were settled back in wartime London, just at a time when Zeppelin raids were on the increase, Muriel was quickly disappointed to find that her husband was as frightened as she was. It was only now they were living together that she really began to discover how complex Rupert was.

Cecil Estall

A much worse blow for Muriel at this time however came in August 1917 when the family received the news that Muriel's brother Cecil Estall had been killed in Flanders, at the Paschendaele offensive. Yet another death in the family owing to the war and a further burden of guilt for Rupert to bear. Muriel was inconsolable and her parents were totally devastated. According to Rupert's son Cecil (named in memory of his late uncle) Cecil Estall's loss led directly to the early death of his father Thomas Estall, just 3 years later, and was the cause of his mother, Cecil's grandmother, 'going right round the bend'. And so it was that the period at Druimgigha that September turned out not to be a holiday at all, but more a kind of mental recuperation and 'coming to terms' for Muriel and Rupert.

On returning to London in the Autumn of 1917, the newlyweds settled into their first home and as has been said, in spite of Muriel's occasional doubts about her husband's behaviour, the first 4 or 5 years of married life for the Gould's were in fact very happy and fulfilling.

On 5 April 1918, Rupert's great aunt, Jane Hilton died. The net value of her estate was £58,535; a substantial legacy, the equivalent of well over £1m in today's money.[105] Dodo and Rupert were among the many

14. The in-laws, Emily and Thomas Estall, with Muriel's brother Cecil. Cecil's death at Paschendaele in 1917 was a devastating blow for them all. (© Sarah Stacey and Simon Stacey, 2005)

beneficiaries and this eased what had been a rather tight financial situation and enabled the young newlyweds to live life to the full for the first few years of the 'roaring' 1920s.

Campden Hill Court

Home was a large flat, No.12 Campden Hill Court, in Kensington, a very select address, where they would live for just over two years. It was here that, on 24 May 1918, their first child was born, Cecil Hilton Monk Gould. The boy was ostensibly looked after by his mother but, as was

15. The beautiful bookplate, shown slightly over full size, that Rupert drew for his mother Dodo in 1918, when her grandson Cecil was born. Amongst the incredibly fine detail are references to the whole family. Dodo's husband, represented by the gramophone, takes a back seat. (© Sarah Stacey and Simon Stacey, 2005)

normal practice at the time in such families, was mostly cared for by the
family's nanny. Alice Bullard, who was employed immediately after
Cecil was born, would be his constant companion for the next few years.

Of the in-laws, it was Rupert's side that had more regular contact
with the young family. Rupert and Muriel would often travel down to
Southsea to stay at Bedworth House with William and Dodo. There were
visits to the Estalls, but these were fraught with difficulty, Muriel's mother
being rude and unkind to Nanny Bullard who ever after loathed her as
much as she liked Dodo. Once Muriel's father died, in 1920, Emily Estall
slipped increasingly into a kind of malign eccentricity. She sold the house in
Evelyn Gardens and proceeded to live a somewhat nomadic existence.
Moving from hotel to hotel, she kept in touch with her daughter, sometimes still managing to be a baleful influence on the young Gould family.[106]

A Fresh Start

The end of the war in November 1918 brought a fresh start for the
young Goulds. Early in 1919 the couple decided to indulge in a motorcar
of their own and bought a smart American two-seater roadster by

16. Muriel and Rupert with Cecil, outside Bedworth House in April 1919.
Regine, the Scripps-Booth motor car had recently been acquired. (© Sarah
Stacey and Simon Stacey, 2005)

Scripps-Booth, which they nicknamed *Regine*.[107] Muriel drove the car as much as Rupert and her photo album includes many pictures she took of the car at Southsea in April 1919.

It was a happy time and things were looking up for the Goulds. Baby Cecil was healthy and growing fast. Muriel was now expecting their second child. Rupert was doing well at the Hydrographic Office, and his job seemed secure. The couple's social life was full and exciting too, and now the war was over Rupert's health had very much improved. During these first few years of marriage the charming and beautiful Muriel and the erudite and urbane Rupert were a popular couple and socially much in demand. Cecil recalled friends of his parents remarking on what a beautiful pair they had been: 'Though my father was strictly the better looking of the two, it would have been my mother who would have drawn most eyes when they came into a room. My aunt once told me that people had stood on chairs in hotels to look at her'.[108]

On 14 October 1919, Lieutenant Gould received the standard promotion to Lieutenant Commander, in this case with the qualification 'retired', as he was no longer on active service. His new boss in the Hydrographic Office, Vice-Admiral F.C. Learmouth, was very happy with his performance, and granted him another 3 years employment as assistant in the department. In addition, with effect from 31 December 1919, Gould was to receive full pay and allowances of rank, plus a 15 per cent bonus.

Although Rupert enjoyed his work at the Admiralty, much of it involved little more than administrative work for the Department and at this time he was beginning to search for further interests to stimulate and occupy his mind. As mentioned in Chapter 1, when Rupert was a child he had read his Uncle Hilton's copy of Britten's *Old Clocks and Watches*, and he had been developing an interest in horology ever since. Britten's book contains a brief description of John Harrison's pioneering work on that indispensable kind of timekeeper that was the Navigator's own, the marine chronometer. Britten's narrative, though sketchy and incomplete, greatly impressed Rupert and in 1919, for his own interest, he began to gather historical information specifically on chronometers. After all, with the Admiralty library and archives outside his office door, he would not have to go far for his sources. Additionally, when spare money allowed (they were relatively inexpensive by today's standards), he would occasionally buy an antique chronometer with the intention of slowly forming a collection, to be the focus of his interest.

He never had enough funds to acquire anything more than a handful of them however.

Meanwhile, as there seemed little prospect of further advancement in the Hydrographic Office beyond his current rank of Lieutenant Commander, Gould's career seemed to be stalling. With this in mind, and with a certain amount of persuasion from all four parents, Rupert and Muriel agreed at this time that he should begin to look elsewhere for a 'proper' career. After much discussion it was settled that he would study for the Bar.[109] They decided that his phenomenal memory and rapid powers of deduction, combined with an eloquent turn of phrase, ideally suited him for the courtroom. He promised he would soon begin study, in his own time, while for the moment continuing his work at the Admiralty.

Waylands

The arrival of the Goulds' second child marked another happy event. On 27 February 1920 Jocelyne Muriel was born and, as Muriel now considered the flat in Campden Hill Court too small, her father bought them a lease on a much larger house. 'Waylands' (named after the Skinners' house in Beckenham), was a large and pretty mid-Victorian villa at 46 Kensington Park Road[110] where they would live happily for two years. It was here that two of the extraordinary timekeepers by John Harrison would soon be subject to Rupert's careful attention.

The purchase of the house was one of the last things Muriel's father would do for his daughter and son-in-law, as Thomas Estall died that summer. Rupert noted in a letter to George C. Williamson in early August 1920 that he was busy travelling around in the Scripps-Booth, dealing with the administration: 'family bereavements have caused me to chase all over England in a small car . . .'.[111] Williamson was a prominent member of a dining club that Rupert had just joined, named The Sette of Odd Volumes. This letter was the first of many Rupert would write to members of the club in years to come.

In February 1921, Rupert's personal file records he was 'granted permission to proceed abroad on 1 March 1921'. This, at last, was the couple's delayed honeymoon to Paris and, leaving the two children in the capable hands of Nanny Bullard, the Goulds took the boat to France to reaffirm the happy relationship.

❧ 4 ❧

John Harrison and the Marine Chronometer

As noted in the last chapter, after 3 years working under the Hydrographer, and being asked to undertake less stimulating administrative work, Rupert had agreed with his family to seek a new career in the legal profession. At about the same time however, he began to take an increasing interest in a subject that was very closely associated with the work in the Hydrographic Office. The subject was of course the marine chronometer, an instrument that, over the previous century, navigators had increasingly relied upon for safe position-finding at sea. Important as the marine chronometer was, its history had never been properly told and, realising he was ideally suited to research and write such a history, Rupert decided he would undertake this project, chiefly in his own time, before leaving the Department.

Rupert's fascination for clocks and watches had been just one facet of his wider interest in all things technical, but now, as his attention turned to the marine chronometer during the early 1920s, it evolved into something much more serious. Gould began what would prove to be the most significant activity of his working life, a fully academic involvement in the world of antiquarian horology. If there was a single achievement of Gould's which posterity might say justified his existence, it would be his contribution in that field, the esoteric world of clocks, watches, sundials and all things relating to time keeping and time telling.

Specifically, two achievements single him out: the writing of the book *The Marine Chronometer, its History and Development* (1923)[112] and the restoration of the great eighteenth century marine timekeepers by John Harrison. A truly Herculean labour, this latter work occupied his time, intermittently, for nearly half his life and was undoubtedly responsible for shaping its course. It could even be argued that his work on these machines was directly responsible for the disastrous events that would nearly destroy him in the years to come. What is it then, about such a subject that is so captivating for some people?

Horology

For an intelligent and enquiring mind, the subject of horology, in its widest sense, is likely to prove, at the very least, a fascination, as there is scarcely a walk of life in which the subject of timekeeping doesn't play a significant, if not a central, role.

The time-centred structures of the natural and civilized world itself, the daily and seasonal routines of life that are the divisions of the day and calendar, are all naturally horological subjects. From a scientific perspective, the theory and practice of horology touches on all forms of scientific endeavour—it has always been especially important in the histories of astronomy and navigation. And as practitioners of one of the first engineering technologies, clockmakers played a vital role in the origins of the Industrial Revolution in the eighteenth century. For anyone with an interest in the history of science, and a fascination with technological evolution, horology has an immense amount to offer. Even today, learning about the science and craft of clock and watch-making introduces many of the fundamentals of physics and mathematics, while to study antiquarian horology is to study the history of furniture and jewellery design. Then again, on a more lateral plane, there can scarcely be a philosopher in history who has not pondered the meaning and nature of time itself. It is perhaps not surprising then that such a broad and rewarding interest would captivate a man like Rupert Gould.

For the real aficionados, the pinnacle of the horologist's art is the marine chronometer. Primarily an instrument of navigation, it has always been the finest made and most accurate of all portable timekeepers and can claim to be the only kind of timekeeper that has actually saved lives—certainly many thousands of them over the years. It is because of this vital role the chronometer played, that the very best scientific and horological minds were directed into their design and construction. The marine chronometer consistently stood among the most advanced pieces of horological technology: it was the 'aristocrat' in the horological pecking order.

The full story of how this exquisitely beautiful instrument originated is too voluminous to be told here; besides, a certain book just referred to tells the story far better than this author ever could. But for the reader without specialist knowledge, it would be helpful to give a short explanation of how and why such a thing as a marine chronometer came into being. A brief description of the Harrison timekeepers is also necessary in order to understand Gould's contribution (there is also a small glossary of terms in Appendix 6 which may be useful as an *aide memoire* when reading the horological parts of this book.)

Longitude

Up until the middle of the eighteenth century, navigators had been unable to determine their position at sea with accuracy and they faced the huge attendant risks of shipwreck or running out of supplies before reaching their destination. It was, as Dava Sobel has described it: 'the greatest scientific problem of the age'.

Knowing one's position on the earth requires two very simple but essential coordinates: rather like using a street map where one thinks in terms of how far one is up/down (one's Latitude) and how far side to side (one's Longitude). The latitude, how far north or south of the equator one is, is relatively easy to find by the height of the Sun at midday or (in the northern hemisphere) by the height of the pole star; sailors had been finding their latitude at sea for centuries. The longitude is a measure of how far *around* the world one has come from home and has no naturally occurring base line like the equator. The crew of a given ship was naturally only concerned with how far round they were from their own particular home base.

Even when in the middle of the ocean, with no land in sight, knowing this longitude position is very simple . . . in theory. The key to knowing how far around the world you are from home is to know, at that very moment, what time it is back home. A comparison with your local time (easily found by checking the position of the Sun) will then tell you the time difference between you and home, and thus how far round the Earth you are from home. The Earth can be divided up, like the segments of an orange, into 24 one-hour time zones, the 24 hours making up the whole 360 degrees round the earth and each hour's time difference equivalent to 15 degrees of longitude.

The great flaw in this 'simple' theory was 'how does the sailor know time back home when he is in the middle of an ocean?' The obvious, and again simple answer is that he takes an accurate clock with him, which he sets to home time before leaving. All he has to do is keep it wound up and running, and he must never reset the hands throughout the voyage.

This clock then provides 'home time', so if, for example, it is midday on board your ship and your 'home time' clock says that at that same moment it is midnight at home, you know immediately there is a twelve hour time-difference and you must be exactly round the other side of the world, 180 degrees of longitude from home.

The principle is indeed simple, but the reality was that in the eighteenth century no one had ever made a clock that could suffer the great rolling and pitching of a ship and the large changes in temperature while still keeping time accurately enough to be of any use. Indeed, most of the scientific community thought such a clock an impossibility. Even the great Sir Isaac Newton considered it so. But the stakes were high, especially after 1714 when the British government offered the huge sum of £20,000 for a solution to the problem, with the prize to be administered by the splendidly titled *Board of Longitude*.

John Harrison

It was this prize, worth about £2 million today, which inspired the self-taught Yorkshire carpenter, John Harrison (1693–1776), to attempt a design for a practical marine clock. History relates that, after a life dedicated to achieving this seemingly impossible task, resulting in the creation of an extraordinary series of five prototype timekeepers, Harrison succeeded in his goal. With the work of a small band of horological pioneers following in his footsteps, the practical marine chronometer became a reality. From the early years of the nineteenth century and through the following century and a half, chronometers served in regular use aboard Navy ships and merchant vessels alike.

It was these fantastically complex and abstruse timekeepers of Harrison's that, after a century and a half of neglect, Gould restored to their former glory during the 1920s and 1930s. In order to convey the magnitude of what Gould achieved with the restoration of these machines, it will be helpful to explain a little of Harrison's development work, and the nature of the timekeepers themselves.

The Government prize of £20,000 was the highest of three sums on offer for varying degrees of accuracy, the full prize only payable for a method that could find the longitude at sea within half a degree. If the solution was to be by timekeeper (and there were other methods since

17. John Harrison (1693–1776) in the engraving by Tassaert (c.1768) after the portrait by Thomas King. Harrison's extraordinary series of five prototype marine timekeepers led to the solution to the longitude problem, and the fourth, H4, was also the first of all precision watches. (© National Maritime Museum, London)

the prize was offered for any solution to the problem), then the timekeeping required to achieve this goal would have to be within 2.8 s a day, a performance considered impossible for any clock at sea and unthinkable for a watch, even under the very best conditions.

At the time, the only precision timekeepers, of any kind, were pendulum clocks. In the 1720s Harrison himself was making such clocks, which he claimed were capable of maintaining an accuracy better than one second in a month, in spite of the fact that they were mostly made of wood. To achieve this high precision, Harrison incorporated several extremely ingenious new ideas, including a mechanism to automatically compensate for the effects of temperature. All clocks and watches, rather like the people who use them, tend to go slow when they experience a rise in temperature. Harrison invented a special form of compensated pendulum, using a grid of brass and steel wires, to ensure his clock kept time, whatever the temperature.

He also designed his clocks to run without the need for any oil, the 'Achilles heel' of clockwork: in the eighteenth century clock oil was derived from animal fat and often quickly deteriorated into a kind of acidic glue. By designing and incorporating bearings that used rolling contact, instead of sliding contact, Harrison's anti-friction bearings cleverly side stepped this problem. No one before Harrison had ever made a mechanical clock to work without oil, and very few have done so since.

Watches on the other hand, were universally dismissed, being seen as jewellery, and not as serious timekeepers. Even the very best pocket watches of the day could only keep time to within about a minute a day and their timekeeping was generally thought of as impossible to improve. So clocks looked like the logical instrument to develop, but a pendulum clock would be of no use at sea, owing to the ship's motion. So Harrison decided to create something based on his precision long case clocks, but made to withstand movement and wide temperature changes.

And so it was that in 1728 John Harrison began to design a series of 'sea clocks', as he called them, which were to become the most celebrated and arguably the most important timekeeping devices ever constructed in the history of mankind. These were the machines that led Harrison to prove, in the face of universal skepticism, that a marine timekeeper was a practical possibility. Harrison's machines led directly to the solution to the longitude problem, immeasurably strengthening the British Royal and merchant navies and saving of countless lives at sea over the following two centuries.

Harrison eventually built five timekeepers, the last two in the form of large watches, and since the 1950s they have generally been referred to as 'HI' to 'H5'. It should be noted that the term 'marine chronometer' was not widely used until after Harrison's death. The word 'timekeeper', however, had very special significance in the eighteenth century. It was only used to describe a portable machine capable of high accuracy.

18. H1. Created between 1730 and 1735, Harrison's first marine timekeeper was intended as a portable version of his precision longcase clock design. It ran without any lubrication, had automatic temperature compensation and was a highly sophisticated but complex mechanism (see plate 4). (D6783 © National Maritime Museum, London)

Harrison's First Marine Timekeeper H1

Built between 1730 and 1735, HI is essentially a portable version of Harrison's precision wooden clocks, except that it runs for only one day (the longcase clocks run for eight) and is considerably larger than the wooden movements, weighing 34 kg and standing 63 cm high. After completion this timekeeper was taken to London where it amazed and excited the scientific community. It was sent on a semi-official trial to Lisbon in 1736 and though its going on the outward journey appears to have been rather poor, on the homeward course it corrected the estimated longitude by some 60 miles and was to impress both the ship's Master and the Board of Longitude. Although not a prize winner, H1 can be considered as the first workable marine timekeeper ever constructed.

Here it would be helpful to mention briefly, in simple terms, the names of the separate parts that go to make up the insides of a clock, the mechanism of which is generally known as the 'movement' (Box 4.1). H1 can be used to demonstrate these parts (See also the Glossary in Appendix 6 for explanations of the terms).

BOX 4.1 The anatomy of a clock mechanism

The Frame

The frame holds all the working parts of the timekeeper together and in H1 this is made in brass, consisting of a series of plates, held apart by pillars.

The Power Source

At the 'bottom end' of a clock movement is the power source, usually a weight or a coiled spring, which is the element that drives the clock. H1 is driven by two mainsprings connected with chains to a 'fusee', a conical pulley invariably found in English spring driven clocks and which ensures a uniform driving force to the wheels of the clock.

Maintaining Power

While the clock is being wound up however, this source of energy is removed from driving the mechanism and so, to ensure the clock continues to run and doesn't lose any time, it is necessary to add a special mechanism called 'maintaining power'. The maintaining power Harrison designed for H1 was totally automatic and this design went on to be used in virtually all chronometers made since.

The Train of Wheels

The wheels in the movement, which feed the energy to the timekeeping element and supply the correct motion to the hands and dial of the clock, are known collectively as the 'train'. Most of the wheels consist of an axle, the 'arbor', with a large toothed wheel and a smaller toothed 'pinion' mounted on it, the wheels meshing with the next pinion in the train. In H1, Harrison made the wheels out of oak, just as he had in his earlier long-case clocks, and made the pinions in the form of little rollers, using *lignum vitae*, a naturally greasy hardwood (*Guaiacum Officinale*), each roller running on a fixed brass pin through its centre.

Escapement and Oscillator

Finally there is the all-important 'escapement and oscillator', a grand-sounding name for the most important part, the 'beating heart' of the movement, the bit that actually measures out the time. It was to be improvements in the escapement and oscillator, made by Harrison and a number of other clock and watchmakers before and after him, which created the 'modem' mechanical watch and clock of the twentieth century. The most common oscillator in a clock is the pendulum and the usual one in a watch is a little oscillating wheel called the 'balance'.

The escapement is the part which feeds in the energy, in the form of little pushes, or 'impulses', to keep the oscillator swinging, whether it be pendulum or balance. Common longcase clocks of the day used one known as the 'anchor escapement', so called because the 'pallets', the parts which receive the impulse from the wheels of the clock and transfer it to the oscillator, look like an inverted anchor.

In the 1720s, when Harrison was beginning to develop his precision timekeepers, all traditional longcase clocks needed regular oiling to work properly, because the anchor escapement they used works with a sliding action and requires lubrication. So, for his precision long case clocks, Harrison invented a new type known as the 'grasshopper escapement'. In this device, the pallets give impulses to the pendulum with direct pushes and without any sliding (and therefore do not need oil) 'kicking' back to their normal position after each impulse, the action looking rather like the legs of a grasshopper.

In H1 'the oscillator' is in fact made up of *two* interlinked bar-balances. These are not wheels but are formed like dumb-bells, connected across their centres with crossed ribbons—a type of frictionless gearing between them. This ensured that any motion affecting one balance was counteracted by that same effect on the other balance. The balances are also linked with steel balance springs, of helical form, at top and bottom, providing H1 with what Harrison called 'artificial gravity' (the 'restoring force', a bit like the effect of gravity on a swinging pendulum, causing it to keep returning to the centre). With this artificial gravity, when the balances were swung apart they would begin to oscillate together, each swing taking one second.

Harrison used the oil-less grasshopper escapement in H1 and in his next two, H2 and H3. The main moving parts in all three timekeepers

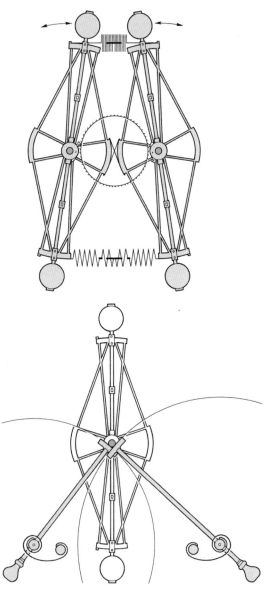

19. H1's balances. The springs connecting them at top and bottom provided what Harrison called his 'artificial gravity'. The wheel in the centre is the grasshopper escape wheel which drives them. The lower drawing shows the pairs of anti-friction rolls upon which each balance runs (see plate 5). (David Penney (www.antiquewatchstore.com). Copyright)

are mounted on anti-friction rollers so they run entirely without oil. Additionally, all the parts were counterbalanced and controlled by springs so that the machines are entirely independent of gravity, an essential for any marine timekeeper.

Harrison also adapted the gridiron principle from his pendulum in order to compensate the timekeepers for changes in temperature. (In the balance-controlled timekeepers, a rise in temperature not only causes the balances to get larger, but also the balance *springs* to become *weaker*, making the clock's tendency to lose even greater). In H1, this compensation mechanism was originally designed to be incorporated into the balances themselves (theoretically the better solution), but had to be altered and put in the main frame of the clock by Harrison, as he couldn't get the earlier design to be reliable.

H2

Encouraged by H1's relatively good performance, Harrison made H2 between 1737 and 1739. Larger and heavier than HI, H2 stands 66 cm high, weighs over 39 kg and is made almost entirely of brass. The only wood he used in this timekeeper is in the lignum vitae parts and pallets. The concept is fundamentally the same as HI's, except that the temperature compensation is of a simplified design and Harrison fitted a device known as a 'remontoir' to H2.

Even in a well-made clock, small errors in the manufacture and meshing of the wheels and pinions will cause variations in the driving force delivered to the escapement. This, in turn, causes variations in time-keeping, and the remontoir is intended to remove these variations. A small spring drives the escapement independently of the main train of wheels. The function of the main train is then simply to wind up this small spring at regular intervals, as and when it needs it. In H2, this rewinding occurs every 3 min 45 s. Thus H2 is rather more complex than H1. As with H1, H2 would originally have been fitted into a glazed case and mounted in large 'gimbals', a kind of large universal-jointed suspension, to ensure it remained horizontal at all times, though the gimbal mounting itself unfortunately does not survive. Harrison discovered a deficiency in the linked bar balances when H2 was moved and, unable to correct for this, he simply set H2 aside. In spite of 2 years hard work he began all over again and launched into construction of his third marine timekeeper.

20. H2. Built over two years, between 1737 and 1739, H2 was a refined version of H1, with a remontoir to ensure a uniform drive to the balances. A fault in the design of the balances caused Harrison to reject the clock and start work on H3 (see plate 6). (D6784 © National Maritime Museum, London)

H3

H3 was supposed to be Harrison's final word in timekeeper design and for years he was convinced it would be the prize winner. Evidently the scientific establishment was too, as in 1749 Harrison was awarded the Royal Society's highest honour, the gold Copley medal, on the strength of his progress, and the Board of Longitude continued to support him with grant money. However, even after 19 years of painstaking labour, H3 was stubbornly refusing to reach the necessary accuracy. Although Harrison learned a great deal from this Herculean endeavour, and incorporated a number of brilliant inventions into H3, its ultimate role was solely to convince him that the solution lay in another design altogether.

24. H3. Harrison spent 19 years working on this immensely complex timekeeper. For many years he was convinced it would win him the longitude prize. It is the only timekeeper to retain its original carrying case (see plate 7). (D6785–5 © National Maritime Museum, London)

H3 stands 59 cm high and weighs 27 kg (43 kg in its case) The balances are wheels instead of dumb-bells and are arranged one above the other. Like the balances in H1 and H2, they are linked together, beat seconds and are driven by the grasshopper escapement. However, the balances are not controlled by helical springs but by one, short spiral spring, which controls the upper balance only. H3 is by far the most complex and difficult to work on. Like H2, it has a remontoir, but this rewinds every 30 s.

The improved temperature compensation is of a historically important design. Instead of using a grid of brass and steel rods, Harrison created a most wonderfully elegant and effective alternative. He fixed together, side by side, two metal blades, one of brass and one of steel. Because the blades are stuck

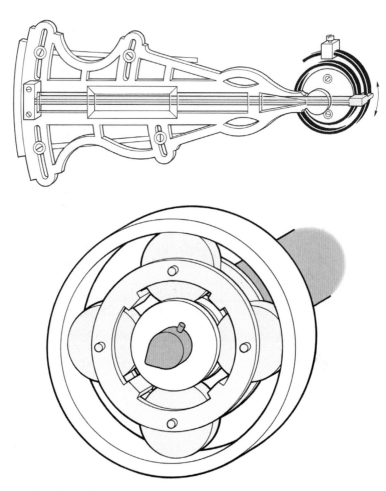

22. The bimetallic strip temperature compensation and the caged roller bearings in H3 are both highly important inventions and both still very much in use today in many technical applications (see plate 8). (David Penney (www.antiquewatchstore.com). Copyright)

23. John Harrison's 'RAS' regulator. Built approximately in parallel with H3, this was Harrison's 'state of the art' fixed pendulum clock. He claimed time-keeping as good as a second in 100 days from it (see plate 9). (D6786–1 © National Maritime Museum, London)

together and because brass expands more than steel, when the temperature rises this 'bimetal', or 'bimetallic strip', will become curved. With one end fixed, the movement of the other end can be used to automatically adjust the timekeeper when the temperature changes. The bimetal is still very much in use today in thermostats, electric kettles, toasters, motorcar direction indicators etc., and is a highly significant invention.

Another important device Harrison created especially for H3 was the 'caged roller bearing'. This was the ultimate evolution of his anti-friction designs and was the predecessor of the caged ball bearing, a device used in virtually every machine made today. The caged roller bearings are employed in H3 in an extraordinarily abstruse mechanism known as an 'isochroniser', designed to ensure that the swings of the balances take the same time whether those swings are large or small.

H3 consists of over 700 parts and, with its remontoir, the isochroniser, the temperature compensation, and all the anti-friction devices, it is the most complex and difficult to understand of all Harrison's timekeepers. Just as it had given Harrison the greatest heartache, it was to give Gould the most trouble during its restoration in the twentieth century.

The 'RAS Regulator'

In parallel with his work on H3, from 1740, Harrison was also busy working on an equivalent 'regulator' (a high accuracy, fixed, pendulum clock) to supersede his wooden precision clocks, now 15 years old. The clock, now known as the 'RAS regulator' (The clock belongs to the Royal Astronomical Society) was, as far as we know, never finally adjusted by Harrison. Nevertheless, he predicted that one day this regulator would be able to keep time with variations no greater than 1 s in 100 days. This was far more accurate than clocks of any kind until the twentieth century, and the RAS regulator is therefore of exceedingly high importance in horological history. Although weight-driven and pendulum-controlled, it is similar to H3 in many ways, employing caged roller bearings, the grasshopper escapement, a 30 second remontoir and working without oil of any kind.

Harrison's Fourth Timekeeper H4

In 1753, Harrison commissioned John Jefferys, a London watchmaker, to make a watch following Harrison's own novel designs. Harrison discovered with this new watch that, if made with certain vital improvements, it had the potential to be an excellent timekeeper. He began to realize, after all this time, that he had been following the wrong path with his earlier experimental marine clocks. Harrison discovered that timepieces with a relatively small, 'high frequency' oscillator (such as a fast beating, watch balance), if made to the correct proportions, are much more stable timekeepers when they are carried about, than the earlier 'portable clocks'. This apparently simple discovery is one of Harrison's great achievements.

So, while ostensibly still working on H3 in the late 1750s, Harrison was in fact busy developing a new watch design, inspired by his success with the Jefferys watch. His fourth timekeeper that resulted from this development was just 13 cm in diameter and weighed 1.45 kg. It is thus completely different from the earlier machines. Both externally and to some extent internally, it looks like a very large, contemporary

24. H4 (left), 1759, 'the most important watch in the world', alongside Kendall's copy, K1, 1769. K1 enabled James Cook to navigate with great accuracy during his second and third voyages of discovery and was dubbed by him his 'trusty friend' (see plates 10 and 11). (B5165 © National Maritime Museum, London)

pocket watch, even to the extent of having pair cases (with an inner case for the movement and a protective outer case around it).

Technically, however, it is different from an ordinary watch in a number of significant ways. Apart from being exceedingly finely constructed, its balance is much larger, although still relatively light, and oscillates at a higher frequency. It swings to and fro no less than five times a second, giving the balance a great deal more stored energy when running, which renders it much less vulnerable to physical disturbance. Temperature change is compensated for by using the bimetallic strip, a smaller version of H3's device. It also contains a miniature remontoir, rewinding every seven and a half seconds, to ensure a constant power source. Creating a remontoir on this small scale demanded the highest accuracy and quality of manufacture and H4 is an absolute *tour de force* of horological design and construction. Naturally this means that, even for an experienced watchmaker, H4 is a tremendous challenge to dismantle and adjust.

A type of verge escapement (the sort used in common watches) was fitted, of a highly modified form, which ensures that the balance is isochronous as in H3's balances. Although Harrison was unable to miniaturize the anti-friction devices, and H4 requires oil on all its bearing surfaces, jewelled bearings were fitted in many places to reduce friction to a minimum. This was not the first time jewelled bearings had been used in watches, but was relatively early for such an extensive use.

H4 was completed in 1759 and was sent by the Board of Longitude on two official trials to the West Indies. The story of H4's glorious triumphs in these trials and the Government's reluctance to payout the prize money are told in excellent detail by Gould in *The Marine Chronometer* and cannot be covered in detail here. Suffice to say that Harrison's timekeepers, as the foundation for the modern marine chronometer, saved countless lives and innumerable ships and cargo, and placed John Harrison as one of the greatest of the eighteenth century's scientific achievers. Indeed, one can also say that H4 was not just the world's first practical marine timekeeper, it was the first of all truly accurate watches, the 'father' of every precision watch which came after.

In 1765 the Board of Longitude reluctantly agreed to give Harrison £10,000, half the prize money, insisting on a full disclosure of the construction of the watch, the details of which were then published in *The Principles of Mr. Harrison's Timekeeper*, in 1767.[113]

The Kendall timekeepers

On the instructions of the Board of Longitude, a copy of H4 was made, in 1769, by Larcum Kendall, that watch being known today as 'K1'. At the same time Harrison and his son William were making a second marine watch, now known as 'H5', in an attempt to meet the Board of Longitude's changing requirements for winning the remaining half of the prize money. Kendall's copy was shown to the Harrisons, father and son, who both agreed that it was even more beautifully made than their own watch; a very remarkable compliment. On its first commissions, K1 distinguished itself in service with the greatest navigator of the day, Captain James Cook, on his second and third pioneering voyages to the South Seas. Cook, who was originally sceptical about its merit, eventually referred to it as his 'trusty friend' and his 'never failing guide', a great 'celebrity endorsement' for Harrison's design.

A particular problem with the H4/K1 solution however was the great complexity and consequent cost of the instrument. K1 cost the Board £500; just 50 such instruments would be the equivalent of the cost of a whole 2nd rate ship of the line for the Royal Navy. So Kendall was commissioned to design and build a simplified version, now known as

'K2'. This watch, made in 1771, was issued on a number of voyages of exploration, the most famous being in 1787, when it joined Lt. William Bligh on the Bounty, the ship that was the subject of the famous mutiny on that voyage. The watch was taken by the mutineers to Pitcairn Island and was only returned to England in 1840, after a series of adventures in South America. A further timekeeper, even more simple in design, was made by Kendall in 1774 and issued to Captain Cook on his third voyage of discovery to the south seas. It is known today as 'K3'.

The 'H' and 'K' prefixes

It was Gould who introduced the 'H' and 'K' abbreviations for some of the timekeepers, but it is curious to note that, although from the outset, in the 1920s, he referred to the Kendall watches as K1, K2 and K3, he didn't use an 'H' prefix for any of the Harrisons until 1939. One wonders if his familiarity with the Royal Navy's K class submarines (he had his own clockwork model of the submarine 'K2') first inspired the use of those terms for the Kendalls. The Harrisons on the other hand were always referred to as 'No.1', 'No.2', etc., until 1939, when hurriedly packing up the timekeepers as the Second World War approached, Gould scribbled the abbreviation 'H4' for the watch. Then, in correspondence with David Evans (1919–1984) of the Royal Greenwich Observatory (RGO) chronometer workshops, the term H4 was used again in 1945. Only in 1952, 4 years after Gould's death, was the term 'H1' formally adopted for use in an article on that timekeeper by D W Fletcher.[114] It was then only in the early 1960s that the others began to be referred to in their abbreviated form, the first reference to the complete series, with the prefixes H1 to H5, being in the catalogue to the National Maritime Museum's exhibition *Four Steps to Longitude*, published in January 1962.[115]

H5

Harrison's second watch, H5, completed in 1770, was technically very similar to H4, but with slight improvements to the escapement and compensation, and with a simplified appearance to the movement and dial (see plate 12). By this date Harrison was desperate for the recognition that his watch had solved the longitude problem. Rightfully feeling he deserved the remaining prize money, Harrison sought the support of

King George III, himself a keen amateur 'natural philosopher' and very interested in advances in watch -making.

The King tested H5, with extremely good results, at his own private observatory in Richmond and then promised the Harrisons his support. This resulted in a Parliamentary debate and the award to Harrison of the remaining prize money, as a bounty from Parliament. This, including expenses, came to £23,065, considerably more than the total prize money, though Harrison still complained that he had been short changed!

It cannot be claimed that Harrison's design for H4 is identical with the modern chronometer, but there is no doubt that those who followed Harrison, and who were responsible for simplifying and standardizing the chronometer, based their ideas extensively on his work (Box 4.2).

The first man to achieve all four features in a successful chronometer was the London watchmaker John Arnold (1736–1799). There is considerable evidence that his work was based on Harrison's ideas, much of it taken from Harrison's published description of H4 (1767). As we have seen, Harrison understood the need for a balance with compensation,

---

BOX 4.2 The critical features of the modern marine chronometer

In simple terms, the critical features of the modern marine chronometer were four:

First, the balance, which oscillates and keeps the time, is on the scale of a watch (as opposed to large clockwork) but is relatively large and runs at a relatively high frequency, with *high energy stored* in it, as it swings.

Second, the balance is *detached*, that is, it is allowed to swing freely and is only impulsed (i.e. given pushes by the escapement to keep it swinging) very briefly, just as the balance passes the middle of its swing.

Third, the swings of the balance are *isochronous*, that is, whether the swings are large or small they take the same time. In the modern chronometer this is ensured by making the balance spring of the correct form.

Fourth, the chronometer is compensated for changes in temperature by incorporating some form of compensating device *in the balance itself*.

It is interesting to note that H4 itself only has one of these features (the first) and yet performs very well. The type of oscillator is thus perhaps the most critical of all these features.

H1 originally had such a device, and he published a statement confirming the principle in 1775, which Arnold later quoted as his source.

The majority within this next generation of chronometer pioneers were English, but the story is by no means wholly that of English achievement. As will be seen when discussing Gould's book *The Marine Chronometer*, one French name, Pierre Le Roy (1717–1785) of Paris, stands out as a major presence in the early history of the chronometer. Another great name in the story is that of the Lancastrian, Thomas Earnshaw (1749–1829), a slightly younger contemporary of John Arnold's. It was Earnshaw who finalized the format and the production system for the marine chronometer, making it truly an article of commerce, and a practical means of safer navigation at sea over the next century and half.[116]

The period of neglect

One of the conditions in qualifying for Board of Longitude prize money was that the Harrison timekeepers became the property of the government. In October 1765, Harrison was instructed to hand over H4 though, after much pleading from Harrison, the Board reluctantly allowed him to keep the three large timekeepers a little longer. Then, on 23 May 1766, these too were taken from him. Harrison's *bête noire*, the Astronomer Royal Nevil Maskelyne turned up, without warning, at Harrison's Red Lion Square home with an un-sprung cart!

Once back at the Observatory, all three machines were subjected by Maskelyne to a timekeeping trial, a completely pointless exercise. These were not the prize-winning timekeepers, neither were they in a fit condition for good running after their rough ride down to Greenwich. Following the inevitably disappointing results, they appear to have been displayed as curiosities in rooms at the Observatory. The astronomer Jean Bernoulli III recorded having seen two of them in 1769 and the artist John Charnock, visiting the Observatory around 1770, depicted H2 in a little watercolour sketch. Soon after this they appear to have been put away in storage, not to see the light of day for over half a century.

The watches H4 and K1 were of course considered as usable chronometers (the term being introduced, in its contemporary sense, by John Arnold in 1780). There is no evidence however that H4 was ever issued for formal navigational purposes, even then being considered an historical relic. After Cook's two south sea voyages, K1 enjoyed a brief but eventful career as a ship's chronometer, sailing with Capt. Arthur Philip in Sirius, part of the 'first fleet' to Australia in 1786–1790. Then, after repairs by Thomas Earnshaw, K1 was lent to Sir John Jervis, later Earl St Vincent and First Lord of the Admiralty, who kept it for nearly ten years.

From the early years of the nineteenth century H4 and K1 were pensioned off and preserved as historic instruments. The evidence is that they were in fact quite well cared for and neither ever got into the kind of state the larger machines were to suffer in later years.

H5 was never government property and remained in the hands of the Harrison family until 1869, when it was sold to the Scottish shipbuilding magnate and collector, Robert Napier (1791–1876) of Gareloch. The Napier collection was sold at auction by Christies in 1877 and H5 was bought by the art dealer W. Boore of the Strand, in London. Boore offered H5 to the Clockmakers Company, but it was turned down on the grounds that H5 was only really a copy of H4 and was of insufficient interest! However, in 1891 the Rev. H.L. Nelthropp finally persuaded the Court of the Company to acquire it, at the princely sum of 100 guineas.[117] H5 has always been the best preserved of all the timekeepers and remains to this day in pristine condition in the care of the Company.

The first three, large timekeepers were to suffer a very different fate. After more than 50 years in poor storage, it was decided by the Astronomer Royal, John Pond (c.1767–1836) in 1824, that the timekeepers should be cleaned. They were sent to John Roger Arnold, son of the celebrated chronometer maker John Arnold, and the principal supplier and repairer of chronometers for the Observatory at the time. As so often happens with complex but non-urgent projects such as these, the cleaning of the three timekeepers seems to have been considered a low priority by Arnold and 10 years passed by without the work getting started. In 1835, Pond was replaced as Astronomer Royal by a 'new broom', the super organized, autocratic George Biddell Airy (1801–1892), who was to reign at the Observatory for an amazingly productive and ordered 46 years.

One of the first things Airy did was to conduct an inventory of the Observatory's instruments and he quickly discovered that the Harrison timekeepers were with Arnold, by now in the partnership of Arnold & Dent. Airy wrote in October 1835 asking for the immediate return of the clocks and the company replied 2 days later asking for a little more time. They reported that H1 was in 'a complete state of decay . . .', H2 was basically sound and H3 was suffering from rust on the steel parts, the glazed case having a broken glass panel, allowing moisture in. They also reported that very little was known about the timekeepers and, as no plans of them appeared to be available, they proposed to have some technical drawings done.

Over five years went by and Airy was obliged to write again requesting a date for the return of the timekeepers. After another slight delay, the clocks finally went home to Greenwich at the end of August 1840. They had been surface cleaned, but were evidently not in working order, though a set of drawings, some of them in very fine colour wash, had been commissioned from the draughtsman, Thomas Bradley, Lecturer of Geometrical Drawing at King's College London (see plate 13). Five of these beautiful plans of H1, H2, and H3 are still preserved in the RGO Archives at Cambridge University Library.[118]

For another 80 years the timekeepers languished in the stores at the Observatory, though for H2, H3, and H4 there was a brief respite in 1891 when they were exhibited at the Naval Exhibition in Chelsea. H1 was by that time considered too badly deteriorated to display, but H2 and H3 were exhibited in their un-restored state. H4 had been cleaned by the noted watch and chronometer maker, James U. Poole, the year before.[119]

Notwithstanding the brief moment of glory for H2, H3, and H4 in 1891, all the timekeepers were out of order and neglected again when, on that fateful day, 5 March 1920, Lt Cdr Rupert T Gould came to visit the Observatory in search of these neglected and largely forgotten masterpieces.

✺ 5 ✺

Research and the First Restorations
1920–1922

As already observed, it was inevitable that Gould, already fascinated by clocks and watches, would develop a great interest in marine chronometers in his professional role as a Navigating Officer. This was one of the essential tools of the trade; something that particular officer would have relied upon every day at sea. And, given its 'ultimate' position in the horological pecking order, it was for Gould something that just had to be fully understood and its story published. After all, virtually nothing had been written about it up to that time and he, with his insatiable appetite for collecting historical data, was the ideal person to do it.

This was something Gould probably needed to do for personal reasons too. His work under the Hydrographer, although sometimes fascinating for him, was hardly challenging for a man with such a lively and creative mind. Here was a chance, before he began studying for the Bar, to make his mark in an historical subject close to his heart. After his problems during First World War, there was probably also a need to prove his personal worth, both to himself and everyone else. Indeed, in the years to come, this need to achieve, regardless of sacrifice, was responsible for Gould's enduring determination to see the Harrison timekeepers restored.

Gould's research for the book The Marine Chronometer had begun in 1919, though his association with the Harrison timekeepers began the following year with his first visit to the Royal Observatory. On 5 February 1920, his boss, the Hydrographer of the Navy, Admiral Learmouth, wrote to the Astronomer Royal, Sir Frank Dyson, introducing 'an Officer serving in this Department . . .'. He attached Gould's synopsis for the book and requested Dyson's help with access to the Observatory's records and 'further sources where information can be obtained on this subject'. Dyson replied positively a few days later and so it was that Rupert Gould wrote his first letter to the Astronomer Royal, on 28 February

1920, requesting to visit the Observatory. The appointed hour they agreed was 11.00 a.m. on Fri. 5 March 1920.[120]

It is probably true to say that sooner or later in almost everyone's life there occurs a 'defining moment', a cliché which nevertheless expresses the phenomenon precisely. For Rupert Gould, one of these had already occurred in 1914 as he fell ill with his first nervous breakdown, but here, undoubtedly, was another, on the morning of 5 March 1920 when he first set eyes on the Harrison timekeepers.

In later years[121] he described this first meeting with the clocks. 'All were dirty, defective and corroded—while No.1, in particular, looked as though it had gone down with the Royal George and had been on the bottom ever since. It was covered—even the wooden portions—with a bluish-green patina.' Apparently H1 had for many years been stored next to the gas ovens used for temperature testing in the chronometer room (now, appropriately enough, the office of the National Maritime Museum's Horology Section at the Observatory) and its corroded state was thought to be due to exposure to the exhaust gases. Gould continued: 'I could not bear to see them in this condition. It seemed to me such a futile, tragic ending to a great adventure. They were the first accurate marine timekeepers ever made—the life work of an original genius who was also an Englishman—and here they were; discarded . . . forgotten . . . buried. Surely they deserved a better fate'.

H1: The First Restoration

A second visit to study the timekeepers on 14 May 1920 confirmed his determination to see something done about their condition. All this work that he was now considering would of course have to be done in his own time, out of office hours.

Ten days later the Goulds' son Cecil would have his second birthday and their daughter Jocelyne, born on 27 February 1920, was just 3 months old. But so affected was Gould by the sight of the Harrisons that time spent with the family and that spent 'off on clouds of Horology', as Cecil later described it, would now openly conflict. On his return to the Campden Hill Court flat that day, Gould wrote directly to Dyson:

will you allow me to put forward a proposal which, I trust, you will not think impertinent? The mechanism of No 1, not being, as are those of Nos 2 and 3, preserved in an airtight case, is far gone in decay, and will, I fear, soon become irreversible. May I then offer my services in the cleaning of it? I would give you a bond in £100, or any other sum you may think necessary, to do it no manner of damage, but to clean and reassemble all parts of it, and to ensure that, for the future, it should not be liable to corrosion; also to provide, at my own expense, an airtight case to replace the present one. I make this offer, in order that I may have the honour of associating myself with the preservation of this unique memento of Harrison's early work. I have some slight qualifications as an amateur mechanic, and have recently cleaned and repaired an old Berthoud chronometer belonging to Commdr Edgell, of the Hydrographic Department, who would, I think, certify to you that I have returned it to him in better condition than I received it.[122]

Gould continued in his letter to Dyson: 'I hope that you will not think me too forward in asking this favour of you, but I do feel, most strongly, that an effort should be made to clean No.1 and to preserve it from further deterioration, especially since it establishes the fact of Harrison's priority over LeRoy's chronometer (preserved at Paris) in an indisputable fashion!'

Pulling out all the stops in his attempt to persuade Dyson to let him take on H1, Gould even resorts to an appeal to the Astronomer Royal's national pride. This is actually quite an extraordinary remark for Gould to have made, as the central conclusion of his magnum opus *The Marine Chronometer*, published just 3 years later, was that the Frenchman Pierre Le Roy was indeed the true father of the modern chronometer.

Gould had, incidentally, joined the British Horological Institute (the BHI), the clockmakers' professional institution, as an ordinary member, directly after seeing the Harrisons for the first time in March 1920. The Astronomer Royal was ex-officio President of the BHI and Gould was thus appealing to a senior member of the horological establishment too.

The Amateur Mechanic

Gould was perfectly frank and open about his lack of qualifications. His description of himself as an amateur mechanic was spot on, he had no qualifications, nor even much experience for that matter, and with no horological training of any kind, he was entirely 'making it up as he went along'. As a Navy Officer, he was in a somewhat privileged position. As most practical horologists were socially his 'inferior', little was ever said to him personally in the trade that he grew to know so well. However, over the years a number of practitioners have expressed the view quite strongly that he was not sufficiently qualified and should not have been allowed to tackle these important pieces.[123]

In any event, the letter to Dyson had the required effect, as 3 days later Dyson replied to 'gratefully accept' Gould's offer, but expected the work to be done at Greenwich. A few days later Gould wrote again asking Dyson to reconsider and let H1 go to the home workshop in Kensington. Gould doesn't mention it, but the family were just about to move again, to 'Waylands', the Kensington Park Road home bought for them by Muriel's father. A great saving of time was Gould's reasoning for having H1 at the house, 'I should be able to give my spare time wholly to it . . .', whereas at Greenwich he could manage only once a week, and that his own 'tools and appliances' would be all around him. The letter reminds the Astronomer Royal that the timekeepers were allowed to go to Arnold & Dents back in the 1830s and remarks that H1's transport will not be difficult '. . . since I have my little car . . .', the tiny Scripps-Booth two-seater, open-top sports car. He promises, nevertheless, to insure H1 against burglary and fire etc.

Dyson evidently agreed, as on 28 May 1920 Gould took away the timekeeper, signing a certificate promising '. . . to clean it thoroughly and reassemble it exactly in its present mechanical condition except for such structural repairs as may be absolutely necessary to prevent further derangement of its mechanism and to return it safely to the Observatory on or before the expiration of one calendar month from this date . . .'. On 10 June he wrote to Dyson with a preliminary but detailed report and broke the news that work was going to take longer than expected. Given its condition, its not surprising that many of the parts of H1 were seized together, though Gould noted '. . . the dials have cleaned up very well and will, I think, make a marvellous difference to the machine's appearance'.

The wooden wheels were in perfect condition except for one, which needed some of the teeth segments gluing back in place. H1 was kept in a locked room on the top floor of the house, but parts were cleaned in the basement 'since the poisonous green dust which rises from them is rather unhealthy for a spare bedroom'.

Work continued for over three months, Gould reporting again on 23 September 1920: 'the difficulties I have had to contend with in dissecting No 1 would fill volumes . . . the work of getting off the scale on the brass-work generally has been very hard . . .'. H1 was now reasonably clean, but was of course nowhere near in working condition. He proposed

25. Waylands, 46 Kensington Park Road, in West London. The mid-Victorian villa was where H1 and H4 were restored for the first time. The house was pulled down in the 1930s to make way for new blocks of flats. (© Sarah Stacey and Simon Stacey, 2005)

26. H1 after and before its cleaning in 1920. Slightly over-stating the case, Gould described it humorously as having 'enough missing parts to fill a bucket'. It would be another twelve years before he restored it to working order. (© Sarah Stacey and Simon Stacey, 2005)

fitting to the back of the timekeeper an engraved plate with a long inscription relating to H1's history. The addition was evidently approved as this plate (now removed but still in the NMM collections) can be seen on some of the contemporary photographs of the cleaned timekeeper.[124] He also stated that he was putting together a lecture for the Royal

Geographical Society (RGS) on *The History of the Chronometer*, a kind of 'progress report' on the research for his book, and he comments, 'the more I write the more I feel I am only scratching the subject'.

The lecture,[125] attended by the Hydrographer and the Astronomer Royal, was given on 13 December 1920 at the RGS headquarters on Kensington Gore. H1, now cleaned and reassembled, was displayed at the lecture alongside K1, (which Dyson had also allowed to be present), and Gould's own chronometer by Berthoud, No 37. The RGS published the lecture in The Geographical Journal[126] and it was immediately produced as an offprint, Gould's first published work on the chronometer.

Not having had the chance to study H2 and H3 in any depth, he was obliged, in discussing Harrison's work, to concentrate on descriptions of H1 and H4, but explained this omission with the extraordinary remark: 'Harrison's second and third machines may be passed over, since they exhibit only detailed improvements over No 1 . . .'. It could not have been long before he must have regretted making such a sweeping statement! H1 stayed on at the RGS, returning home 10 days later.[127] By this time Gould had had a display case manufacturer (Hoskins of Old St.) make a new case for H1, instructing that it should be 'lined with dark blue plush'.

The Restoration of H4

In May, Gould brought Miss Dorothy Cayley, of the Admiralty's Photographic Section, down to Greenwich to take pictures of the timekeepers for the forthcoming book.[128] Like his wife Muriel, Gould was himself a very capable photographer and did take many of his own photographs when necessary, but preferred to use professionals when the images were for publication.

While overseeing the photography, Gould was able to inspect H4 and K1 closely for the first time and noted that while K1 was in sound condition H4 was considerably out of adjustment. Emboldened by his success with the cleaning of H1, Gould now wondered whether he might be able to tackle H4. This was a much more delicate operation and bold indeed for one without any training as a watchmaker. Nevertheless, writing to Dyson on 17 May, Gould pointed out that he had recently rectified the adjustments in H5 at the Clockmakers Company at the Guildhall (no details of this work appear to have survived), and now wondered if he could clean H4 on the same terms as he had done H1?

The work would be much quicker as the watch was smaller and better preserved, and he guaranteed it would be returned in good going order. He suggested that if this were done, a display case could then be made to show off the movement, just as H5 was at the Clockmakers. He continued that: 'there must be many working watchmakers who would do it far better than I could. On the other hand, I have made a special study of its mechanism . . . and think I know as much about its design and construction as I do about any modern chronometer'. He had evidently studied the available literature on H4 closely and, incidentally, had also studied K2, the Bounty timekeeper, in October 1920 at the Royal United Services Institution at the Banqueting House in Whitehall, just over the road from the Admiralty. He made a fine drawing of K2's movement at that time, a beautiful piece of draughtsmanship reproduced as a print by the Hydrographic Department that year.[129]

Understanding the design of a complex mechanism, and being able to illustrate it beautifully, is all very well, but having the manual dexterity to take it all apart, and reconstruct it safely, is another matter entirely. But Gould was confident and, although he had virtually no experience in practical watch making, he does indeed seem to have had a natural dexterity when called upon to really concentrate.

The Astronomer Royal had faith in him anyway and, naturally grateful for all Gould's work on H1, Dyson replied gladly accepting the offer to clean H4. Dyson informed him he had had a nice table made for Harrison's newly cleaned No 1 and was hoping that 'we are gradually getting rid of any grievances his ghost may have against the Royal Observatory'.[130]

Family life

Harrison's grievances may have been diminishing, but Rupert's family were gradually finding some of their own, related to the growing amounts of time he was dedicating to Harrison, and not to them. During the months of May, June, and July 1921, Muriel took the children on holiday to Frinton, Rupert joining them at various times for a break at the seaside, on one occasion with his old friend Val Hirst. But the breaks with the family were short—when the Harrison work allowed—which was not as often as Muriel would have liked, and from this time onwards she began to become rather fed up with his absence.

After official approval from the Admiralty, Gould was able to collect H4 personally from the Observatory on 29 July 1921, signing a receipt

Bobbie Bunny Cecil.

27. Happy times. Rupert with the children, Jocelyne (affectionately known as 'Bobbie' at the time), and Cecil, on holiday at Frinton in the summer of 1921. (© Sarah Stacey and Simon Stacey, 2005)

promising to return it within 3 months. Naturally he was thrilled to have H4 in his hands. Writing to the polar historian, H.R. Mill (ostensibly on arctic history) Gould noted excitedly: 'I have just received from the Royal Observatory Harrison's prize-winning chronometer No 4 for cleaning and repair and am as proud as a dog with two tails'.[131]

The Austrian horologist Heinrich Otto, an instructor at the BHI and later something of a hero-worshipper of Gould's, recalled that he visited Gould at Waylands at this time and was shown H4 which (Otto observed rather critically) Gould kept in his desk drawer.

Six weeks later, on 8 September 1921, after concentrated dismantling and study of the watch, carried out at his desk at Waylands, Gould reported enthusiastically to Greenwich on the quality and design of its mechanism. He noted: 'it would puzzle a good Clerkenwell workman to turn out a duplicate of the third wheel, with a deep rim as thin as paper and 120 internal teeth'. The third wheel in H4 is indeed an amazing piece of work and we are unclear, even today, how Harrison made it. Gould himself found some of H4 puzzling: 'It took me three days to learn the trick of getting the hands off-I more than once believed they were welded on—and about as long to dissect the remontoir'.[132]

Cecil & Rupert.
Harrison ho: 4.

28. 'As proud as a dog with two tails'. Rupert and Cecil in the back garden of Waylands in the Summer of 1921 with H4, about to undergo restoration. (© Sarah Stacey and Simon Stacey, 2005)

Apart from a broken mainspring, Gould pronounced H4 to be generally in fine condition but complained that the previous overhaul by J.U. Poole had been done badly, the watch over-oiled and the escapement adjusted incorrectly: Gould was not at all impressed with Poole the horologist. The broken mainspring could, Gould said, be replaced for 10 shillings, as long as Dyson approved, which he did by letter on 19 September.

Controversially, H4's original mainspring was then broken up by Gould into a number of short (approximately 30 mm long) pieces and distributed widely to collectors and museums as relics. The Science Museum, for example, approved the acquisition of such a piece (now in the NMM collection), and over the years others have been donated back to the Observatory. The National Maritime Museum collections now hold four pieces of the original spring.[133]

Although Gould was evidently aware of the need to record the work on these important machines, and his letters to Dyson were most informative on what he was doing, he had not, at this stage really begun the systematic recording of the work he undertook. It was only later he started compiling detailed notebooks. As discussed later, the work done to H1 in 1920 was only very sketchily noted, and he has left us no notes at all on the cleaning of H4 in 1921.

Research for the book

On the same day as Dyson was writing to Gould, the 8 September 1921, the Goulds were on holiday at Druimgigha in Scotland and Rupert was actually writing to Dyson. His aim was to persuade the Astronomer Royal that he should be allowed to take the original Board of Longitude archives home to work on in Kensington. He explained that he needed to hasten the writing of his book *The Marine Chronometer* for two reasons: 'Ditisheim is preparing something of the same kind, and may forestall me. The other is, that next year I must seriously tackle my reading for the Bar, and that in consequence I shall have much less spare time than I have now'.[134] 'Ditisheim' was a reference to Gould's friend and contemporary, Paul Ditisheim (1868–1945), the distinguished Swiss watchmaker who had already helped with material for Gould's new book.[135] The request to Dyson to take the archives home was duly granted, and work on H4, and writing the book, continued in tandem, all in his 'spare time' at home, as usual.

The Mudge Timekeeper 'Blue'

Other horological matters kept Gould's mind occupied at this period too. On 29 October 1921, he wrote to Frank Dyson asking if he would provide a Foreword to the new book, but took the opportunity to raise a further matter with the Astronomer Royal, who happened also to be Master of the Clockmakers Company (the London Livery Company) that year.

At the time, the London Fine Art dealers, Hurcombs had for sale a highly important marine timekeeper made in 1777 by Harrison's contemporary, the celebrated watchmaker Thomas Mudge (1715–1794). Known as 'Blue', to distinguish it from its pair 'Green' (named after the colours of their shark-skin covered cases), the timekeeper was in very fine condition and Gould and his friend, the distinguished engineer and horological collector Sir David Salomons (1851–1925), felt strongly that it should be in the Company's collection. So keen was Gould to see this happen that he generously offered, as an incentive, to donate his own chronometers by Earnshaw and Berthoud to the Company if they agreed to buy Blue, 'but any action should be immediate or an option obtained, as otherwise 'Blue' may pass into the hands of a private purchaser'.

As well as agreeing to write the Foreword, Dyson promised in his reply to propose the offer to the Court of the Company. Alas, it was refused and Blue did indeed 'pass into the hands . . .' of another purchaser, ending up in the Staatlicher Mathematisch-Physikalischer Salon in Dresden. This was a tremendous opportunity missed, and doubtless regretted to this day by the Clockmakers Company, but Gould had tried his best.

Meanwhile, professional work at the Hydrographic Office continued to fill the days. In September 1921, Gould compiled a comprehensive set of notes for Ernest Shackleton (1874–1922) for the ShackletonRowett Expedition that set off in SS *Quest* for the Antarctic soon afterwards.[136] November 1921 saw Gould filing his internal report based on his extensive research on the history of the International Dateline. This report was published by The Hydrographic Office in later years, though without any mention of Gould's authorship.[137]

The move to Epsom

When Gould wrote to Dyson reporting on H4's condition, he didn't mention that the family were soon to move house again. In early 1922 the Goulds left Kensington for a small, detached house in Epsom in Surrey. Muriel had been the instigator, always impetuous and unable to settle, she decided that living in London was too expensive, though the house at Kensington Park Road had been a poor investment and they suffered a considerable loss on the sale of the lease.

The Cottage, in Lynwood Avenue, Epsom, is (it is still there) a detached four-bedroom house standing on the corner plot at the elbow of this quiet *cul-de-sac*. It was probably new, or very recently built, when the Goulds moved in. At the end of the garden, with access from Lynwood Avenue from the side, is the garage, though another of the financial cutbacks on moving was the sale of the Scripps-Booth, and the Goulds no longer possessed a motor car. Later however, the garage was to be used as a workshop for restoring H2 and beginning work on H3. In the best-known photograph of Gould, alongside H3 and with the balance of H2 on his lap (see p. 138), he is shown sitting in the garden, the side of the garage/workshop behind him.

As noted before, being struck by lightning was one of a number of phobias that preyed on Rupert's conscious and subconscious mind during his life. One of the first things he did on moving to the Lynwood Avenue house was the fitting of lightning conductors, still to be seen on the chimneys of the house today.

Four months went by and on 21 January 1922 H4 was still with Gould at Epsom, but he was able to report that it was all cleaned and running. Remarkably, given his small experience in working on chronometers,

29. The Cottage, Lynwood Avenue, Epsom *c.*1922. The work on H4 was completed here and H2 was restored in the garage (behind the camera) between 1923 and 1924. (© Sarah Stacey and Simon Stacey, 2005)

he seems to have had little difficulty in dismantling, cleaning and assembling this highly complex and delicate watch. On completion however, he found that H4 had a very large losing rate and he needed Dyson's approval for an additional 'stiffening piece' to be attached to the tail of the balance spring to bring the watch to time. This would be fitted 'reversibly' so that no permanent change was made.

Once completed, he proposed the construction of an airtight display case for both H4 and K1. As part of this overhaul, Gould had taken the rather unusual step of lacquering the brass parts of H4's movement and he suggested displaying the watch open, and K1 closed, so both the inside and the outside of the watches could be seen. With K1's case polished, and an ivorine label produced, the little museum exhibit would be complete and this is what Dyson agreed to soon after, when H4 was returned to Greenwich.

In fact the mounting for H4, with its movement open, was to stay in use long after this time. Years later, when the timekeepers were displayed in the new National Maritime Museum in 1937, the mount was used, and it remained with H4 right up until 1993 when the timekeepers were redisplayed in honour of Harrison's tercentenary. The mount is now preserved in the museum's reserve collection, as an associated part of H4.

☙ 6 ☙

The Magnum Opus 1921–1923

Gould's 'spare time' in the years 1922 and early 1923 were spent completing work on *The Marine Chronometer, its History and Development*. Research had begun in 1919 and, in his Preface, the author lists the comprehensive catalogue of sources he studied and credits the many people he received help from.

It is remarkable how thorough and painstaking Gould's research was. Even today, when researching some obscure backwater in the horological world, in the hopes of discovering something really new to reveal to one's colleagues in antiquarian horology, one often finds that R.T. Gould has been there before and has already noted, if not published on, the subject.

Twenty-four chronometer makers were sent a synopsis of the book with a request for help. Sadly, but predictably, in this period of decline in the industry, less than half even bothered to reply and only eight were able to provide anything useful. By 1920, with the post-First World War depression and the already rapidly diminishing British watch and clock-making trade, those chronometer makers who hadn't already gone into receivership were struggling to survive or were diversifying into other manufactories.

Gould's friend Frank Mercer was one of the few English makers to reply with help and it is perhaps no coincidence that three of the other seven companies who responded were French. France was the other great player in the history of the chronometer and, as a nation, has always been supportive and proud of its scientific achievement. In researching museum collections Gould took the trouble to visit Paris too, and inspect Pierre Le Roy's celebrated marine timekeeper, his *montre marine*, along with those of Ferdinand Berthoud, at the Conservatoire des Arts et Métiers.

Paul Chamberlain

It was during this research that he first encountered the distinguished American horological collector and historian Paul Mellen Chamberlain (1865–1940). Although twenty-five years his senior, Chamberlain and he got on extremely well, and were firm friends thereafter. Gould sold Chamberlain his Berthoud chronometer (No 37) in July 1922, after just a year or so of ownership. By the early 1920s the Goulds had spent the majority of the money that had been bequeathed to them from the family. In need of cash, Gould was obliged to withdraw the Berthoud chronometer from the Science Museum displays where it had been on loan since the previous year.[138]

A series of preparatory notes, got up in his typical DIY bindings by Gould himself, were intended as 'dry-runs' for the book. In June 1921 he made up a volume which he entitled *Photographs Illustrating the Evolution of the Marine Chronometer From its earliest Beginnings to the Present Day (1660–1921) With some notes by R. T. Gould*.[139] This included comprehensive illustrations of the Harrison timekeepers, especially H1 before and after its recent cleaning, and a number of the early French timekeepers, including Le Roy's montre marine and a number of Berthoud's. 'The Modern Chronometer' is epitomized by an example of a chronometer by Kullberg and a selection of special compensation balances are shown.

Two bound-up volumes of detailed notes, made by Gould during his preparation for the book, are also still extant, as are a series of interesting observations on the second volume, made by the distinguished amateur horological historian, David Torrens (1897–1967).[140] The writing of the book itself was done by Gould directly onto a typewriter. Typically confident in his knowledge, he made no hand written scripts first and just transferred the text straight from his head onto the page.

As for the final result, Gould's *Marine Chronometer* was a long way ahead of its time. As has been said, in the 1920s, very little of any kind had been written on the subject of clocks and watches, and what there was tended to be dry and matter of fact in style. Nothing had been written specifically on the marine chronometer at that time, yet it is generally accepted in the horological world today that, in spite of a number of more recent books attempting to cover similar ground, we still have no finer discussion of the subject. The book is not only packed from cover to cover with clearly and logically presented facts about the subject, the author writes in an informal, even 'gossipy' and lucid style, making a potentially dry topic much more accessible.

The very first sentence in the Introduction asks, intriguingly: 'Would the reader be good enough, for a few minutes, to imagine he is Christopher Columbus?'. With this engaging start, grabbing the reader's attention, he goes on to describe the navigational difficulties faced by Columbus and

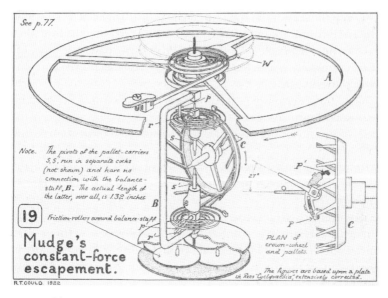

See p. 77.

Note. *The pivots of the pallet-carriers S,S', run in separate cocks (not shown) and have no connection with the balance-staff, B. The actual length of the latter, over all, is 1·32 inches.*

19

Friction-rollers around balance-staff

**Mudge's
constant-force
escapement.**

R.T.GOULD. 1922

W
A
P
C
P'
27°
O
P'
F
C
B

PLAN of crown-wheel and pallets.

The figures are based upon a plate in Rees' Cyclopædia, extensively corrected.

30. Gould's perspective view of Mudge's extremely complex, 'constant force' escapement. He considered this one of the better drawings, and is one of only two he signed in the book. (© Sarah Stacey and Simon Stacey, 2005)

his contemporaries and, without having to wade through a lecture on navigation, the reader easily assimilates the function of a chronometer.

As Dava Sobel has echoed in her book *Longitude*, written over seventy years later, the story of how this esoteric instrument came into being is actually dramatic and exciting. In his own way Gould imparts every bit as much intrigue and pathos in its telling, while covering the technical aspects of the subject in comprehensive detail.

The technical aspects of the book are also illuminated with his own fine and characterful pen-and-ink technical drawings, masterpieces of their kind. Illustrations such as the perspective view of Mudge's constant force escapement would have been exceptionally difficult to draw (even when based on an earlier engraving, as this was). Drawings such as this make a highly complex and abstruse device admirably clear for those wishing to study the technicalities closely. Typically, the drawing was done rather at the last minute and Gould was, in fact, not entirely happy with all of the illustrations. He noted in one of his annotated copies of the book that he felt the Mudge drawing was one of the more successful illustrations in the book.[140a]

Given the date of the book, the chapter on 'Early Efforts . . .' covers the pre-Harrison attempts to discover a means of solving the Longitude problem in remarkable depth. Having said that, this part of the story has been extensively researched in more recent years and is the part that most shows its age.

Gould was obviously in the best possible position to write on John Harrison's contribution and the two full chapters on his work were the basis of just about everything which has appeared in the horological press since.

Ironically, if there is one flaw in the content of the book, at least in the view of this biographer, it is that, in spite of his huge admiration for Harrison (the frontispiece for the whole book is Gould's own print of Harrison engraved by Tassaert in 1768), the author still fails to recognize the depth of Harrison's contribution to the development of the modern chronometer. Instead he concludes emphatically, in a later chapter in the book, that the true father of the modern chronometer was Pierre Le Roy of Paris.

He states: 'The exact genesis of many great inventions is hotly debated. Whether Heron, De Caus, the Marquess of Worcester, or Savery invented the steam engine; who first printed from moveable type; who invented the mariner's compass . . . But there can be no doubt at all that the inventor of the modern chronometer is Pierre Le Roy'. Of Harrison's and Le Roy's machines, he says: 'The difference in their machines is fundamental–Harrsion built a wonderful house on the sand; but Le Roy dug down to the rock.' And he concludes of Le Roy: 'He stands alone, the father of the chronometer as we know it'.

But modern research suggests that Gould's lionising of him, and damning Harrison with faint praise in the process, was much too simplistic. In 1766, in a paper to the French Academie des Sciences, Le Roy proposed three of the four features for a successful chronometer (nos 2–4 discussed in Chapter 4) and might seem therefore to justify Gould's placing of him as the real 'Father of the Marine Chronometer'. However, the reality is that Le Roy's actual understanding of the principles was not complete and the all-important fourth feature, the scale and frequency of the oscillator, was missing.

This is not the place for in-depth discussions of chronometrical theory; suffice to say that Le Roy's contribution was much too heavily emphasized by Gould, a legacy of misinterpretation which is still with us today, and a flaw in an otherwise outstanding book.[141]

After bringing the chronometer up to date, with chapters on the work of the other great pioneers from France and England, the second half of the book, logically enough entitled Part II, is subtitled 'The Later Development of the Chronometer'. In fact it is really a more technical discussion of the various improvements, made in the nineteenth and early twentieth centuries, to squeeze the very best performance out of what was already a fine design. Thus, the second half is aimed more at the specialist and provides a very readable but technical reference work, rather than a continuing narrative of the chronometer's story.

The Chronometer in 1923

By the time *The Marine Chronometer* was written, the instrument itself was supplemented with Wireless Telegraphy as a means of determining GMT at sea. The chronometer was far from obsolete though, and was not to be wholly superseded for another half a century. Gould ends the text of the book on a realistic but upbeat note:

'A similar fate [obsolescence] one day is probably in store for the chronometer as a means of finding longitude at sea. When it comes there need be no repining, for nothing is permanent and nothing is indispensable. But whenever it comes . . . it will close the history of one of the most determined and successful attempts to solve a mechanical difficulty that Man has ever made'.

It is true that in 1923, when Gould's book was published, the chronometer was still very much in use on ship board, but ironically that period in the early 1920s marked the beginning of the end for the chronometer-making industry in England. Demand for new instruments had simply dried up and that same year, 1923, the Astronomer Royal at Greenwich began receiving letters from impoverished craftsmen pleading for more orders for chronometers. The problem was that the timekeepers they made were just too good. Even one hundred-year-old instruments were still providing good service on Navy ships and, with the First World War over, the demand simply vanished. It was at this time that the majority of craftsmen in this noble profession retired or sought alternative employment, Gould's book ironically seeming in retrospect to be a kind of valedictory.

Of course, no publication of such magnitude can be absolutely perfect, and in addition to the author's conclusion concerning Le Roy, there are a number of curious little flaws in the production of the book.

The references to the plate numbering goes adrift in the text; there are two pages 39; two pages 82; and no less than three pages 234! Overall though the book remains a masterpiece of its kind and stands, arguably, as the finest of all the horological publications of the twentieth century.

Over the years, the book has been 'looted' for quotes and illustrations. Professor Usher of Harvard University used no less than eight figures and several pieces of text in 1929 in one of his publications, and Professor Wolf of London University, in his *History of Science* of 1936 used two figures without an acknowledgement of any kind. Commenting on these contraventions in a letter to Professor Stewart in 1936 Gould, always short of money, decided to pursue the second for payment, noting: 'I think I shall try and twist his tail'.[142]

A pointer to Gould's own view of the book's overall value was that he dedicated it to one of his great idols, Admiral David Beatty, under whom, it will be recalled, he had served in 1908. Having said this, Gould, like a disappointed parent, was always critical of detail within the book and was keen to update it even before it saw publication.

Annotated Copies

Receiving an advance copy (to be referred to as 'Copy 'A')[143] in March, he immediately began annotations for a proposed second edition. Not content with this, he then asked for a full set of unbound pages from the publishers and, in April 1923, had them bound up and interleaved throughout with plain pages, for longer annotations. That copy is now referred to as 'Copy B'.[144]

These annotated copies make fascinating reading. Almost every one of the hundreds of annotations is initialled and dated by Gould, sometimes even with a note of where he was at the time. His criticisms are not confined to his own work either. As for the Astronomer Royal's Foreword, he remarked: 'As originally drafted by the late Sir Frank Dyson, this Foreword was rather wide of the mark, and (in one or two places) flatly contradicted the text. I revised it, I'm afraid, with an unsparing hand. He accepted all my amendments. RTG25-X-40'.[145] Of his acknowledgement in the Preface to the Hon. Horace Woodhouse, for reading the whole text, he noted scathingly: 'No, only a portion of the first half—and he might as well have left it alone'.[146]

Good as the first edition was, the annotations in these books are so interesting and so comprehensive they would have elevated the second

53   JOHN HARRISON.

The fusee contains Harrison's maintaining power, fitted in a very similar manner to that employed by him in his previous machines. The maintaining spring is very large*, and will keep the machine going for eleven minutes, although the operation of winding only takes a few seconds.

Harrison's foresight is well exemplified by his fitting a frictional brake, acting upon the rim of the balance, as shown in Plate XII. Without this, the watch, if it were ever allowed to run down, would stop with the remontoire down too, and, after rewinding, it could not be restarted, however much motion were given to the balance, unless the detent of the remontoire *chain* were first unlocked—a difficult and delicate operation for anyone but a skilled watchmaker. The brake gets over this difficulty. Operated by the last turn of the fusee chain as this unwinds from the fusee, it stops the timekeeper, *with the remontoire wound*, half an hour before the mainspring can run down.

The mainspring† is mounted in a " resting barrel "—that is to say, it does not directly rotate the barrel on which the fusee chain is wound, but fits inside a stationary barrel fixed to the plate nearest the dial, generally termed the " pillar-plate." The outer end of the spring is attached to this barrel, and the inner to an arbor upon which is mounted the barrel for the fusee chain. The resting barrel can be rotated to adjust the initial tension of the spring, and is held by a ratchet and click. This construction was imitated by Kendall and by Mudge‡, but its practical advantages are slight. However, it enables the barrel, being rigidly attached to a pivoted arbor, to be better supported and less affected by side strains than the ordinary construction, in which the arbor is fixed and passes through two holes in the barrel and its cover, constituting the bearings on which the pull of the chain is exerted.

The watch beats five to the second, a slight recoil being perceptible at each beat, and goes for thirty hours. The finish of the movement is very good, particularly for a man not trained as a watchmaker. The plates are of brass, polished, but not gilt. The pivot holes are jewelled as far as the third wheel—that is to say, those of the balance staff, detent, contrate wheel, fly, fifth, fourth, and third wheels. The jewels are rubies, and the end stones diamonds.

* It is not of the modern " split-ring " pattern, but a large spiral spring, considerably bigger than the mainspring of an ordinary watch.

† This spring, as well as those of the remontoire and maintainer, was made by Maberly, a famous London spring maker. The balance spring was made and tempered by Harrison himself.

When I cleaned No. 4 recently, I found the mainspring, which was lettered " I.M.—E.C." and dated " 2.13.(sic)1760," broken. I was able, however, to obtain a duplicate, made for me by Messrs. Cotton, of Clerkenwell.

‡ It is interesting to note that there are many points of resemblance between Harrison's resting-barrel and that invented by Mr. Lewis Donne for going-barrel watches.

[Handwritten marginal note at foot:] This is not quite correct. The fitting -which is complicated- differs, somewhat from that in the previous machines. These also differ, in this respect, among themselves. Thus, No 1's main-trainer had [originally] no stop; and those of No 2 and No 3 are fitted in the pinion of the 2nd wheel, not in the fusee. RG 5-XI-40

31. A page from Gould's annotated *Marine Chronometer*, 'Copy A'. The inclusion of all the notes into a second edition would have created an extraordinarily fine book, but such an edition has not yet seen the light of day (Courtesy of Charles Allix. © Sarah Stacey and Simon Stacey, 2005) edition to a standard beyond anything available today. Gould knew it would take years to complete though; in a flippant note to Dyson in January 1924, he remarked that the second edition would probably appear 'about 1950', sadly unaware that he would have been dead 2 years by that date.

Copy B was sold by Gould to the noted horological bookseller and antiquarian horologist, Malcolm Gardner (1896–1960) in June 1947 (retained by Gould until his death when it was sent to Gardner by Gould's literary agents, Curtis Brown). Gardner made strenuous efforts to get a second edition published in 1948, but in 1952 he was still struggling with the publishers, NAG Press, who were apparently too busy to deal with the matter, and it never appeared. In 1960 Holland Press produced a straight reprint of the first edition (including all the typos etc, but minus the dedication) and produced further reprints, several during the 1970s, a paper-back edition appearing in the United States in 1987.

In 1989 the Antique Collectors' Club (ACC, Woodbridge) produced another reprint, this time reset, including the dedication and with additional illustrations. But this poor new printing only compounded the old errors with incorrect captions and missing plates. At the time of writing (2005) ACC were advertising a new edition, due out in 2006.

In the 1960s the Harrison biographer Colonel Humphrey Quill, who at that time owned copy A, borrowed copy B from its owner and carefully transcribed the hundreds of annotations from both A and B into his own copy of the book, creating what is now referred to as 'Copy Q'.[147]

According to one source, there was a third copy of this important book, annotated by Gould himself, which, for the sake of completeness, could be named 'Copy C'. The horological illustrator Maurice Aimer (d.1986) claimed to own it, describing it as 'a personal copy of Gould's with many notes and corrections he added after publication.'[148] At the sale of Aimer's effects,[149] a copy of the book, marked in Gould's hand: 'Personal copy, uncorrected, R.T.Gould' was sold, and which might be what Aimer was referring to. Although it has no annotations on the pages, it is possible that this copy had papers inserted with corrections, but if so, these are all now missing.

The majority of the original typescript was destroyed by Gould, but a few chapters and the Appendices and Index were sold to Malcolm Gardner in 1936, on one of the many occasions during the latter part of his life when Gould was short of money. These were sold by Gardner in 1958.[150]

Marketing and Reviews

Published by J.D. Potter, the British Admiralty's Agent for charts, the book had been promoted by the publishers since June 1922[151] and received excellent reviews on its arrival in bookshops in April 1923. The *Nautical Magazine* echoed the Astronomer Royal that 'Gould's practical skill is happily associated with the zeal of the antiquary'.[152] The *Mariner's Mirror* (for whom Gould redrew the renaissance-style front cover in January the following year and which is still in use today) enthused: 'This is a book which will rejoice the heart of every reader . . . even those without an interest in the technical niceties of horology.'[153]

In the two whole columns dedicated to the book in the *Times Literary Supplement*[154] the reviewer was unable to find a single adverse comment to make, claiming: 'The book represents not only an astonishing piece of industry, but of scientific work of a high order. If the Royal Society had the agreeable habit of crowning works of merit, like some of its literary colleagues, this book would emphatically merit that distinction'. The British Horological Institute's own Instructor, Thomas D. Wright, reviewed the book for the *Horological Journal*,[155] finding very little in the text to make adverse comment on. The only criticism he made of a drawing (that of Cole's 'double overcoil' spring) turned out, after a letter of explanation from Gould in the next issue of the HJ, to be a misunderstanding on Wright's part.

There were notable literary fans too. Rudyard Kipling, writing from *Bateman's*, his house in Sussex, sent two letters to Gould in June 1923 extolling the virtues of the book and commenting on the use of chronometers at sea.[156]

In October, Gould's old friend Arthur Hinks, Editor of the Royal Geographical Society's *Geographical Journal*, was working on a review for that organ and wrote to Gould to point out a couple of typographical errors. Gould noted in his reply: 'I am glad to hear that you are reviewing my book. My previous reviews have all been sugar and water. Give me some acid for a change'.[157] Hinks could only answer, modestly, that 'I cannot pretend to have done your book justice, but as I could not find anybody with sufficient technical knowledge to do it properly I did the best I could—certainly without sugar and water and certainly not with acid. The liquor will at any rate be found dry'. The two and a half page review was uniformly positive, describing it as a 'capital book' and apologizing for the delay in reviewing it.[158]

But Gould did finally get a little of the acid he sought. A rather late, and rather mixed review appeared in *Nature* in March 1924,[159] the reviewer's initials, appropriately enough, being 'RAS'. Appropriately, because this was in fact Ralph A. Sampson (1866–1939), a.k.a. the Astronomer Royal for Scotland, and a Fellow of the Royal Astronomical Society (RAS).

Not knowing Gould, and making the mistake of not thoroughly reading the book under review, Sampson chose, very rashly, to dismiss Harrison's work: 'To some he appears . . . incurably clumsy. His making clocks of wood, his complications, his retrograde inventions like the grasshopper escapement and the gridiron pendulum, when the all-but-perfect thing was already in existence in Graham's dead-beat escapement and compensation by a jar of mercury, are to his debit.' Digging himself in even deeper, he concluded: 'all Harrison's ideas were of the nature of misdirections'.

This was not quite the species of acid Gould had been seeking but, as it turned out, he was in no fit state to respond immediately anyway. By early 1924 Rupert had plummeted into another period of depression which, off and on, would stay with him for another year and a half. In April 1924, Hinks replied robustly to Sampson on Gould's behalf, in a letter to *Nature*, a somewhat dismissive reply from Sampson appearing alongside.[160] Finally, in June, as his depression eased, Gould himself wrote to *Nature*, pointedly 'I should like, if it is not too late, to have an opportunity of placing on record a considered opinion . . .', explaining, blow by blow, where the reviewer had misinterpreted Harrison's achievement.[161]

7

Horology: The Obsession

The seeds of the new bout of depression had of course partly been sewn over the previous 4 years while writing the book. During this time Rupert had also been managing to keep up a full time job at the Admiralty while trying to bring up a young family, at the same time as restoring H1, followed by H4. The huge stress of this manic overwork, much of it done in the early hours of the morning, had naturally taken its toll. Writing to his Swiss friend, the chronometer maker Paul Ditisheim, in January 1923, he noted: 'after four years of the hardest work I ever had in my life I have at last a chance to rest . . .'.[162]

Similarly, in a letter to Admiral Parry the ex-Hydrographer, he observed: 'the whole of my spare time has been absorbed by my book on the chronometer which has kept me for the last two years working on an average until one or two o'clock in the morning. In fact, my wife says that she is going to apply for a divorce on the grounds of desertion, as I hardly ever see her or the family except at breakfast. . . .'; words which, though said partly in jest, would prove sadly prophetic.[163]

Still, the horological bug would not let go and in February 1923 Gould was discussing the possibility of a lecture and article on chronometers for the Royal Society of Arts, either based on the RGS lecture 2 years previously, or on chronometer improvements funded by the Society.[164] Neither lecture nor article happened, probably owing to his illness and, as discussed later, it would only be in 1928 that Gould finally appeared at the Society, to lecture on the history of the typewriter.

William Monk Gould

The weeks immediately following the book's release were not only a time of exhaustion for Gould, but were tinged with great sadness. Writing to Edward Heawood, Librarian at the Royal Geographical Society on 18 April, Gould remarked, 'For the last month I have been condemned to the suffering of watching my father growing gradually weaker from heart disease, which culminated in his death ten days ago.'[165]

After the publication of the book, Gould spent the summer and early autumn of 1923 recovering from nervous exhaustion. He wrote to Paul Ditisheim again on 23 August from Druimgigha explaining that he was 'over-worked and mentally tired-out'.[166] At the time, Ditisheim would have been sympathetic, as coincidentally, he was having similar problems. In a few months time Ditisheim suffered the same fate, having his own breakdown brought on by overwork, inducing him to leave his native Switzerland for Paris to start a new career.[167] In his letter Gould described to Ditisheim his father's little shooting lodge, Druimgigha, in the wilds of Scotland, saying it was perfect for a rest but reminded him terribly of his father. He spent three weeks there preparing for a return to some kind of normality.

A Crossroads

Here was another crossroads. Which direction he took now would be critical, and there were now new and very real worries about his job and salary. A scheme had recently been introduced in the Hydrographic Office that, while benefiting most retired officers, almost halved Gould's salary from £850 to £450 a year. Writing again to Admiral Parry for help, Gould stated 'this makes me consider very seriously whether I ought, not only in my own interest, but in those of my family, to stay any longer here. I have had an offer to take charge of an electric clock factory; but I declined it, since I don't think I am fitted for business, and I *do* think I can do this job better than anyone else can'.[168]

The reference to the possibility of a job in an electric clock factory is most interesting. The offer was probably made by an acquaintance of Gould's, Frank Hope Jones (1868–1950), founder and owner of the Synchronome Company and a central figure in the development of electric clocks. It is fascinating to imagine how Gould would have got on in the commercial world, especially working alongside Hope Jones, another very pedantic personality. In all probability Gould was right, it would have been a disastrous move.

Another option was to consider a literary career. In his letter, Gould informed Parry he had been asked by the publishers Methuen to write a *History of Geographic Discovery*, and Potters were interested in his continuing the *History of the Hydrographic Department*. They were also considering asking him to do biographies of John Harrison and the polar explorer James Clark Ross (1800–1862). As it turned out, none of these options was taken as it seems his salary was increased again. Parry had succeeded in exerting pressure on his behalf and the Admiralty Board agreed to see his anomalous case resolved. This must have been a great relief for Gould, but posterity is undoubtedly the poorer without a Gould biography of Ross, and especially one of Harrison.

So, which way to turn at this crossroads? The book was written and H1 and H4 were preserved for the future. Even though H1 was not complete and working, posterity would have considered Gould's contribution not only generous and talented, but also very remarkable. He could—logically, he should—have decided to follow his family's wishes and study for the Bar. Perhaps he might also have invested a little more time in resting, while consolidating the family unit. After all, they too had made sacrifices to see the book published and the timekeepers cleaned.

H2 . . . *and* H3

However, in October we find him embarking on a new round of clock restorations in the evenings and, incredibly, the work he took on this time was considerably more onerous. In discovering the timekeepers and realizing their significance, Gould had struck gold. The fatal attraction these timekeepers held for him was simply too great, the full and detailed story of their incredible creation and trials (by no means completely told in *The Marine Chronometer*) was just too exciting.

Driven by an apparently unstoppable determination, Gould was not content this time to accept just one machine for cleaning, he asked for both H2 *and* H3. With Dyson's continued blessing, he collected them from the Observatory on Wednesday, 10 October 1923. Now without his little Scripps-Booth motor car, Gould asked a colleague, Cdr. Phillips from Plans Division at the Admiralty, to help by providing his Ford (a 'shooting brake') to transport them to Epsom.

Gould wrote to Paul Ditisheim at the time, and explained this apparently masochistic decision to load himself with another heavy burden when he should be relaxing. Employing slightly doubtful logic, Rupert stated that as it was writing that had caused his nervous trouble, he was now returning to mechanical work instead, considering it a form of therapy.

The Notebooks and the Restorations

Building on what he had already learned in work on H1 and H4, and in writing the book, Gould was determined now that the whole series of timekeepers should be restored to good working condition. Increasingly aware of the historical importance of the pieces, he was also now determined that these projects should be methodically and carefully recorded. From this point onwards all the work he did was fully written up in notebooks, every aspect covered in minute detail, including dozens of beautiful sketches and descriptions, all to be preserved as part of the history of the timekeepers.

By the end of Gould's life there were eighteen of these books, covering half a lifetime's work and including data on every one of the machines.[169] They make fascinating reading. Some of the later books are bound up carbon copies of his correspondence (almost always typed of course) relating to the restorations. Often letters would be illustrated with his painstaking pen and ink sketches to explain problems or describe parts needing repairs. In these cases, Gould would not only do these drawings for the top copy, but also would then sit down and make an identical, and equally fine, drawing for his own copy of the letter.

The earlier notebooks are in manuscript and cover the restorations on a day-to-day basis. Written in diary style, the prose is careful yet familiar, but decidedly self-conscious, with frequent references to his own feelings or circumstances. The overriding impression one has is that Gould is acutely aware of the importance, and perhaps the personal price paid, in these Herculean tasks. With his constant asides and occasional, rather pretentious, Latin expressions thrown in, he tries hard to convey the sheer unrelenting toil involved. Looking through the pages and pages of notes one gets a real flavour of the months spent, for example, cleaning the multitude of small pieces of brass, and the great frustration when parts are broken or the timekeeper fails to work, followed by the huge relief and satisfaction when the pieces go together to create a working machine again.

The work was almost all done in the weekday evenings, sometimes into the small hours, and went in phases, occasionally entering the periods of illness. The handwriting itself is an interesting pointer to Gould's state of mind during these times. Depending on his mood, it is sometimes barely recognizable as by the same hand.[170]

Other literary tasks would also interrupt the restoration work; sometimes proof reading of his publications, sometimes translations of horological texts. Gould was fully conversant with Latin from his school days and could get by in a number of foreign languages, but spoke none fluently. Paul Ditisheim once said that Gould spoke French like the eighteenth century chronometer-maker Ferdinand Berthoud! But Gould was quite prepared to work his way through many thousands of words of Spanish or German using a dictionary, and what knowledge of the language he had.

The Quality of the Work

As for the restorations themselves, Gould was the first to admit he was hardly the obvious choice to tackle them and the 'amateur mechanic' has been mentioned already. Without any horological training or qualifications whatever, he came to the subject as an intelligent novice and *dilettante*. Thus his approach was a curious combination of the scientific and logical, with the amateurish and unorthodox. For example, the newly extended ends of the broken pallets in H2 were attached by binding-on with linen thread, and the balance connecting wires first fitted by Gould were made from florist's wire. One can imagine what a professional antiquarian horologist would say today.

The publication of *The Marine Chronometer* did at least establish Gould as a serious academic from a literary point of view, but even here Rupert would occasionally have what might be described as a lapse of good judgement. In October 1923 he published in the *Horological Journal* details of a peculiar addition he had made to a longcase clock. The trunk door of the clock had a pendulum 'lenticle', that is, a little window through which the pendulum bob could be seen passing as the clock ran. Gould decided, in a fit of facetious humour, to paint a *face* on the pendulum bob and designed a system of wires and strings so that the face *winked* at the viewer as the pendulum swung. Not exactly guaranteed to ensure the professionals took his work seriously.

In retrospect, poor security in Gould's workshop was a potential threat to the Harrison timekeepers too. Keeping H4 in his desk drawer rather than in a safe was certainly a risk. In later years a leaking roof and, on one occasion, the outbreak of a fire in the workshop would both prove to be potential threats to the timekeepers during the restorations. Similarly, his practice of taking many small parts of the timekeepers, all neatly packed in envelopes, on holiday with him to Norfolk, so that he could clean them in his hotel room during the night hours, would also be unthinkable today.

A Systematic Approach

On the other hand, Gould's logical, systematic approach to the restorations was vital to their success. He always took far longer to consider a technical problem than any professional could afford to. His constant companion was his curly, 'Sherlock Holmes' style pipe, and technical problems were often described as needing a 'one-pipe' or a 'two-pipe' solution, depending on their difficulty. The design and construction of specially made torque meters and special tools for determining the inaccuracies of the wheelwork, (see Fig. 48, p. 242), was very much ahead of its time and something the conventional chronometer makers would have had too little imagination even to contemplate.

It could also be argued that a scientifically independent mind, without preconceived horological ideas, would be perfect for interpreting the deeply unconventional thinking of John Harrison. Anyone, qualified or not, would have to 'feel their way' with these unique machines. It has to be said however, that Gould did struggle to understand them sometimes. It is probable, for example, that he never fully understood the grasshopper escapement, nor how the temperature compensation worked on H1 and H2, though he managed to get both to work well.

Then again, one or two of Gould's other restoration procedures would today be totally condemned as botchery. Without the experience of a professional clockmaker, Gould had the dual handicaps of little horological experience and only average manual dexterity.

Clock and watchmakers rely on their insight—born of experience—rather more than even they themselves realize. Without it, dismantling and the interpretation of missing parts of a mechanism is a tortuous, lengthy business. And it goes without saying that with such delicate yet complex machines, a high order of manual dexterity is a great asset. It is true that Gould tackled the abstruse H4 successfully and he certainly could manage good manual dexterity when he was able to concentrate and really put his mind to it.

Breakages

On the larger timekeepers though he did not always achieve this. It may be that he considered them rather more 'agricultural' than H4, and though he prepared himself mentally for the extremely delicate work in that watch, he hadn't fully taken on board the fragility of some parts of the larger timekeepers. Small parts, usually the brittle, high-tin bronze pieces, would sometimes get broken during the course of the restoration work and these would then have to be repaired or remade. This was usually (in the later restorations) carried out by R.J. Hopgood, Gould's trusty instrument maker in Blackheath, always ready to help out when needed.

Often parts of the timekeeper would be scratched with code letters or punch-marked with numbers to ensure they returned to their correct respective places. On one or two occasions, where parts were particularly difficult to get together, Gould would resort to cutting or enlarging holes in the frame to improve access or visibility. At one point in the restoration of H1, some of the riveted pillars were simply pulled out of the plate in order to clean them properly and had to be fixed back in by screws instead.

In short, one cannot say that the work was carried out to the exacting standards of twenty-first century conservation practice. But he wasn't working in the twenty-first century but in the 1920s, and he may well have been right in remarking that without his attentions the timekeepers may not have survived at all.

Books 1 and 2

Strangely enough, the first comprehensive data on the restorations begins in Gould's book No. 2. Book No. 1 is really a nineteenth century copy Gould had acquired of one of Harrison's manuscripts (*An Explanation of my Watch . . . of 1763*). Having a few blank pages at the back, the book was then used by Gould, turning it back to front, to make notes on the first cleaning of H1 in 1920. These only cover a dozen or so pages and actually say very little beyond noting the arrangement of the plates and the wheels.[171] Half a dozen further pages relate to work done on H1 in the 1930s, but the majority of this work was covered in the later books.

Book No 2 therefore starts with the journey in Commander Phillip's Ford on Wednesday 10 October 1923, when H2 and H3 were delivered to the brick built garage which was Gould's workshop at the Cottage in Epsom. Ever scrupulous in his detail, Gould records that the time-keepers arrived at precisely 7.15 p.m. that evening and work started the same night.

8

H2 is Restored 1923–1925

No sooner had Phillips left the Cottage at Epsom than Gould got down to work. Both H2 and H3 were in their original glazed cases and he began by getting them both out of these, remarking that H3 was more corroded than H2 but 'Both have suffered from the red-lead used to lute their cases'. Of both he said: 'I see little to do mechanically, except the pallets & balance connections, but the stripping and cleaning will be a long job'. The entry ends 'Got to bed 1am, & to sleep about 4'.

The following day saw work at the Hydrographic Office as usual, and the evening at home was then rounded off with 2 hrs work in the garage workshop. The dismantling of H2 itself began at 10.00 p.m., and stopped at midnight. The next evening, work continued at 11.00 p.m., finishing at 12.20 a.m. and, after the weekend, Monday evening saw Gould carry on from 10.30 p.m. until 12.30 a.m. All parts were carefully stowed in marked envelopes and pages of little drawings in the notebook indicated the relative positions of parts and how they interact.

After six evenings of this, Gould wrote to the Astronomer Royal on 16 October summarizing the condition of H2 as found. Basically the machine was mechanically sound, with only a few steel parts rusty but the pallets of the escapement both had the acting ends broken off and the connecting wires for the balances were either missing or broken.

Having planned the making of new pallets and repairs to the connecting wires, Gould wrote to the Astronomer Royal with dimensions for both H2 and H3, with a view to new display cases being made for them (and rendering their original travelling cases somewhat redundant, sadly). An estimate of £27/10/00[172] for the two, arrived a week later at the Observatory from A Wescott (later to be an employee of the Observatory) and the cases were ordered.

Breguet celebrations

Between 22 and 27 October 1923 Gould was in Paris, representing the British Horological Institute at the celebrations to mark the centenary of the death of the great French watchmaker Abraham Louis Breguet (1748–1823). After the official opening, the programme for the celebrations included a weeklong congress at the Paris Observatory, a reception at the Hotel de Ville (City Hall), and a special exhibition of Breguet's work at the Musée Galleria (typically, Gould made careful notes and diagrams of the layout and details of the exhibits in his pocketbook).[173] There were also visits to the Eiffel Tower broadcasting station and the Breguet aviation works, which would no doubt have fascinated him. The celebrations ended with the *Seances Solennelle*, attended by the *Ministre de la Marine* and *l'Amiral Attaché à la Presidence*, at which event Gould was invited to wear his full dress uniform. Cecil remembers being sent a postcard from Paris by his mother at this time, so it is likely Muriel accompanied her husband on the visit. During their stay Rupert planned to visit Paul Ditisheim, one of his horological friends whom Muriel liked too, so there was also socializing on the trip.

Back in England in November, Gould was invited by The British Horological Institute to give his first lecture to the Institute. It was entitled *Early Marine Chronometers*, and was held at their headquarters at Northampton Square in Clerkenwell on the evening of the 21st.[174] He was given permission to display H4 at the meeting and, during the lecture, he referred to the work currently under way on H2 and H3, producing one of the balances of H2 to show members.

Work at the Admiralty during the day was continuing at the usual pace too. In late 1923 Gould was off again, this time to Oslo, in Norway. He was representing the Hydrographic Department in giving advice during international discussions on the extent of Norwegian territorial waters. The Norwegian experts were particularly happy with Gould's help, and a message of gratitude was sent to the Hydrographer, fortuitously arriving just as a new three-year term for Gould's employment in the Department was up for consideration.

Thankfully, the Hydrographer kept him in post and all seemed well on the professional front. Things at home were not so rosy however. Muriel certainly didn't see the new Harrison restoration project on H2, with H3 waiting in the wings—and all still carried out in his home time—as beneficial to a settled future at home with the family. It was becoming clear that any thoughts of his studying for the Bar had gone out of the window and that a regular stable relationship with wife and children was not exactly the highest priority for this particular husband. Things would not get better over the next 4 years either, and Muriel was to discover just what these projects really did entail. They were turning into a lifetime's work, and she began to wonder where she and the children fitted into these plans.

At the beginning of December 1923, on his return from Norway, Rupert got back to work on H2 and on 20 December he was able to report to Sir Frank Dyson that he had H2 temporarily working, by winding the remontoir, though he had not yet cleaned the movement. The note-book records on 18th: 'I feel I have not worked in vain'.

The Grasshopper Escapement

Gould's letter to Dyson included a technical remark about the grasshopper escapement that, surprisingly, was not only erroneous but was to mislead horologists for many years to come. He stated that if the drive on the wheels were temporarily to be held up while the balances were running 'there is a point, at the middle of the swing, when the escape wheel is free, and if the force of the mainspring comes on it at that point the wheel will run and, probably, carry away the pallets'. It is true that, *in extremis* this situation can occur on the grasshopper escapement as fitted to the pendulum clocks, like the Royal Astronomical Society's clock. It is also true in the marine timekeepers if the escapement is not correctly adjusted, but the grasshopper escapement cannot do so in the marine timekeepers when it is correctly set up. In January 1924 Gould continued experimenting to discover the correct form for the pallets and stated that he was 'coming to understand the grasshopper escapement, little by little . . .'. Perhaps he eventually came to realize his error in the earlier assessment.

Gould now began the initial cleaning of parts of H2, and several late nights at Epsom saw progress, after which, he assured Dyson, he had: 'no doubt that when returned the machine will remain in going order for many years to come'. On 13 January he records that he was about to embark on 33,000 words of Spanish translation, but this too came to a halt on 10 February owing to an 'opportune breakdown of my typewriter', giving him a Sunday to start cleaning the plates of H2.

However, the entries stopped here for a while, as yet more projects intervened. An article had been promised for the *Geographical Journal* on an area of deep sea in the Antarctic known as the 'Ross Deep', and as well as getting this done, Gould also then undertook the restoration of the mechanical orrery belonging to the Royal United Service Institution.[175]

Overwork

Not surprisingly, this bout of overwork now induced another period of illness, beginning with severe nervous depression and insomnia starting in early March, followed by 'flu and bronchitis; the illness finally easing in the middle of May.

Muriel had little time for her husband's nervous depressions, partly brought on, she knew, by his overwork, and Rupert resorted to staying with his mother in Southsea. He spent over 2 months there, half of which time he spent in bed and for 6 weeks of which he was 'unable to do anything'.[176] In a letter to Paul Ditisheim sent from Southsea, he explained tactfully that: 'there was more room here to be ill...' and mentioned that he had had a visit there from his friend the American horologist Paul Chamberlain.[177]

Work on H2 finally started again on 25 July 1924 'after a long interval, due to illness and other work...' as recorded in notebook No 2. The smaller brass parts were now cleaned by Gould in a solution of soft

32. Rupert T. Gould at Epsom in 1924, with one of H2's balances on his knee, and H3 on the table. He is seated in front of the garage at the Cottage, where the restoration work was carried out. (© Sarah Stacey and Simon Stacey, 2005)

soap and ammonia (familiarly known in the horological world as 'clockmakers' soup'). Gould, using this method for the first time, remarked, 'effect is good, but not equal to polishing . . .'

The July issue of the *Horological Journal* contained a note in the Astronomer Royal's annual report mentioning that Gould was restoring H2 and H3, and in the August issue Gould sent in a photograph of himself at Epsom with H3 and the balance of H2, that he hoped 'may be of interest'. The photograph had probably been taken a few weeks before. In the same issue of the Journal, Gould demonstrated his occasional (but decided) tendency to be ascerbic when communicating with people who were prone to muddled thinking. A letter by him was critical of an article in the Journal, on compensation balances, which had employed over-simplistic, misleading science. Responding in the next month's issue, the author, gratified at comment from 'so eminent an horologist . . .' replied that it was necessary to keep the explanations simple, even if not quite correct, in order for the readership to understand. This further muddled thinking caused Gould to reply very sharply, defending the intelligence of the HJ readership and ending sarcastically 'I have no claim to be considered . . . 'an eminent horologist'. If I were one I should not have taken any notice of his article'.

Notebook No 2's entry for 3 August 1924 contains a summary of the work done on H2 so far. Most of the cleaning was now complete and most parts had been polished with metal polish and rinsed in petrol, and only required lacquering. In a letter written that day to William Bowyer, the Astronomer Royal's assistant, Gould says 'the cleaning has been a tremendous success and the parts look like new . . .'. He intended to make a better repair of the pallets and then build a wooden cradle in which to support the movement while reassembling.

H2's original glazed case

Unfortunately he decided at that stage not to put H2 back into its original glazed case, which was still languishing, un-restored, in the garage. Gould explained: 'It is a most clumsy affair and has practically to be built up round the machine. I am quite certain some damage to the latter would be done in the process . . .'. His letter to Bowyer arranged for the new display case to have the mounting pillars positioned correctly and he provided a list of the main mechanical defects he found and repaired in H2.[178] On 23 August the machine was ready to be reassembled and, listing the order in which he believes this should be done, Gould ponders 'I wonder how soon I shall have it together? I have pinned up Bradley's drawings opposite the bench. They are extraordinarily useful'.

During the reassembly that followed, two of the anti-friction segment centring springs were broken and had to be repaired. By 29 August the machine was running again but Gould could not keep it running, even with the remontoir set at its maximum output. After a number of experiments with the torque required to run the balances he realized that there was a problem with excessive energy consumption by the balances, probably caused by the connecting wires being too strong and too tight. These were replaced and the problem was reduced but he still decided to have new, stronger, remontoir springs made. This should not have been necessary as it is unlikely the original remontoir springs had lost much of their strength, but as they do appear to have been slightly rusty this is a possible explanation.

Another breakage

The next task was to fit the parts of the temperature compensation, a job complicated by Gould's making 'a bungle of it' and breaking two pivots, which then had to be repaired. Much further experimenting with the escapement and balances saw that long day's work complete, Gould remarking 'But for the breakage of the compensation pieces I should say I had done a good day's work.'

The London firm of Usher & Cole, watch and chronometer makers, were sent the remontoir springs on 1 September, in order to make copies but 30 per cent stronger. They telephoned on 4th to say that as the spring makers they usually used had failed to make good ones, they were now making them themselves. By that time Walters (a local handyman in Epsom) had repivoted another broken part and delivered the compensation pieces later that evening, so the rest of the time that day was spent putting the compensation assembly together; 'quite a good evenings work' Gould recorded in the notebook.

The following day the remontoir springs arrived and, after experimentation to ensure they provided sufficient torque to drive the balances, H2 was running 'under its own steam' for the first time, though there were still a few problems to be overcome and the machine was still consuming much too much energy. Gould could not yet get the maintaining power to wind the remontoir springs at the setting needed to run the balances, and he hadn't yet managed to get the fusee chain attached to the barrel; a temporary gut line attachment was connecting it. Nevertheless, he now experimented to bring H2 roughly to correct rate by counting beats 'I will now light a pipe & count 180 swings . . .'

The entry for 6 September 1924 finds Gould 'thinking over various points . . .'. He was on the right track as he was wondering about possible frictional losses in the balances and elsewhere in the machine, and how they can be reduced. Connecting up the fusee chain was also on his mind and he decided that as the machine will have to come apart again to change the set up on the maintaining power, he should start again on this problem.

Getting the fusee chain to hook on the barrel may seem, to a professional horologist, a fairly straightforward job, but with a machine weighing over 100 lbs and the barrel underneath, just lifting the movement up to reach the barrel is itself not easy. Having planned the next dismantling in detail, writing down the complete course of action, he began at 8.30 p.m. and as midnight saw the new day arrive, Gould had had the movement apart, adjusted and complete again. After a scare because it would not run he found the simple cause, one of the connecting wires on the balances was off its arc, and H2 was now ticking again. Full of optimism, after a few hours' sleep, and breakfast, Gould started work later that morning, Sunday 7 September.

Another disappointment

He was in for another disappointment. The notebook states wearily: 'As has been truly said, when we get too uppish, Fate always leaves the dustpan on the stairs for us. Having fitted the axial wires, & successfully wound the machine, I fitted the hour wheel & found it didn't meet the teeth of the minute wheel pinion. I examined, & found the arbor of the minute wheel had snapped. I must have jarred it with the cover when taking that off. Words fail me.'

With the help of Walters the machine was taken apart again and the broken arbor was sent, along with two pages of typed and illustrated instructions, to the professional chronometer makers, Usher & Cole, in London, for a replacement to be made. The job took just 2 days, though the new arbor was made in steel rather than bronze. At 9.00 p.m. on 9 September H2 was again complete and started, though the next morning it was found to have stopped in the early hours, both balance spring connecting wires having broken by their securing screws. Splicing these together temporarily, the machine was set going again and at 9.45 p.m. Gould noted: 'Joy! I return to find that the machine is going strongly, and has been going all day!' though the machine was still running considerably too slow.

Further disappointment

He was not quite out of the woods yet though, as the balance spring wires were still giving trouble. Two days later, at 10.45 p.m. Gould was 'Disappointed. Machine will not go. I have translations to do so must stop' and the next evening, still depressed at the machines recalcitrance in spite of all his efforts, he writes 'I return to my work, with a wild wind raging & a lowering sky. A fit setting!'

However, after much experimenting and with new balance spring wires fitted, the next 10 days saw H2 running, keeping reasonable time and coming close to completion. On the 19th, Gould 'took a squeeze of the winding square . . .', presumably in plasticine, and sent this and the pipe surrounding the square off to have a crank key made. Last minute problems with the escapement tripping, and the remontoir fly sticking, necessitated another partial stripping of the machine again on 21 September and the next 2 days were spent bringing it to time for the last time at Epsom.

Work on H3 begins

Undeterred at the time and trouble H2 had taken up, and seemingly oblivious to the growing irritation of his wife, Gould now completely stripped H3, eager to start the next project while he was bringing H2 finally to time. Writing to the Astronomer Royal on 22 September 1924, he announced the completion of H2 and his intention to start the full restoration of the third timekeeper.

H3 was Harrison's most complex and mysterious, and some would say his most beautiful creation. It is in some ways the Holy Grail of Harrison's life's work. Just as H3 had caused Harrison the greatest problems and stress, so it was to be with Gould. The circumstances of this restoration would prove to be as complicated and convoluted as the story of the machine itself.

In his letter to Dyson in September, Gould noted (of H3 versus H2): 'there is a wide difference in the condition of the surfaces. No 2 had oxidised in a slow and gentlemanly fashion and the surfaces of the plates etc. polished fairly easily—although in some spots there were deep-seated stains which I was unable to remove. But No.3 has oxidized much further. The brass has become a sort of deep chocolate colour . . .'

His proposal was that the larger of these brass parts be sent off to the professional metal polishers, Bucks of Westbourne Grove, leaving the other bits stored in the garage at Epsom. H3, now wholly dismantled, entered what is probably the most vulnerable period in its whole existence.

H2 returns home

The evening of the 23 September 1924 saw Gould making stretchers for carrying H2 and, with his neighbour 'Palmer's' assistance, it was lifted down onto a table and the balances lashed, in readiness for transport. 'For a variety of reasons' he had decided to transport it to his office at the Admiralty for a few days before seeing it taken down to Greenwich.[179]

The following day the lorry arrived from the Admiralty and H2 (and 'most of H3', for sending to Bucks for polishing) was loaded, though the lorry was found to be too small for the (un-restored) original cases. With the Bradley drawings and Gould's own mezzotint of Harrison also put on board, the consignment endured 'a bumpy journey . . .', arriving at the Admiralty in Whitehall at Noon. Gould and his colleague Collings carried H2 up to Gould's office without damage but Gould then broke both connecting wires when adjusting them. With the experience of fixing these parts several times now, he took just 40 min to replace them, having the timekeeper running by 12.45.

Over the next few days, in late September 1924, it was brought to time and appears to have settled down to reliable running. Naturally enough, after such a tough project, Gould felt his achievement warranted some form of public acknowledgement and arranged a press release, notices then appearing in the *Daily Mail* and the *Daily Telegraph*, amongst others.[180] He also wrote a short article for the *Nautical Magazine*, including two drawings he did especially for it, showing H2 in its original glazed carrying case and of the movement only.[181]

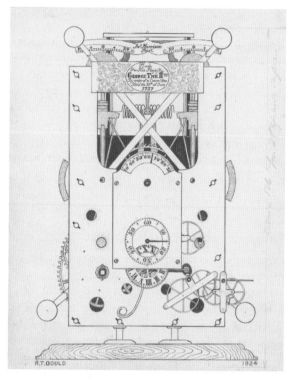

33. H2, as drawn by Gould for an article in the *Nautical Magazine* in 1924. Gould also loaned the drawing to his friend Paul Ditisheim for use in his book *Pierre Le Roy et la Chronometrie* in 1940. (© Sarah Stacey and Simon Stacey, 2005)

The RAS Regulator

His appetite for horology unsated by H2's splendid refurbishment, Gould immediately started planning for the completion of H3. Many of the parts were however on their way to Bucks in Westbourne Grove to be polished and, horologically greedy as ever, and with no work to do on H3, Gould now began negotiations to take on the restoration of the Royal Astronomical Society's regulator by Harrison as well.

In November, submitting his bill for expenses to Greenwich for work on H2,[182] Gould continued in his covering letter: 'I had a look the other day at the RAS Harrison clock. I can't make out what Cottingham's been up to [E.J. Cottingham had overhauled the clock in 1911]. The driving weight is absurdly too large, and the remontoire winds with a most fearful bang and consequently is always liable to trip. I hope to put matters right, so that the clock can be kept going. There is no reason why it should not go well for a long time. The movement is almost a facsimile of that of No. 3, with a pendulum substituted for the balances'. He consequently wrote to the RAS offering his services, an offer which was gratefully accepted by the President and Secretary of the Society.

However, even Rupert's enthusiasm for horology eventually had its limits, and the overwork now began to make itself felt. At the end of 1924 Gould began a steady decline into another severe depression.

H2 goes to Wembley

As for H2, after its return to the Observatory in late 1924, it was decided (probably in the spring of 1925 but the actual date is uncertain) it should be exhibited, in going order, at the British Empire Exhibition at Wembley. This had opened on 23 April 1924 and the timekeeper was a 'late entry' in the show. But the Exhibition went on until October 1925 and there was still plenty of time for visitors to appreciate this marvel of technology and the wonderful restoration of it by Gould. The watchmaker and BHI instructor, Heinrich Otto, always an admirer of Gould's, displayed a copy of Gould's book *The Marine Chronometer* throughout the exhibition, as a tribute to Gould's continuing work on chronometers, though he recalled that the man himself, now in the throes of nervous depression, never visited the show at Wembley.[183]

While at Wembley H2, in its magnificent restored condition', came to the notice of the Director of the Science Museum, H.G. Lyons. Impressed and excited at this interesting potential exhibit, Lyons wrote to Dyson in the autumn of 1925, enquiring about a loan of H2 with the watches H4 and K1. Dyson replied positively, suggesting a five-year loan of H2 after the Wembley exhibition had ended on 2 November, but decided against lending the watches.[184]

The Astronomer Royal was unsure if Gould would be well enough to escort H2 and in the event it seems it was Dyson's assistant, William Bowyer, who set it up at the Science Museum. Gould did however write a detailed, 23 page description of the timekeeper, with the usual fine pen-and-ink illustrations, for the museum early the following year.[185]

H2 was professionally photographed at the Museum soon after the loan began and these were sold by the Science Museum as postcards marked 'Copyright Science Museum'. Over the years, this has led to the occasional misunderstanding by the public that the timekeeper was, at the time, the property of the Science Museum. H2 was to remain on loan to the Museum, on display and in running condition, with its rate of going recorded at various times, for over ten years.

With H2 now nicely settled in to a highly appropriate home, where its intricate splendour (and Gould's restoration) could be admired by every visitor, one might imagine Rupert was content. But he was approaching another crossroads in life and was in no fit state to make sensible decisions. With H1 cleaned, H4 overhauled and H2 restored and running, here was an ideal opportunity to rest his overworked system as his mental health was now deteriorating again. It might also have been a chance to repair the split which was daily opening in his marriage. Unfortunately for Rupert though, even if he had tried, it would by now have been too little too late, as Muriel was also now at her wits end and the period that followed would mark the beginning of the end for their marriage.

9

The Sette of Odd Volumes

The period after the completion of H2's restoration proved catastrophic for Gould. Over the previous few years, several factors had contributed to the growing calamity, but the seeds of one were planted back in 1920, and to explain, it is necessary to look at how Rupert Gould's social life had been developing. The fashionable parties attended with Muriel after they were first married were great fun and were enjoyed by both, but they didn't provide the intellectual stimulus that Rupert needed, and he longed for more challenging conversation.

Ralph Straus

On the evening of Tuesday 25 May 1920, Gould was invited to attend a dinner by his friend, the writer and journalist Ralph Straus. The venue was the select and fashionable Imperial Restaurant in Regent St, in London's West End, two doors down from the Café Royal. Gould was one of two guests Straus was introducing to his literary dining club, The Sette of Odd Volumes, with the intention of proposing Gould for membership, should he enjoy the occasion. He did, and remained a member for the rest of his life.[186]

It is not certain how Gould and Straus met, but, as noted in Chapter 3, it is possible it may have been in connection with Rupert's psychological illness during the war, as Straus had looked after servicemen with shell shock. In 1920 Rupert drew the illustration for the dust jacket of Straus's new novel *Pengard Awake!*, the subject of which, John Pengard, was a schizophrenic. In the story, the narrator, who appears to have been modelled partly on Straus himself, was President of a literary dining club called The First Folio Club. Apparently drawing his ideas from life, Straus clearly had The Sette of Odd Volumes in mind, of which he had been President throughout the war years, 1913–1919, and it is possible Pengard's character was loosely based on Rupert Gould.[187]

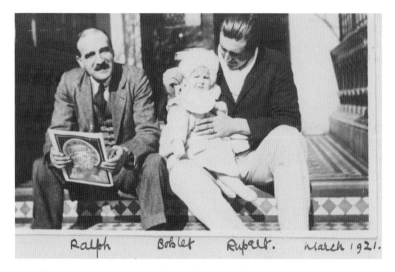

Ralph   Bobley   Rupert.   March 1921.

34. The author and journalist Ralph Straus, with Rupert and young Jocelyne at the front door of Waylands in 1921. Straus introduced Gould to The Sette of Odd Volumes in 1920. (© Sarah Stacey and Simon Stacey, 2005)

Conviviality and Mutual Admiration

The Sette of Odd Volumes, which survives to this day (2005), was founded in 1878 by the noted antiquarian bookseller, Bernard Quaritch. It was one of a number of literary dining clubs flourishing from the end of the nineteenth century, one of the more famous perhaps being The Savage Club. All had particular *raisons d'être*, The Sette of Odd Volumes' simple mission being 'Conviviality and Mutual Admiration'.

The underlying aim of members was to enjoy good food and drink, along with an interchange of ideas and researches from an extremely broad range of disciplines, while not taking themselves *too* seriously. The Sette's rules and procedures could be very odd indeed, even 'pythonesque' on occasions. For example: 'Rule XVI: There shall be no rule XVI' and 'Rule XVIII: No O.V. shall talk unasked on any subject he understands', though one of the more sensible ones (Rule X) was that 'discussions about Anthropology, Religion and Politics shall be put down by the President'.

Publication was another activity of the Sette's. The little books, usually resulting from papers given at dinners, were called 'opuscula' and were finely produced on hand made paper in small limited editions, usually of odd number, of course, and only with the odd numbers used.

As to membership, Quaritch had originally decided the Sette should be confined to twenty-one 'brothers', 'this being the number of volumes of the Variorum Shakespeare of 1821',[188] but supplemental 'O.V.'s were eventually allowed, up to another twenty-one. Brothers adopted names appropriate to their main interest or profession; Gould assumed the title 'Hydrographer' (having first, very sensibly, got permission from his boss, the real one!)[189] and the use of these names was obligatory at dinners and meetings.[190]

The Sette included a Chairman/President (to be addressed as 'His Oddshippe') and the other usual officers, as well as a Master of Ceremonies to ensure everything went according to plan at dinners. During Gould's time these were usually held in private rooms hired for the occasion at one of the better class of London restaurant. Up until 1926 the venue was The Imperial Restaurant in Regent Street, whose proprietor was, coincidentally but happily, called Auguste Oddenino; the Sette naturally approving of the restaurant's more familiar name, 'Oddenino's'. Briefly (in 1926 and 1927) they met at Gatti's in the Royal Adelaide Gallery, round the corner from Charing Cross station, and from 1927 until after the Second World War, they took rooms at the Savoy Hotel on the Strand.

Facetious Humour

As was usual for the period, the dress code was always white tie. Cecil, who was a guest of his father's at one of these Odd Volumes dinners, recalled that proceedings were conducted with a sort of facetious, donnish humour, blended with an enforced, rather pretentious formality. There was a small amount of associated paraphernalia, including an elaborate President's chair (surviving today at the Savile Club in London) and an electric bell for marking 'time's up' during after dinner speeches. This latter object was presented by the scientist Sylvanus Thompson (1851–1916) who had, some years previously, been a member of the Sette (Brother 'Magnetiser') and its President in 1904.

One of the stranger and less hospitable traditions of the Sette was that guests would, where possible, be introduced in such a way as to embarrass them. For example, Rupert Gould noted[191] that on one occasion Sir Samuel Hoare, of the highly important and influential banking dynasty, was a guest at an Odd Volumes dinner and was introduced by his host, Sir Maurice Healy ('Prattler' to the Sette), with the pronouncement 'My guest this evening practices in one of the oldest professions in the world . . .'

The dinner before Gould's introduction had seen Ralph Straus introducing Osbert and Sacheverell Sitwell to the Sette. Alas, history does not relate what was said. Cecil however, remembered very well being introduced by Rupert who observed that unfortunately Cecil had not inherited his father's extremely good looks ('My son gets his looks from his mother'). This, although only intended as a joke (and untrue anyway, Cecil was very like his father) did not go down well with Cecil, whose personality was very different indeed from Rupert's, and he stated frankly that he disliked the whole evening.[192]

For Rupert however, The Sette of Odd Volumes appealed immensely. At that period the membership consisted of some very distinguished and important figures, from all walks of artistic and scientific, intellectual life and Gould, with his wide knowledge and interests, duly discovered a rich source of 'Mutual Admiration'.

A look at a few of Gould's new circle of friends will provide a clue to the intense distraction he discovered here, and will introduce many of his social contacts during the 1920s and 1930s.

Oddly Distinguished

Contemporary brothers of Gould's with literary interests included Oscar Wilde's son Vyvyan Holland (1886–1967), a particularly close friend of Rupert's, who was appropriately named 'Idler' in the Sette. Wilde himself had been a guest at one of the Sette's dinners in 1889 and had been a childhood friend of Sir Edward Sullivan, who was 'Bookbinder' to the Sette in Gould's day.

Another member was Wilde's associate and friend John Lane (1854–1925), 'Bibliographer', founder of The Bodley Head and publisher of *The Yellow Book*, with illustrations by Aubrey Beardsley which so influenced Gould in his early drawing.[193] The young writer Alec Waugh (1898–1981), 'Corinthian' and later Secretary to the Sette, was just 22 years old when he joined in 1920, the same year as Gould. Five years before, the 17 year-old Waugh had left Sherbourne School under a cloud and 2 years later wrote his first novel, *The Loom of Youth*, based on his school experiences, causing a sensation by its frank treatment of the subject of homosexuality, and quickly becoming a best seller.[194]

Artists and Antiquarians

Brothers with more specifically artistic credentials included John Hassall (1868–1948), who was brother 'Limner'. Hassall, known as 'The Poster King' was the well-known cartoonist and pioneer of modern poster design. He was creator, *inter alia*, of the famous poster 'Skegness is so bracing' and was tutor of many later 'names' including H.M. Bateman. Charles Holme (1848–1923) 'Pilgrim', founder of the fine and applied art magazine *The Studio* in 1893 (also employing Beardsley for his illustration) had lived with William Morris who influenced much of his work.

Edward T. Reed (1860–1933), caricaturist for *Punch* magazine, and Frederick Townsend (1868–1920), the same magazine's Art Editor were both members of the Sette. So was David Low (1891–1963), 'Exaggerator', the much-loved cartoonist for the *Evening Standard* and *Daily Herald*, and creator of the famous 'Colonel Blimp'. The artist Sir George Frampton (1860–1928) was 'Sculptor' (works included the Peter Pan in Kensington Gardens) and George C Williamson (1858–1942), a highly respected antiquarian, was 'Horologer'. Williamson is especially remembered today by antiquarian horologists for his enormous catalogue of the Pierpont Morgan collection of watches (1912).

The Sette's Menus

'The Sette's' dinners always entailed the production of specially printed menus, the covers of which were often enlivened with caricatures of relevant members speaking that evening. These were usually drawn during this period by Hassall or Low, but often with very good caricatures by Rupert Gould. As well as having Sette names, the brothers were given emblematic 'cartouches' by which they could sign themselves on letters using a rubber stamp, and Gould was responsible for the design of most of these during the 1920s and 30s.

Publishing

Philip Allan (1884–1973) 'Hermit', and John Gilbert Lockhart (1891–1960) 'Investigator', co-directors of the publishing house Philip Allan & Co, produced the first three of Gould's books on scientific mysteries: *Oddities* (dedicated to The Sette) in 1928, *Enigmas* in 1929 and *The Case for the Sea Serpent* in 1930. Among the journalists in the Sette was the influential Sir Max Pemberton (1863–1950) 'Hack', Director of Northcliffe Newspapers.

E.B. Noel (1879–1928), 'Paulmier', was Sports Editor of *The Times*, Secretary of the Queen's Club and an expert on the history of tennis (like Noel, Gould had an absolute passion for the sport and, in due course, would have a chance to demonstrate his knowledge of the rules of the game when serving as an Umpire at Wimbledon). Another sporting friend, Lt. Col Henry Wakelam (1893–1963) 'Commentator', was a sports writer and BBC radio commentator (tennis and rugby).

The celebrated writer and politician A.P. Herbert (1890–1971) 'Jester' joined The Sette in 1922 (he too joined the staff of *Punch* a couple of years later). His many amusing and thought-provoking works included a pioneering novel *The Secret Battle* (1919) on the sensitive subject of desertion in wartime, a fine book on sundials and a short but highly intelligent discussion on the subject of the adoption of Summer Time in the United Kingdom.[195]

Scientific brothers

The scientific and technical brotherhood at this time included Harry Wimperis (1876–1960) 'Astrologer' who, as Director of the Air Ministry Laboratories in the mid-1930s oversaw the development of Britain's radar defence system. The surgeon Sir William Arbuthnot Lane (1856–1943) was 'Chirurgeon'[196] and Sir John Scott Keltie (1840–1927, geographer and Fellow and Vice President of the Royal Geographical Society (of which Gould, and a number of other brothers were also fellows) was 'Chorographer'.

Sir Maurice Craig 'Thought Reader', psychiatrist and author of the text-book on nervous exhaustion, may well have had a particular interest in Brother Hydrographer! Another close friend of Gould's was the noted biologist and microscopist Edward Heron Allen (1861–1943), 'Necromancer', one of the Sette's long-standing members, but one who had resigned in 1890 after an altercation with its founder,[197] and who had rejoined in 1925.

Talents in the Sette also included military men such as the highly decorated General Sir Walter Pipon Braithwaite (1865–1945) 'Peltast', who was ADC to the King (George V) and Admiral Sir Reginald Hall (1870–1943) 'Inquisitor' who, among many interesting roles, was the

Director of Intelligence at the Admiralty during WW1.[198] 'Sciolist' to the Sette was Ivor Stewart-Liberty (1887–1952), Director of the famous Liberty's department store in Regent Street, London. His uncle, the founder A.L. Liberty had also been a member of the Sette. Several other members were MPs, a number were High Court Judges and QCs, and at least one was Lord Mayor of London.

So, when Ralph Straus introduced Gould to the Sette in May 1920, the intellectual level would immediately have been apparent, and attractive, to Gould, who found the many and varied kinds of expertise in the Sette both stimulating and educational.

But the 'Admiration', specified in the Sette's mission statement, was certainly 'Mutual'. The average age of the brotherhood was somewhere in their early 60s and the good looking, confident and erudite 30-year-old Rupert Gould would have been an ideal addition to the Sette. Within 2 years Brother Hydrographer was promoted to 'Master of Ceremonies', a role he was amply qualified for.

35. Photograph of an Odd Volumes dinner at Oddenino's Restuarant in late 1924. Gould, as 'His Oddshippe', stands in front of the President's chair, with Ralph Straus at his right. (© The Sette of Odd Volumes)

H4 Miniature

Gould went one step further in this role and created a special badge of office, echoing his own particular passion: John Harrison. Taking what must have already been a fine, late eighteenth century, pocket watch movement by the noted London watchmaker, John Holmes, Gould had a new enamel dial made for it, in imitation (on a smaller scale) of that on H4. He then had it fitted to a contemporary silver case, and wore this 'H4 miniature' round his neck at the dinners.[199]

The 'conviviality' Gould found here was no less rewarding. The witty and facetious repartee blended very effectively with good food and fine wines, both of which Gould appreciated as much as the debate, and over the next 18 years Brother Hydrographer was a prominent member of the Sette.

36. Gould's badge of office as Master of Ceremonies. An eighteenth-century watch movement was adapted as a miniature 'H4', and worn at every dinner. The watch is now missing. (© Sarah Stacey and Simon Stacey, 2005)

37. A caricature by Gould of himself (at the back, wearing his badge of office) with three other brothers, from left to right Ralph Straus, Edward Heron Allen and Vyvyan Holland. The drawing was for a menu insert, probably on the Jubilee and Ladies' night on 24 April 1928. (© Sarah Stacey and Simon Stacey, 2005)

In 1920, Muriel probably had no qualms about his joining. Although women were excluded except on 'Ladies Nights', this was the norm for such clubs and membership did after all give Rupert the intellectual stimulation he craved. It was not long though before qualms must have arisen, especially as a result of the 'fine wines' available on these occasions. Gould's excessive consumption of alcohol would soon become a real issue for her, and a commonplace for members of the Sette. A typically facetious remark in the OV menu of 24 January 1922 (only a year or so after Rupert had joined) hints jovially: 'A discussion will follow the paper, after which Hys Oddshippe's health (and, not improbably, the Hydrographer) will be drunk'.

Looking at the relationship between Rupert and Muriel, it would seem that this period, during the year 1921, marks a turning point in the marriage. The names Muriel chose in her photo album, when putting titles to specific photographs, may perhaps not be a particularly scientific means of determining her attitude to her husband, but it could be significant that up until 1921, all images of Rupert were marked with the family's affectionate name for her husband: 'Bunny', or 'Buns', but from early 1921 onwards she always simply used 'Rupert'.

The first dinner Brother Hydrogapher attended (i.e. as a member of the Sette) was on 26 October 1920, and by way of introducing him, the menu cover featured a sinister and distinctly Beardsley-esque drawing by Gould of 'The King Orgulous'. This was done back in 1909 when Rupert was serving on HMS *Queen*.

The following dinner, on Tuesday 23 November 1920, at Oddenino's as usual, saw Gould's more formal introduction to the Sette. He gave the talk that evening on 'doubtful islands' (a subject he had been studying at work in the Hydrographic Dept.). The menu cover was drawn by him in the form of a self-portrait, a closely detailed study full of symbolism, and covering just about every one of his many interests. Every detail in this drawing contains a reference to one of his preoccupations, the miniscule scale of the drawing itself one of his particular talents. For a brief summary of Gould's activities at the time, this one image is a remarkable document.

This drawing then was Gould's own introduction to the Sette that evening. His lecture told the extraordinary story of professional sailors and navigators who, during voyages of discovery in previous centuries had, with all due diligence, observed, measured, carefully illustrated and

BOX 9.1

Yᵉ 365ᵗʰ Meeting of Yᵉ Sette of Odd Volumes
held at Yᵉ Imperial Restaurant (Oddenino's)
on Tuesday, yᵉ 23ʳᵈ daie of November, 1920

Yᵉ Hydrographer *laboriously compileth, not without sustenance, hys odde*
Dissertatio de Insulis Insolentibus

Hys Oddshippe
Bro. Horace C. Beck, Visionary
in yᵉ Chaire

38. Gould's self portrait for the OV Menu on 23 November 1920. Among
the subtle details are many references to his interests (see plate 14).
(© Sarah Stacey and Simon Stacey, 2005)

On the stand, an isogonic globe, showing lines of equal angular magnetic declination. Variations of magnetic declination across the world were for many years believed to be a possible means of determining 'The Longitude' at sea.

The shelves behind the globe contain several books pertinent to Gould's work. The ships logs for Sir John Franklin's ship HMS *Erebus* in the years 1847 and 1848 would have made fascinating reading had they existed. Unfortunately as we now know, both *Erebus* and *Terror*, the ships from that ill-fated Arctic expedition in search of the North West Passage of 1845, had become ice-bound at King William Island in 1846, the last men from the party dying on the Island over the following two or three years. On the other hand, the logs of the ships *Resolution*, *Discovery* and *Adventure* are very real indeed, being the record of James Cook's second and third great voyages of discovery to the South Seas in 1772–1775 and 1776–1779. Both voyages proved to be highly successful testing grounds for K1, Larcum Kendall's copy of John Harrison's prize winning timekeeper, H4 (the official account of Cook's 1st, 2nd, and 3rd voyages, further along the shelves, provides the full background to that story).

The next book on the shelves, 'The Scandalous . . .'is probably a complimentary reference to *The Scandalous Mr Waldo*, a novel by his friend Ralph Straus, who had introduced him to the Sette. A book on Gould's shelves entitled 'Horlogerie' needs little explanation, though the identity of the author, seeming to begin 'N de L'En . . .'is uncertain. The likely explanation seems to be a 'tongue-in-cheek', punning reference to the Swiss clockmaker Enderlin, who published a piece on pendulums in Antoine Thiout's *Traite de l'Horlogerie* in 1741. The large tome or box file next to it, titled 'Opuscula Vagula Blandula' (brief, pleasing little books), with the large letters OV, is a complimentary reference to the Sette's own publications. 'The Siege of the South Pole' refers to the book of that name by his friend the Antarctic historian Dr Hugh Mill, the book on James Weddell's Voyages adding a further important reference to Antarctic discovery.

The History of the Marine Chronometer was of course not yet written, but work on H1, shown behind his left shoulder, was very much under way, and Gould had had a preliminary look at H4 shown now booked out from Captain Hardy (Gould's colleague in the Department, Ernest C. Hardy, who had been Assistant Hydrographer to Parry), to the current Assistant, Captain Douglas. No doubt the Sette's young 'Brother Hydrographer' was already totally committed to Harrison and horology. Under Gould's arm, three volumes are just visible. The first 'of A.G.Pym' is a reference to the book '*The Narrative of A. Gordon Pym*', a horror story set on the South Seas by Edgar Allen Poe, one of Gould's favourite authors; the second is the 'Log of *Williams* 1820', the record by the English navigator William Smith, discovered by Gould himself and the evidence for Britain's claim to the first discovery of land in the Antarctic, the South Shetland Isles. The third is 'Captain van der Decaen's Log', which would appear to be a reference to Captain-General Charles Decaen who, as Napoleon's Governor of the Ile de France (Mauritius) in 1802, arrested Matthew Flinders R.N. on his arrival at the island after his pioneering circumnavigation of Australia, detaining Flinders there for 6 years.

On the table in front of Gould are charts relating to his favourite Hydrographical subject, the Antarctic regions, with the South Shetlands centre stage at the back, but with many references to 'doubtful islands' scattered about. A chart by Antarctic explorer James C Ross and one of Bouvet Island in the South Atlantic (the remotest island in the world) continue the theme. The chart of the Aurora Isles of 1794, upon which Gould is working, emphasises the subject of the talk that night 'Doubtful Islands'. The text of the lecture is shown, in tiny script (his signature in the full scale drawing is just 3 mm in length) under Gould's favourite typewriter, 'The Blick'[200]. The telephone nearby was another technological instrument which held a great fascination for Rupert.

The work trays at the front of his desk contain departmental H files, which were generally notifications of new hydrographical information from official external sources, such as the Hydrographic Offices of other countries. This was reliable information which was to be used to update charts and which was considered 'most urgent' by the Department, though evidently not by Cdr Gould! 'M papers' refers to a series of notes and reports from foreign ships or private individuals (sometimes also external publications), with similar information for updating charts.

Gould had no illusions about his addictions and the table also contains clues to favourite pastimes of his, including cigar smoking, though a few years later a pipe would be his constant companion instead. He was equally frank about his love of alcohol, including wine, champagne and beer, though his love of spirits was apparently supposed to be a secret. These were evidently drunk neat, as all kinds of water, including that in the soda siphon, are shown in the OUT tray! (© Sarah Stacey and Simon Stacey, 2005) charted islands, which in reality never existed (in many cases being confused with icebergs, cloud banks, etc.).

Later, the antiquarian horologist Malcolm Gardner, admiring this drawing greatly, arranged to have it produced as a single print, with the title 'Good Master Hydrographer, Chart me the Unknown Seas', a quote taken from within the menu. One of the two guests Gould invited that evening was Capt. H. P. Douglas, soon to be Hydrographer of the Navy and Gould's boss. The text inside the menu states, in typical OV form: 'The Guests having been introduced with the strict OV regard for truth, and callous disregard for the Guests' feelings . . .'. One wonders if Gould dared to risk insulting Douglas. There was certainly no love lost between them in the years to come.[201]

Harrison Lecture

In the following years, the Sette's attendance registers show Gould at the majority of dinners, often with guests. In April 1921 his father, W. Monk Gould appears in the register, and on 22 November that year Gould was speaking again, this time on his favourite subject: John Harrison, the menu illustrated on the cover with a reproduction of the likeness of Harrison engraved by Tassaert after the portrait by King. H1's preliminary clean had been completed the previous year and H4 was now at the new Epsom home with its overhaul nearly at an end. A short talk, probably based on what he had said to the RGS the previous December, would have been very straightforward. Guests of Gould's on the occasion were the Astronomer Royal Frank Dyson, the Admiralty librarian W.G. Perrin and, appropriately enough, Frank Mercer of the famous chronometer manufacturers Thomas Mercers of St. Albans, who, in a few years time, would be helping Gould in the next restoration of H1.

One or two other dinners at this period are worthy of note, while on the subject. The meeting on 27 June 1922 was a ladies night with a musical theme (The President at the time was the musician and composer Frederick Keel, 'Singer') and on this single occasion Muriel came as one of three lady guests of Gould's.[202] On 28 November 1922 Gould was speaking again, this time on the subject of 'Jeremiah Horrox, Astronomer', with the Astronomer Royal Frank Dyson attending again as a guest. This talk was published as the Sette's Opusculum No. 75 (133 copies, dedicated to his brother Harry, lost in The First World War, and dated 1923) the first of two Gould was to author.

The Yellow Book was the subject of John Lane's address to the Sette on 22 May 1923, a meeting Gould surprisingly appears not to have attended (at least he did not sign the attendance register), in spite of being the author of the menu's cover. He did however attend many of the dinners later that year and one or two in 1924. That year, 26 February saw him taking three guests—a considerable expense for him—to hear E.B. Noel speak on 'An Odd Survey of the Royal Game of Tennis'.[203]

Trouble begins

As told in the previous chapter, in the Spring of 1924 Gould slipped into another period of depression and life at the Epsom home was becoming increasingly stressful for everyone. Cecil recalled, even as a six year old, being very aware of the tensions in the house. Nanny Bullard had recently left and, as a result of an advert placed in *The Times* by Muriel,[204] a new and very severe governess, one Miss Farrow, had been appointed to look after the children. Within only a few weeks however, she was sacked, after some extremely tearful scenes. Cecil noted, of the period that followed: 'I was soon to discover in different and even more intimate circumstances, it is very unsettling for a young child to witness emotional scenes between grown-ups'.[205]

Recuperating from his depression, Gould had stayed at his mother's in Southsea where, it will be recalled, he was visited by the American horologist and friend Paul Chamberlain. On Gould's recovery, Chamberlain attended an O.V. dinner with him, on 27 May 1924, to hear Sir Maurice Craig ('Thought-Reader') on the ominous subject of 'Crime and Insanity'.

Still recovering from his depression during the Summer of 1924, Gould stayed at Epsom while Muriel and the children took holidays on Hayling Island a few miles east along the coast from Portsmouth. Muriel's photograph album from this period contains only pictures of the children, mostly happily playing on the beach, but Rupert was not with them apart from occasional visits. He now needed to get on with work on H2, a job he would be involved in until the autumn when the restoration was finally finished. During that autumn, (on 18 November 1924) Gould's old friend and colleague, Lt. Cdr Archibald Colquhoun Bell, 'Brother Investigator' to the Sette of Odd Volumes (until 1927), spoke on 'The Writing of Naval History'. Gould naturally attended, and drew the menu cover. A few months before, Gould and Bell had been jointly

39. Gould's close friend the horologist and collector Paul Chamberlain. The two antiquarian horologists greatly respected each other's learning and publications and were frequent correspondents. (David Penney (www. antiquewatchstore.com). Copyright)

responsible for the new cover of the *Mariner's Mirror*, Bell contributing the calligraphy for the title, Gould the illustration.[206]

The Absent President

The year 1925 was to be Gould's turn as President of the Sette of Odd Volumes, and he was installed as 'His Oddshippe' at the dinner on 16 December 1924. But, as mentioned in the last chapter, trouble was now on the horizon, as he was gradually becoming ill again with nervous depression. The pressure of work at the Admiralty had been as constant as ever and work in the evenings had not eased up. The stress of H2's completion earlier that autumn, was now combined with the additional work on articles for the *Nautical Magazine* (on Pioneer Chronometers) and *The Geographical Journal* on 'The First Sighting of the Antarctic Continent'.[207] The prospect of the new Presidency of the Sette probably also contributed to the stress he felt, as, perhaps, did the growing social unrest in Britain at the time. What the exact combination of factors was is difficult to say, but his mental health deteriorated steadily over the next few months.

The first two Sette meetings of the year were attended by Gould as President, and the printed menu for the third, on 24 March 1925, bills 'His Oddshippe' to Chair the dinner as well as giving the talk: 'A True Account of the Life History of the Sea-Serpent'. This was a subject with which R.T. Gould would soon become intimately associated and upon which he would research and write in very great depth, this lecture being his very first thoughts on that interesting and controversial matter. Although the attendance register does not contain his signature, it would appear he *was* there and gave the talk, as it appeared the following year as Opusculum No.80, dedicated, 'with sympathy and affection to all surviving specimens of *MEGOPHIAS MEGOPHIAS*'[208] and stating it was read at the 24 March meeting.[209]

Rupert is also billed as chairing the next dinner, on 28 April 1925, where A.P. Herbert gave the address 'Concerning the Distinction between Heaviness and Weight', but again Gould's signature does not appear and he was almost certainly not present. What is certain is that he did not then appear again at O.V. dinners at all during his Presidency. The Vice President, the surgeon Owen Lankester, 'Leech', being billed on menus to take his place. Gould was expected to appear at the official hand-over to the new President on 27 October 1925 and John Hassall illustrated the menu with caricatures of Gould and Lankester, but again Gould appears not to have been present as there is no entry in the register. In fact his signature only appears again in the register for dinners at the latter end of 1926.[210]

❦ 10 ❦

Separation 1925–1927

As told in Chapter 8, the reason for his absence was that in May 1925 Rupert was experiencing another severe bout of depression, and was in fact approaching another full nervous breakdown. According to Muriel's account, Rupert had occasionally suffered from nervous paroxysms, usually exacerbated by alcohol, since the beginning of their marriage, and that this culminated in April 1925, at the Cottage in Epsom, with him threatening suicide, followed, a few weeks later, by total collapse, at which point he was sent to a private hospital for nervous diseases.

What can one say of these repeating depressions that so scarred Rupert's life? It has been suggested that this kind of depression is often associated with those of strongly artistic or literary talents and that Gould's would seem to be a splendid case in point. Recent published studies of severe and manic-depression suggest that, for example, the ability to work incredibly hard over sustained periods without sleep, followed by a period of deep introspection, often associated with alcohol abuse, might even be part of the creative process itself and the cyclic repeating illness may in fact be inseparable from the ability to produce really meaningful, positive work.[211]

It is tempting then to consider Rupert Gould's affliction as an example of this, but there are however doubts about such theories on the supposed link between artistic temperament and depression. It has been pointed out that depression affects 25 per cent of the whole population, and that with such a large proportion, a significant number are bound to be of strong literary or artistic talents, and that, by definition, they are the cases most likely to be publicised. In fact many professionals in the United Kingdom think that mental illness *inhibits* true creativity and endeavour. Therefore this biography must avoid any attempt at such an analysis.[212]

Gradual recovery

Though severe, this second breakdown of Gould's did not last as long as that he sustained during the First World War. After a few weeks he was on the way to recovery, and we get a glimpse of what was happening in early June 1925 as Arthur Hinks, the Secretary of the Royal Geographical Society (RGS), wrote to Muriel asking how Rupert was. Hinks had recommended Gould as someone who might (in his spare time!) be able to catalogue the large collection of watches in the Fitzwilliam Museum in Cambridge, and now wondered if this was likely in the near future. Muriel replied courteously that Rupert was 'making good progress and hopes to be back at the Admiralty in 6 weeks from now, but if he had intended doing the cataloguing "on leave", he would not be having any for some months, I expect'.[213]

Leaving Rupert to recuperate in hospital, Muriel now took the children off for the holidays in Hampshire where they had taken part of a large rectory for the summer. Their son Cecil remembered 'my father came down at weekends and behaved strangely. He read the lesson in church one Sunday at the Rector's suggestion, and then, when he was asked to do it again, I heard him say "No. I found it had a bad effect on my nerves" '.

It was during these holidays that Muriel, having had enough of Rupert's overwork on his own projects while neglecting the family, decided that she and the children would leave Epsom and her husband for good.

A difficult husband

In fairness to Muriel, it is not too difficult to see why life with Rupert would have been frustrating and unsatisfactory. The product of a tradi-tional upper-middle class Victorian family, Gould was essentially an old-fashioned husband for whom the need for hard work was instilled from birth and infidelity was just about the only sin. Though in fact, as their parents had feared, the couple had little in common, Rupert genuinely did love his wife. He just didn't recognize that times had changed and that Muriel, along with so many other women at the time, were no longer prepared to accept the simple role of dutiful wife and mother, fill-ing in for an absent husband until he deigned to make an appearance. It must be said that even today there are plenty of wives and partners who *are* happy to accept a relationship on this basis, but that is their choice and Muriel was not of that mind. The short answer is simply that Rupert and Muriel were, as their parents had suspected, quite unsuited.

Rupert could also be infuriatingly pedantic which must have irritated Muriel. For example, Cecil recalled that his mother, the inveterate romantic, once remarked to Rupert how beautifully blue the sky was, only to be told that it *wasn't* really blue, it only *appears* to be blue! Her husband's intellectual pursuits would also have had a fairly limited appeal to Muriel, and, what with his spending much of the time alone in the workshop, it seems likely she became bored with life at home after the first few years of marriage.

Muriel's statement about suing for a divorce after Rupert had spent so many night hours writing his book, *The Marine Chronometer* (mentioned in Rupert's letter to Admiral Parry) had no doubt been entirely serious. But such was Rupert's lack of understanding that he evidently saw it as simply a flippant remark. But she meant it, and now, after two further years in which it was clear he was not going to change, the time had come for action.

So, taking the children, Muriel left Epsom while Rupert was still recuperating in hospital. For the first few weeks they moved in with her mother, the baleful and troublesome Emily Estall, in London and, when the holidays were over, into a small hotel in Queen's Gate. Cecil, who was now just 7 years old, was sent to board at his preparatory school, Kingswood House, at Epsom, and Muriel arranged to have the Cottage at Lynwood Avenue put up for sale. Muriel's decision to send her young son Cecil to boarding school was tough on him as he was only just seven and was a shy and sensitive boy. In later years Cecil admitted it was very difficult for him, it was something he would never forget and, as he later found out, it was something Muriel would feel guilty about for the rest of her life.

And so, on Gould's discharge from hospital in July 1925, he found no one at the Epsom house, only parts of H3, in pieces (the other parts, due to go for polishing, had gone back to Greenwich when H2 returned home). While not accepting the separation, there was nothing he could do in the short term apart from establishing exactly where his family was. So, after arranging for the boxes of H3 parts to return to Greenwich, Gould, still weak and needing convalescence, went to stay with his mother again.

Downside

In fact, in a sense it was Dodo who came to him because during the summer of 1925, now that her husband had passed away, Dodo had moved from the old house at Southsea and was now much closer to

Epsom. Her intention, ironically under the circumstances, was to be closer to the Gould family. The home she chose was a large and (by modern standards) rather grand, late Victorian house called *Downside*, at 41 Wood-field Lane, Ashtead, about two miles west of Epsom. In fact Dodo had never really liked the house in Southsea and as she was now rather more comfortably off thanks to her legacies, income from her mother's trust fund and careful investments of her own, she decided to live in the style to which she had been accustomed as a child. As well as two live-in maids, Dodo employed a gardener to tend the large plot, and a 'chauffeur', called Partner, who in fact had a more general responsibility as Dodo's houseman. This comfortable, organised and relatively calm environment allowed Gould to recover for the next few months and settle back into regular work at the Admiralty during the day.

Ever the glutton for punishment though, Rupert, not able to continue with H3, managed to find other horological 'fixes'. In February 1926 he visited the British Library to study a most interesting and important proposal for the Longitude by Jeremy Thacker, *The Longitudes Examin'd*

40. Downside, Dodo's house in Ashtead, from a 1950s photograph when the house was a school. Gould would spend the happiest years of his adult life living here between 1928 and 1938. (Copyright holder untraced.)

(1714). He then carefully transcribed the full text of the pamphlet, making his own pen-and-ink facsimile copies of the elaborate engravings in the work to ensure the copy was complete. Adding a page of his own carefully considered and erudite commentary to it, he sent a copy to Paul Ditisheim in Paris for his research in chronometry.[214]

In the following month Gould then agreed to catalogue the large watch collection at the Fitzwilliam Museum in Cambridge. In fact the idea had been around for some time. Gould's friend Arthur P Hinks (Secretary to the RGS) had written to the Fitzwilliam's Director, S.C Cockerell in November and December 1923 recommending him. Hinks observed: 'I should be surprised if there is any subject which Gould does not know anything about...' adding gratuitously that '... the people in the British Museum know absolutely nothing about their astrolabes and sundials and other beautiful things they have, and Lewis Evans [whose fine collection of scientific instruments had opened at the Old Ashmolean Building in Oxford the previous year] knows very little about his from the technical point of view'.[215]

The Fitzwilliam Watches

The Fitzwilliam collection of antique watches, consisted at the time basically of two separate collections formed respectively by Messrs Smart and Perceval. It numbered in total some 190 items and was a fine and interesting, if eclectic, mix from many periods and several countries of origin. At that period, with very little at all published on antique watches, it would have been something of a challenge for any antiquarian to catalogue fully without a good deal of research.

Putting his trust in Hinks' recommendation, Cockerell agreed that Gould should do the work, and the 5 days of 13 to 17 April 1926 saw Gould in Cambridge studying the Perceval watch collection. Cockerell must have approved of the result as on 3 June Gould wrote to him from Downside thanking him for his appreciative note, and apologising for the delay in reply, pleading ' the recent disturbances as an excuse'.[216] The 'recent disturbances', part of what is today known as The General Strike, were the result of a cause which had been fermenting among working people in Britain since the end of the Great War, but which had been steadily escalating since the beginning of 1926.[217] Gould, with his mortal fear of civil unrest and revolution, was of course particularly worried by the General Strike, and this did nothing to help his mental condition. In his letter to Cockerell, he stated that the catalogue could have been much better had he not been 'pressed for time and not very well'.

Nevertheless, Cockerell was obviously very satisfied, as Gould was paid a fee and was apparently sent a further *tranche* of the collection of watches for cataloguing at Downside, returned to Cambridge by post a few days later. Gould only catalogued those he considered worthy of retention, recommending some to be disposed of. All this took considerable time and trouble. The resulting catalogue, sent on 28 August, was all neatly typed up, Gould remarking cautiously: 'not, of course, at the Admiralty, but privately'. He was always careful to separate his own projects from his daytime work, for fear of a conflict of interest and accusations of furthering his own projects in the Hydrographer's time. The catalogue was provided with 'Prefatory Notes' which he considered 'constitute a short essay on the history of watch-making'. Gould went on to ask about the likelihood of the collection being published one day, expressing the hope that his notes 'may be useful', and probably hoping he might be asked to provide the full text.

One watch, purporting to be by the celebrated German renaissance watchmaker Peter Henlein, was condemned by Gould as a fake but, for a second opinion, he suggested the well-known horological dealer Percy Webster (1862–1938). Actually, Webster was not a very good choice, as today he is rather well known for his own 'imaginative' restorations and fakes.

G.H. Baillie

Cockerell did have a published catalogue in mind, but decided first to seek a review of Gould's work. Sensibly he did not approach Webster, but chose to ask the independent antiquarian horologist Granville Hugh Baillie (1873–1951), a researcher of the highest calibre who would publish the seminal work *Watches, their History Decoration and Mechanism* in 1929. Baillie sent in a learned critique of Gould's typescript, with many detailed but fair corrections. Cockerell now had to tactfully broach the subject with Gould, who was not in the habit of 'being corrected' about anything horological.

Although Baillie had been studying antique watches for many years (almost since before Gould was born) he was not generally well known and Gould had not heard of him. Needless to say, Gould was initially frosty, writing in reply to Cockerell on 6 December, 'I cannot say that I concur with him in his opinion of the Perceval watches'.[218] However, when Gould received the detailed comments, he asked to be put in contact with Baillie, evidently recognising the work of a serious and erudite researcher. They finally met in the Spring of 1927, an interesting occasion for antiquarian horologists to contemplate, as these were arguably the two finest horological historians of the twentieth century. Gould hoped 'that, in consultation we shall be able to evolve something satisfactory to both of us and also to you.' But the catalogue never appeared, probably because Baillie was busy researching for his own publication and Gould was, by early 1927, in the throes of more serious trouble.

Leatherhead

Some months after Muriel had walked out, back in 1925, mother-in-law, Dodo, assuming the role of peacemaker, had invited her, with the children, to stay at Downside. She was trying to persuade the couple to give the relationship another try. And Dodo was prepared to put more than just sound advice and encouragement into the proposal. As the Epsom house had long since been sold and the mortgage repaid, the reunited family would be homeless, so Dodo was prepared to buy another house, to be leased to them, in order to get things onto a stable footing again.

It was some time before a reconciliation succeeded, but eventually it seems, swayed by Dodo's generosity and good intentions, Muriel agreed to try again with Rupert, so a house was sought. The house Dodo found was in Kingston Road, Leatherhead, a couple of miles from Downside. Cecil remembered it clearly, describing it as 'ghastly', standing, as it did, between a gas works and a home for what was euphemistically called 'fallen women'.

So the family were reunited in the summer of 1926 at 'Glenside', 98 Kingston Road, a 1920s detached house, somewhat larger than the Cottage at Epsom, and with a 'lean-to' on the side, ideal, as it turned out, for a makeshift workshop. Here at Leatherhead everyone hoped life would return to some kind of normality, but sadly it was not to be, and this time not all of the problems in the relationship could be placed at Rupert's feet. Muriel had returned to the new family home with the children, but with someone else in the wings.

Vivian Gurney

Back in 1922, soon after the Goulds had moved to Lynwood Avenue, Cecil had been sent to have dancing lessons at the fashionable establishment of one Miss Vivian Gurney in Epsom. Cecil recalled her being quite terrifying (he described her as one of the 'dragons' of his childhood) and it seems the dancing lessons did not last long.[219] It transpired however, that although the classes had ceased by 1926, Muriel had kept in touch with Vivian.

Miss Gurney had apparently become very attached to Muriel and now spent an increasing amount of time staying at the new house. Cecil recalled his mother and Vivian Gurney then going on a short holiday together to Frinton and returning with a pet dog, a bulldog, 'apparently their joint property'. Needless to say Rupert had put two and two together and was deeply unhappy about the new relationship. The subject itself was currently topical though, with celebrities like Vita Sackville West becoming well known for their 'proclivities'[220] and the publication, and immediate banning in 1928, of Radclyffe Hall's pioneering novel *The Well of Loneliness* which dealt with the issue of love between women.[221] In the same year, the *New Statesman* described lesbianism critically as a 'completely widespread social phenomenon', blaming it on the actions of the Suffragettes![222]

With the topical subject of women's sexuality and the hotly debated issue of votes for women (all women above the age of 21 were finally given the vote the following year, in April 1927), Rupert must have been very much aware that feminism and women's rights were currently top of the agenda. In a sense, the story of much of Rupert's life is a story of the influence of strong women. First, his mother of course; Dodo was the dominant figure in his childhood, but then his wife, her mother, and then his wife's partner, all had a powerful and unwelcome effect on the course of his life.

Predictably, tensions between Rupert and the couple, Vivian and Muriel, quickly got worse. Vivian rented a bungalow for the rest of the summer of 1926 at Fairlight, on the south coast close to Hastings, and Muriel joined her there, with the children, for the whole summer.

In spite of his unhappiness at these marital developments, Gould's mental health seems to have entered a more stable period at this stage. He continued his work at the Admiralty during the week while living at

Glenside and going down to Fairlight at weekends to be with the family. Of course, he bitterly resented Miss Gurney's presence but he soon realised he would have to get used to it: she was to remain among the family for some time to come.

H3 at Leatherhead

Meanwhile, with the main cataloguing of the Fitzwilliam watches done by the autumn of 1926, and with Muriel's affections directed elsewhere, Rupert's desire for another 'fix' of Horology and Harrison surfaced again. To reconcile this, he managed to persuade the Astronomer Royal to part with H3 for the second time. On 6 July 1926, Dodo arrived at the Observatory on his behalf, with a letter of introduction, to collect the parts of the timekeeper, thirty of which had by this time been cleaned and polished by Bucks in West London.

Gould set up a workshop in the lean-to at Glenside to continue the restoration work; his son Cecil clearly remembered seeing H3 completely in pieces, spread out on tables in the house. For several months though, little was done to the timekeeper. Muriel and Miss Gurney were unlikely to take much care of the pieces if they got in the way, and H3 had entered another period of great danger. Complex objects such as this really are vulnerable when in a wholly dismantled state, as Rupert certainly knew: he had himself experienced an extremely embarrassing situation of this kind a few years before. Cecil recalled the incident, which had been caused by his mother at the Epsom house. Rupert had agreed to overhaul a large English chiming clock for the collector George Matthey (the same man who intended to propose to Muriel just before she and Rupert met) and had left a small number of the pieces of this clock on the dining table to inspect later in the day. Unfortunately Muriel appeared and 'tidied up', throwing the pieces away, assuming they were 'rubbish' Apparently, Matthey reacted in a very gentlemanly fashion, and agreed to say no more about it, sending the clock to a professional clockmaker to be restored.[223]

Progress on H3 may have been slow, but Rupert's plans for Harrison work were, as usual, racing ahead. In spite of H3 still being in pieces, the Royal Astronomical Society's regulator was still on his mind and, at the same time, he was dreaming about the completion of H1, writing to the Observatory in September asking for technical details on the barrels and fusee, so he could mull over its further restoration.

But there was little actual restoration done, and H3 remained, completely in pieces, on the tables.[224]

Although there was little practical horology going on in Gould's life at this time, he did receive a token of respect from the horological world. In October 1926 he accepted an approach by the Council of the British Horological Institute to join them. His election was approved on 20 October and was announced in the Horological Journal the following month.[225] This would have been a welcome sign for him of the esteem in which he was held by the professionals, and it meant regular contact and stimulating technical discussions with notable horologists such as Frank Mercer and Frank Hope-Jones.

One of the reasons work at home was slow was probably that at this time his work at the Admiralty during the day was particularly important and challenging: Gould was one of two joint secretaries (he representing the Admiralty and another representing the Dominions Office) at a confidential conference in October and November 1926. Held at the Dominions Office, the highly sensitive subject under discussion was British Policy in the Antarctic, the aim being to consolidate British claims on that continent. Gould's responsibilities included summarising the data concerning the history and charting of the ice barriers and recording and summarising the deliberations of these secret meetings, the whole eventually being printed for government circulation. This work was later complemented with a hugely detailed 75-page typescript report by Gould, dated September 1927, on *Notes on the South Sandwich Islands*.

Trouble at Glenside

From the autumn of 1926 Miss Gurney had been virtually living at the Leatherhead home, and with family relations becoming increasingly strained, there was little incentive to stay at home and try to repair the rift. From October 1926 Gould began to attend dinners at the Sette of Odd Volumes again. This was now meeting at Gatti's, in the Royal Adelaide Galleries in King William Street, a slightly less expensive venue than the Imperial.[226]

On 25 January 1927 Gould was himself back on form, reading a paper to the Sette on 'The Mystery of Sir John Franklin'. As mentioned in Chapter 3, this concerned the fascinating but tragic story of Franklin's ill-fated expedition of 1845 in search of the Northwest Passage. With such an interesting topic, and one close to Gould's heart, it is not surprising to find it attended by many guests of his. These included the polar explorer Apsley Cherry Garrard (1886–1959) a member of Scott's Antarctic expedition of 1910–1913; the Admiralty Librarian and erstwhile Editor of the SNR's *Mariners' Mirror*, W. G. Perrin; the Secretary of the RGS, A. R. Hinks; and Major L. T. Burwash from the Canadian Department of the Interior. As mentioned, Burwash would soon make an exploratory flight over King William Island in the Arctic, in search of Franklin's ships, and this was research Gould was currently engaged in at work. Admiralty chart No. 5101 of 'King William Island, with the various positions in which relics of the Arctic Expedition under Sir John Franklin have been found', was published in May 1927, following Major Burwash's visit.[227]

The RAS Regulator

By early 1927, Gould was beginning to start practical restoration work again at the Glenside lean-to workshop but, incredibly, he did not choose to complete H3. John Harrison's immensely complicated and interesting timekeeper H3 had been in pieces, laid out on tables in the house, virtually untouched, for several months. Surely the completion of H3 should now have been a priority? But no, instead he decided to take up the question of restoring the Royal Astronomical Society's ('RAS') regulator again.

In late 1924, just as he had begun work on stripping H3 at Epsom, he had also offered to restore the RAS Harrison regulator, an offer that had been gratefully accepted by the RAS Council, but which he had been unable to honour as his mental health had begun to deteriorate. Now, in early 1927, just as he could have restarted work on H3, he again proposed work on the RAS regulator.

It may be that he felt they should be restored *in tandem*, as they were so very similar in concept. But to take on two such big and complex tasks was hardly wise, especially in such an unstable domestic and emotional environment. But take it on he did, and the RAS regulator and H3, both in pieces on tables at Kingston Road over the next few months, would now be witness to the most turbulent and difficult period of Gould's emotional life. As the restoration of these timekeepers took place during this complicated period in his life, the restoration work itself has been separately described in a later chapter (Chapter 13), hopefully keeping the two matters separate and thus easier to follow.

The Final Break-up

As for the life in the family home at Kingston Road, things now deteriorated rapidly, ending in a kind of personal catastrophe for Rupert. According to Muriel, things came to a head in the Spring of 1927, when Rupert's excessive drinking had become progressively more frequent, and on Saturday 21 May, when he was drunk again, she decided she had had enough. She walked out for the second and final time: there would be no going back now.

In the contradictory statements that followed it is difficult, with so many years elapsed, to be certain exactly where the truth lies. It would seem probable however, that in spite of the new home and fresh start, Rupert, deeply unhappy with Vivian Gurney's presence, was indeed drinking more.

Naturally, when Muriel left it was with Miss Gurney and they took Jocelyne with them. Cecil (who was still at boarding school at this time) stated that his mother left a note for Rupert informing him of what had happened. If what Muriel said was true, then Rupert was probably unconscious (she described him as 'drunk and incapable') when they walked out. With nowhere else to go, a hotel was the only course and Muriel, Miss Gurney and Jocelyne temporarily checked in at the Langham in London. They used false names to prevent Rupert from tracing them, something he attempted, without success, as soon as he found the note.

The Petition

Actually, things were even worse than he would have supposed on reading the note. Not content with leaving him, Muriel, undoubtedly encouraged by Miss Gurney—'playing Iago' as Rupert described her—went the following week (Tuesday, 24 May 1927) to her solicitors,[228] and had the children made Wards in Chancery, preventing Rupert from claiming custody of them.

She went further, at the same time petitioning for judicial separation. The 24th was also Cecil's ninth birthday and he recalled that day at his boarding school getting a visit from the maid at the Leatherhead house to see him with his birthday present from his father, the maid appearing rather flustered. He knew why; in fact he had already received a note earlier from his mother, without birthday greetings he noted (it being coincidental the note arrived that day), but instead bringing him up to date with what had been happening. The children had no illusions about either parent being models of their kind and Cecil recalled that Muriel could sometimes be just as selfish as Rupert, in her own way.

Unable to find his wife and daughter, Gould could only carry on alone in the house and wait for news. He continued his dining at the Sette too, and on the evening of that Tuesday when Muriel had started proceedings at her solicitors, Rupert attended the monthly dinner (without a guest, but in company with the architect Edwin Lutyens, a guest of another brother). He listened to Maurice Healy, 'Prattler', speak (appropriately enough) on 'Irish Wine'. The neatly ordered signatures in the attendance register for that evening include Gould's, carefully written, but upside down.

Gould vs Gould

Muriel would probably have preferred a divorce to judicial separation, but this would have been much more difficult. At the time, for a spouse to guarantee divorce, proof was necessary that the partner had been guilty of both cruelty *and* adultery; cruelty alone was insufficient. It was not until 1937 that the law was changed to make this possible.

In fact, although it didn't help Muriel, the law had recently improved, becoming much fairer for women. Before 1923, for a woman to be granted a divorce from her husband, she had to prove cruelty as well as adultery, while for the husband, adultery *alone* would have been sufficient. But one of the few things Muriel could not level at Rupert was infidelity, so a petition on grounds of cruelty, resulting in a judicial separation, was the only option. Unlike a clean-cut divorce, judicial separation was a rather unsatisfactory kind of 'twilight condition' in which the couple were officially parted but were unable to remarry, and the change in the law in 1937 was undoubtedly welcomed in most broken marriages.[229]

It would seem from the case papers which, incredibly, have survived, that Muriel's solicitors advised her in the first instance to avoid anything sensational (which may attract the attention of the Press) by keeping the allegations rather general. The relevant paragraph of the first petition thus states rather blandly that: 'the Respondent is a man of intemperate habits and since the said marriage has treated your Petitioner with great unkindness and cruelty and has used threats towards her'. An affidavit, verifying the petition, was signed by her the same day. At that first meeting, her solicitor did however make notes, also surviving among the papers, and these much more specific allegations tell a sorry tale of intemperance and bad behaviour.[230] A penned note at the foot of this page simply reports: 'Respondent Denies'.

The case was to be heard in the High Court of Justice, within the Probate, Divorce and Admiralty Division, naturally Divorce being the relevant branch in this case.[231] There was a preliminary hearing of the evidence on 17 June, when both sides' solicitors presented their cases. Gould used the celebrated firm of Lewis and Lewis of Ely Place, founded by Sir George Lewis, and solicitors to the rich and famous.[232] Presumably Rupert engaged such an established and high profile firm hoping for a better chance of getting the petition quashed.

The gloves come off

The result of the hearing was unfortunately an order demanding further clarification and details of the alleged cruelty. It seems Muriel's solicitors now advised her that if she was to succeed she would have to bring up every small complaint and make as much of it as she possibly could. This resulted in more specific and lurid allegations, sworn by her on 20 July.[233] These can be summarised as of three kinds: excessive drinking leading to threatening and abusive behaviour; unreasonable sexual demands upon her; and mental fits and paroxysms terrifying her, all of which, it was alleged, led to her mental strain and insecurity in the home. The claim of excessive drinking included a specific reference to the Sette, in that it was stated that Gould's drinking was particularly bad on the last Tuesday of every month: the usual date for the Sette's dinners.

Considering the evidence that has come down to us, including the opinions of both children, Cecil and Jocelyne, it would appear that all Muriel's allegations against Rupert had a basis in truth, but all were exaggerated to the point where they were misleadingly damning.

Cecil and Jocelyne frankly admitted[234] that at the time they only really heard and sympathised with their mother's viewpoint. They were, after all, closer to her emotionally and were under her care, so heard her argument on the matter (no doubt backed by Miss Gurney) on a daily basis. In subsequent years, both children realised that perhaps their father also had had reasonable grievances and that the hearing had been weighted heavily against him, but neither child was old enough to fully appreciate this at the time.

A simple statement of answer by Gould, through his solicitors, was submitted on 29 July 1927, saying that he was 'not guilty of cruelty as alleged in the said petition, or at all' and asking for the petition to be rejected. Both parties now awaited a date for the hearing, due sometime in the autumn.

Meanwhile, it was clear there was now no hope for the marriage, in the short term at least, and Rupert left the house at Kingston Road for the last time in July 1927, packing up all the pieces of H3 and the RAS regulator, amongst everything else, and taking all he had over to Downside to live with his mother, who was now obliged to sell Glenside again.

Muriel's new life with Miss Gurney continued to establish itself during the Summer of 1927. The children were now living with them at Vivian Gurney's flat in St George's Square Victoria, and from here the couple took the children on summer holidays to Bournemouth that year.

41. Vivian Gurney with Jocelyne and Cecil on holiday in 1927. Muriel and Vivian would remain together for a turbulent ten years. (© Sarah Stacey and Simon Stacey, 2005)

Gould, meanwhile, attempted to maintain a social life, attending every O.V. dinner throughout the Summer and Autumn. As well as his Franklin research at the Admiralty during the day, he took up his 'Harrisonian' tasks at Downside with new vigour during this time too, the RAS regulator fully occupying his evenings and weekends for many months (as discussed in Chapter 13).

The High Court

In preparation for the hearing, the Court ordered on the 12 September that each side should make a full copy of the statement/response available to the other side, for information, and on 20 September Muriel applied for interim costs to be funded by her husband, pending the outcome.[235]

The case was finally heard over three days, Friday 18th, and Tuesday 22nd and Wednesday 23 November 1927, before the Honorable Sir Alexander Dingwall Bateson at the Royal Courts of Justice in the Strand.[236] In case a witness was needed, Muriel had made the decision to bring Cecil, at 9 years old, out from boarding school and he remembered waiting nervously outside the Courtroom in those grand and awe-inspiring buildings on the Strand.

Happily, he was not called to give evidence. Rupert did however need a character witness and his mother Dodo stood by him on the day, though the evidence given by her daughter-in-law must have been very depressing for her. In her own way, Dodo had a lot at stake too. Apart from not wishing to see her son dragged through the courts in this terrible way, she too might suffer from the inevitable opprobrium. In her retirement—she was now 67 years old—Dodo was engaged in a number of prestigious charitable works, not least of which was membership of the Executive Committee of the 'Queen's Institute for District Nursing'. This committee was actually due to meet a couple of weeks later, and the whole affair must have been very distressing and embarrassing for her.

At least they were spared a presence in the public gallery. As soon as Muriel began to give her lurid evidence it quickly became clear to the judge that the matters being discussed were of a very explicit nature, and the hearing was moved *in camera*. No doubt Dodo's evidence of Rupert's good character was strong, but it was insufficient to convince Justice Bateson, and on the last day of the hearing his judgement was clear cut: a decree of Judicial Separation as of that same day. He remarked that 'a man may be good at his work, but this does not necessarily make him a good husband . . .', and he 'condemned the said Respondent in the costs incurred . . .'.[237] The final bill for these, over and above that already paid, was £413/1/8, equivalent to approximately £12,250 today (2005),[238] a devastating sum and one Rupert didn't have a hope of paying without longstanding debt.

Press interest

As if this wasn't bad enough, word of the case had aroused interest by the Press, and *The Daily Mail* and *The Times*, amongst others, decided to cover the case. Unfortunately they had also managed to get a smattering of some of the more titillating details before the hearing was closed to them.

By coincidence, a change in the law the previous year was supposed to have reined in the Press's tendency to reveal every bit of salacious detail in cases like this.[239] Nevertheless, the following day, a piece appeared in The *Mail*, headed 'Husband's Cruelty', giving over half a column to the story and including some of the sordid details.[240] The piece reported that medical evidence had been provided on both sides, Rupert's relating to his earlier nervous breakdown and Muriel's relating to the strain she had been put under coping with her husband. The *Mail's* reporter, who gratuitously added the heading 'Gagging Story', noted: 'Mr Justice Bateson, in his judgement, said in his opinion the wife must succeed. He considered her evidence was true and that her allegations were true'.

In fact, Rupert did not deny that some of these had truth in them, 'but only to a very limited extent' as he put it. An interesting point reported in *The Times* was that according to Muriel's evidence, she had tried for a reconciliation and that 'Mr Gould's refusal no doubt put a great strain upon the wife and affected her health adversely and did her injury'. Whether Muriel really did try for a full reconciliation is unlikely now to be proved, but there seems to be nothing surviving to support such a claim. As there was no co-respondent on Gould's side, it is difficult to see why he should have simply 'refused' such an approach, unless of course, it came with the proviso that Vivian Gurney had to remain with Muriel, an issue which Gould naturally could not possibly have raised in court.[241]

THURSDAY,   The Daily Mail   NOVEMBER 24, 1927.

HUSBAND'S CRUELTY WARNED BY JUDGE.

WIFE WHO WANTED TO BURY THE PAST.

DIVORCE DIV.—Mr. Justice Bateson.

Mrs. Muriel Hilda Gould, of St. George's-square, Westminster, S.W. petitioned for a decree of judicial separation against her husband, Lieut.-Commander Rupert Thomas Gould, R.N. (retired), at present engaged in the hydrographic department at the Admiralty, alleging cruelty, which affected her health. The cruelty was denied.

There was a body of medical evidence on both sides, and Commander Gould called his mother, Mrs. Agnes Gould, of Downside, Ashtead, Surrey, as a witness.

Mr. Justice Bateson, in his judgment, said in his opinion the wife must succeed. He considered that her evidence was true and that her allegations were true without a witness.

It was acknowledged that Commander Gould did his work perfectly satisfactorily, but a man might be a perfectly sound and nervous breakdown, but he had recovered sufficiently to justify his getting married.

GAGGING STORY.

Mr. and Mrs. Gould were married in 1917. Previously the husband had had a serious illness.

After the marriage there were some difficulty, and something was said by the husband about gagging his wife. Their two stories differed, but he believed Mrs. Gould. At any rate, there was talk about it.

In 1918 a son was born, and after that the husband, who had the reputation till

DEALINGS IN NINE CULTURE PEARLS.

KING'S BENCH DIV.—Mr. Justice Swift and a Common Jury.

Mr. Jack Brompton, of Chancery-lane, W.C., dealer in precious stones, sued Messrs. Carter Bros., jewellers and dealers in precious stones, of 36, Gerrard-street, for the return of nine culture pearls or, in their default, £135, and damages for detention. Carter Bros. did not admit that the value of the pearls was £135, and denied that the pearls were the property of them or that they were his property.

Serjeant Sullivan, K.C., said that in November 1926 a Mr. Stambois took on approval from Mr. Brompton two culture pearls and a reconstructed ruby, leaving a diamond ring as security. On November 8 Mr. Stambois paid £55, of which £45 was in respect of the pearls. He then received the nine pearls, the ring again being given as security. On November 15 Carter Bros. wrote to Mr. Brompton asking for the return of the money as the pearls were not what they "were represented to be."

Two days later, Mr. Brompton was arrested, it being alleged that he attempted to steal the ring. ... Bow-street police Court. After Mr. Stambois had been examined, the case collapsed and Mr. Brompton was discharged.

Mr. Brompton, cross-examined by Sir Walter Greaves-Lord, K.C., for the defence, said he gave the pearls to Stambois as culture pearls.

The defendant, Mr. Walter Ernest Carter, said he parted with the nine pearls to Mr. Stambois, who gave him £55 of the £65 which he had paid for the two pearls.

MENACING A WITNESS.

SHIPOWNER LOSES LIBERTY.

JUDGE ON CONTEMPT OF COURT.

From OUR SPECIAL CORRESPONDENT.
Cardiff, Wednesday.

The hearing of the charges at Glamorgan Assizes, Cardiff, against Watkin James Williams, the Cardiff shipowner, of offences in connection with the loss of the steamship Eastway and 23 lives, and the alleged habitual overloading of the steamship Tideway, was interrupted by Mr. Justice Wright to-day to deal with a complaint against Williams.

Captain James Agnew, the former master of the Tideway, who was giving evidence, but on rising in the court yesterday returned to the witness box yesterday.

Sir Douglas Hogg, the Attorney-General: When you left the witness box yesterday did the prisoner say anything to you? Yes. He said he would ruin me before he had finished with me.

Mr. Justice Wright: That is a very serious thing to say to a witness giving evidence.

Sir Douglas: So serious that I have taken the ... I get instructions from ... who heard ...

Mr. Justice Wright decided to hear these witnesses at once, and two solicitor's clerks and Mr. W. H. Coombs, managing director of the Navigators and General Insurance Company, said that they heard Williams, as he went to the clerk of the

42. The Daily Mail, 24 November 1927. The column headed 'Husband's Cruelty' covered the story in some detail. The subheading about the wife wanting to bury the past would seem quite misleading, given the actual situation.

Whether or not Gould wished for the marriage to continue, it is clear his attitude was unchanged as far as the Sette was concerned. He attended every dinner during the autumn of 1927 (since October these were held at the Savoy Hotel), including that held on the evening of 22 November, right in the middle of the Court hearing, even illustrating the menu cover for that dinner.[242]

The Consequences

With no family and no home, Gould's only option was to remain at his mother's house in Ashtead. The judgement of the Court was devastating for him, though it was to be several weeks before the full import of what had happened would completely register as there were two further terrible blows awaiting him.

Rupert had long been concerned that his post at the Hydrographic Office was something of 'a luxury' to the department. He had predicted 4 years earlier in a letter to Admiral Parry that 'if the question of reducing the staff crops up, I am quite certain I shall be the first to go', continuing philosophically: 'I am in some ways a fatalist, and I believe that when the right time comes for me to leave here I shall not be allowed, even if I wanted to, to remain . . .'.[243] And so it was.

H. Percy Douglas

For some years the Goulds had been on friendly social terms with Rupert's colleague in the Hydrographic Office, H. Percy Douglas (1876–1939) and his wife. By 1927, Douglas (by now Vice Admiral) was Hydrographer and was Gould's boss. Douglas had been finding Gould, with his pedantic attitude, increasingly irritating to work with and the two cordially disliked each other. Muriel on the other hand, had always got on well with the Douglases, and during the crisis in the Gould's marriage, they were very supportive to her. So, the messy and shameful marriage separation now gave Douglas the perfect excuse: Rupert was summarily 'released' from his employment in the Hydrographic Office.

Admiral Douglas was not alone in feeling that Muriel had been ill treated by her husband. Rupert's erstwhile friend and colleague, Cdr. Archibald Colquhoun Bell, also now supported Muriel. Bell was about 4 years Gould's senior and the two families had been close for some years, so Bell and his wife would have been familiar with both sides of the

43. Vice Admiral H. Percy Douglas. Douglas was Hydrographer of the Navy in 1927 and Gould's boss in the Department. He disliked Gould and did not find his dismissal from the Department difficult. (UK Hydrographic Office (www.ukho.gov.uk))

argument in the Gould *vs* Gould affair. Witnessing Rupert's obsession and his activities in the Sette (Bell had joined in 1924), and hearing Muriel's complaints from home, they too were now critical of Rupert.

Bell resigned from the Sette of Odd Volumes at this time, probably as a result of the affair. Soon afterwards, John Gilbert Lockhart, a Director of the publishing house Philip Allen, and a friend who would stand by Gould throughout the court case, was admitted to the Sette[244] and he assumed Bell's old Sette name of 'Investigator'. Perhaps Rupert suggested it as an appropriate and symbolic way of settling the loss of his friendship with Bell.

Val Hirst

Commander Bell was not the only friend to desert Gould now. In later years Rupert stated that he had suffered a triple blow, from his wife, his

Val . Rupert

44. Rupert's closest friend, Val Hirst, with Rupert at Frinton in 1921. They never spoke again when Hirst sided with Muriel over the judicial separation in 1927. Gould was particularly sad about the split. (© Sarah Stacey & Simon Stacey, 2005)

chiefs and a very close friend. This was Val Hirst, one of his dearest friends, who now also sided with Muriel, something which Rupert was terribly upset about.[245]

The court had originally ruled that Rupert would at least have access to his children for half of their holidays but, on appeal, Muriel had managed to persuade the court that this should be reduced to one quarter, on the extraordinary grounds that 'the children didn't like their father'. Of course Gould also had to pay alimony to Muriel, eventually set at 15/- (Fifteen shillings) a week, as he now had virtually no income save his small Navy pension.

It didn't seem possible but somehow, at a stroke, Gould had managed to lose everything: his wife, custody of his children, his dearest friend, his home and now his job. At thirty-seven years old, Rupert Gould's world had simply collapsed.

❦ 11 ❦

Oddities and Enigmas 1928–1929

With the separation now official, Muriel and the children settled in to a new life with Vivian Gurney. Miss Gurney had now taken a larger flat in Kensington, the top two floors of 62 Courtfield Gardens, to house the new family. As for Rupert, he had no choice but to remain at Ashtead, and live permanently with his mother, Dodo; he had no other home. Having lost his job his only income was now his tiny Navy pension, of which Muriel took over half in alimony and there were the huge costs of the court to pay, so no chance of affording a lease on his own home.

It was a terribly sad and bitter time for Gould; the consequences of the case had far-reaching and humiliating effects on his life. Some years later he wrote: 'For about three years I was like a man who has been permanently stunned, with no ambition and very little interest in life. I had lost all my ideals, and I felt that something had broken inside me and would never be mended . . . [but] . . . the one thing I set myself to do was to avoid self-pity'.[246]

Every aspect of the break up had been acrimonious, not least with his ex-employers. Explaining his situation to colleagues at the Scott Polar Research Institute the following year, he wrote: 'After eleven and a half years in the Hydrographic Department I was compelled to resign my appointment in consequence of the publicity attaching to an action for separation which my wife brought against me. As I was in the right, and had done my best to prove it (otherwise I would never have fought the case, but consented to a separation by deed and kept my job) I felt this high handed and unjust action very keenly. *Inter alia*, my wife is a close friend of Mrs Douglas, the wife of the present Hydrographer. You will see from this that relations between the Admiralty (particularly the Hydrographic Department) and myself are not very amicable at present . . .'.[247]

Adding insult to injury, the Hydrographer now made efforts to remove Gould's association from work published by the Department. One of the pieces of work Gould was very pleased with while at the

Hydrographic Office was the preparation of the new edition of *The Antarctic Pilot*, the Admiralty's most complete and up-to-date sailing directions for the whole of the south Polar Regions. After a delay of three years, this was finally published in 1930 without a single reference to his contribution, and a number of highly relevant articles by Gould were simply omitted from the Bibliography published within it.

It is perhaps difficult to understand today the profound shock and disdain which was commonly felt by much of society when a case like the Goulds' came to the public notice. No doubt many people's expressions of disapproval were quite hypocritical—their own behaviour may well have been far worse behind closed doors. The real problem was not that such things happened, but that they should be discussed and broadcast publicly, and for this Rupert Gould would sadly pay a high price; he would soon discover that all round him doors of opportunity for a new start were closing in his face.

Keeping busy

In coming to terms with his situation, horology was always a great consolation for Gould of course, and work on one of the great Harrison masterpieces, the regulator belonging to the RAS, began again almost immediately at Downside, most of the work being done once Gould had set up his workshop in the large attic room which Dodo had allowed him to take over. This was eventually followed by the completion of H3's restoration and finally the reconstruction of H1; work which continued, alongside Gould's other activities and writing, virtually constantly from his arrival at Downside in 1927 until the mid-1930s. Just as with our description of the work on the RAS clock, this further restoration work on H3 and H1 will be considered in later chapters, to keep the narrative as simple as possible, but it should be remembered that the practical horology was constantly going on in tandem with his writing.

Once Rupert had begun to come to terms with his new situation, research, writing and lecturing on his many interests also began again. Thankfully, not all his friends deserted him after the judicial separation, and Frank C. Bowen, who ran a publishing company specializing in naval history, provided Rupert with desk space in the company's Westminster offices until he had set up his library and workshop at Downside. One of the first major events in this new phase in Gould's life was to give a lecture to the British Horological Institute in December 1927 on 'Horological Curiosities'. But even here at the BHI it seems there were knock-on effects of the court case and, owing to 'the publicity', Gould soon felt it would be best if, for the time being, he resigned from membership of the Council of the Institute. This he did in June 1928, less than two years after his election.

Money was also now a real issue. Fortunately he had paid for life membership of the BHI, but he was unable to continue paying the fees for Fellowship of the Royal Geographical Society and was struck off the list of Fellows. And it was soon clear that the Society's hierarchy were shedding no tears over this particular loss of a Fellow.

Rupert had submitted three reviews for inclusion in the *Geographical Journal*, and on publication he discovered that, while other reviewers were credited by their initials, his were anonymous. Challenging the RGS Secretary, his erstwhile friend Arthur Hinks on the matter, he queried whether this decision might be 'connected with my wife's recent action . . .'. The reply by Hinks confirmed, with weasel words, that he had considered it might be better if 'for the time being' anything submitted by Gould were not identified as such. Some years later, in a letter to Hugh Mill,[248] Gould noted that as a result of the court action he had been, and still was, *persona non grata* with the hierachy of the Society 'and while I was unable to alter that attitude, I was not in the least disposed to endure it', and that he discontinued contributions to the Journal and meetings.

His interest in Hydrography was unaffected of course, and in February he published a letter in *The Times* on the currently sensitive political issue of the ownership of Bouvet and Thompson Islands in the south Atlantic. Having made a special study of those islands when at the Admiralty he was uniquely qualified to comment on such matters.[249]

Typewriters

Thankfully, the public opprobrium Gould was experiencing did not apparently prevent his continuing association with some of the larger institutions. Another big lecture he gave at this time was to the Royal Society of Arts, on 21 March 1928,[250] and was on the subject of 'The Modern Typewriter and its Probable Future Development'. It was a subject which had interested Gould for years and was something he had been studying recently. In September 1927 he had managed to find time to write notes for the Science Museum's curator, Mr Hartley, on their collection of typewriters, one of the largest available at the time.

He had also agreed to loan the Museum two typewriters from his own collection, which he had recently begun to form (this did not require large funds; old typewriters were not expensive to buy, and many were actually given to him). Among the distinguished members of the audience for the lecture were several doyens of the business machine world, and also the electric clock manufacturer, Frank Hope-Jones. The text was subsequently published in the *Transactions of the Society of Arts*, and was apparently even translated into Japanese! The research for this lecture began Gould's life-long fascination for this most useful piece of office equipment and he soon formed a very large collection of them, ultimately (1944) numbering some seventy instruments. In spite of his intense interest in this machine though, he never mastered its use beyond what he termed the 'hunt and peck' technique.

Lecturing

By this time Gould was becoming an accomplished lecturer, and giving talks, whether for small informal groups or large official bodies, held no fears for him. He had amassed a considerable number of lantern slides on the various subjects of his expertise, and he was regularly asked to speak to local literary groups or institutes. He was generally a patient speaker, but noted to his correspondent Professor Stewart that he could not tolerate people coughing while he spoke. On occasions he also found effusive praise a little too much, describing one group of enthusiastic ladies as 'gushing boneheads'! His collection of lantern slides and glass plate negatives survives and is now preserved in the collections of the British Horological Institute at Upton Hall in Nottinghamshire.

Income

Keeping busy with his lectures and studies at Downside was however, not enough: one of the first things Rupert needed to do was find a way of earning a living.

On his arrival at the house, the ever resourceful and practical Dodo must have sat him straight down and discussed his future. Although wholly supportive of her son in his adversity, Dodo was well aware that there

45. Rupert T. Gould with his mother Dodo, at the front door of Downside in 1930. The household was run to Dodo's strict timetable, but life was comfortable at Downside, in spite of Rupert always being short of money. (© Sarah Stacey and Simon Stacey, 2005)

had been some truth in Muriel's accusations that he tended to drink too much. There was no way she would have been prepared to fund a continuation of that kind of indulgence. So, while being provided with a home and stability, Rupert was expected to provide for himself. His meagre pension, part of which went every week to Muriel as alimony, would certainly not.

The difficulty was that with such unfortunate publicity over the court case, finding suitable paid employment would be almost impossible. This unfortunate period in Gould's life also happened to occur at the worst possible time in the British (and the world's) economy. Within months the world was facing the biggest depression in living memory; the stock market collapse in 1929 was followed by the 'slump' and the country entered a period of severe economic depression and high unemployment in which Gould stood little chance. But it was vital he did something remunerative, the trial had totally sapped his (already low) self-esteem and he desperately needed to earn his keep and prove his value to himself and society.

This was not the first time he had dealt with this situation. After his failure to cope in 1914, part of the successful cure for his low self-esteem had been writing (*The Marine Chronometer*), and it was in writing that he sought solace and achievement again. Over the next few years he would write a series of remarkable books on the subject of scientific mysteries, a rather different field from those he had specialised in, up to now. He explained his motivation in a letter to Professor Stewart: 'I wrote *Oddities* and *Enigmas* to show, as much as anything else, that I was not the cruel, besotted [alcoholic?] wretch that interested persons had made me out to be . . .'.[251]

Scientific Mysteries

Ever since boyhood Rupert had been fascinated by reports of unexplained phenomena and mysterious occurrences. With his sharp, analytical mind, he found that, in reading such accounts, he was capable of dissecting the most complex and confusing of cases. He took real pleasure in logically sifting ideas, studying the data, discarding the 'red herrings' and organizing the significant facts—as has been said, he would have made a fine barrister. Nevertheless he was, for those days, deeply sceptical when dealing with fantastic or pseudo-religious explanations for such events or occurrences. He always preferred to take a cold, hard, logical approach to the problem, and often therefore had to be content with ordering the evidence and leaving others to draw what conclusions they chose. And this new genre, made virtually his own by Gould, appeared at just the right time.

After the First World War almost everyone in Britain had known someone who had perished as part of that conflict, and the inter-war years were marked by a growing interest in spiritualism and the paranormal. These books of Gould's were thus widely read and quickly developed a cult of their own. Most readers were keenly interested in these curious tales, though it must be said they were not always taken quite as seriously as they deserved. Robert Graves remarked in *The Long Weekend*,[252] his controversial but interesting book about this period: 'The Press exploited 'borderland' cases between science and mysticism, hard fact and prodigy. The usual line taken was to print the hard facts of a case but without spoiling the story for those who liked prodigies: Lieutenant-Commander F [sic] Gould, author of *The Case for the Sea Serpent* and similar 'believe it or not' books, was the best-known journalist of this borderland'.

In a similar vein, George Orwell noted of the novelist Charles Reade: 'What is the attraction of Reade? At the bottom it is the same charm one finds in R. Austin Freeman's detective Stories or Lieutenant-Commander Gould's collections of curiosities—the charm of useless knowledge'.[253] It would be interesting to hear what Gould thought of this, if he read it! These mystery books of his may not have been of profound importance in the history of literature, but they are both charming and deeply fascinating, and were hardly based on 'useless knowledge'. Rupert was, incidentally, a friend of the writer R. Austin Freeman, mentioned by Orwell. One of Freeman's later 'whodunnits', on a horological theme, *Mr Polton Explains*—also written in 1940—was dedicated to Gould, 'To . . . the distinguished horologist, this story of a simple clockmaker is dedicated by his old friend, the Author'.[254]

The books

So it was this sphere of his interest, unsolved scientific mysteries, that he took up, partly to aid his recovery, and during the first 6 years at Downside he wrote a series of exceptionally fine books on these subjects, works which were both ground-breaking and seminal. In the majority of the cases he described, the mysteries had never been properly discussed before. As will be seen, it is remarkable, over seventy years later, how (forgetting the somewhat barbed remarks of the likes of Graves and Orwell during his lifetime), many of his essays require little criticism or correction, and have been cited by just about every author who has subsequently written on those subjects.

To this day, students of all things paranormal or 'Fortean'[255] all generally hold Gould in very high esteem for his logical and dispassionate recording and analysis of evidence. Since his death, there have been a number of subsequent reprints, in various forms, of all these books of

Gould's at various times, mostly produced in the United States (see the Bibliography, Appendix 1).

His accessible and engaging style of writing, apparent in his earlier work, *The Marine Chronometer*, is evident in these books too, and for many readers it was Gould who first got them interested in these subjects. The writer and publisher Mike Dash, for 18 years a contributing editor of Fortean Times and publisher of that journal between 1993–2000, who has kindly helped in assessing Gould's contribution in this field, recalls 'the sheer range of Gould's selections was a wonderful stimulus to curiosity and the imagination, and his footnotes (which hinted at an even vaster range of arcane knowledge) were if anything even more enticing than the text'.

Eric Russell, author of *Great World Mysteries* (1957) described him as 'a shrewd if somewhat irascible author who devoted much of his time to analyzing such puzzles . . . and—to judge from his many letters to me—gathering odd-shaped scraps of writing paper from heaven alone knows where' (the odd reference to the 'scraps of paper' comes from Gould's practice of trimming his letters to exact length, saving small pieces to be used for shorter notes. Shortage of money over the years led to a number of parsimonious little foibles such as this.).

Loren Coleman, who has written extensively on the subject of sea serpents (one of Gould's favourite topics), especially appreciated Gould's 'wonder-filled combination of a Victorian sense of adventure and a rugged British love for the no-nonsense examination of the evidence. For Gould, this translated into a great respect for detail and open-mindedness'.

Naturally, with most of the mysteries Gould discusses, the information is based on reports from human observers and he took considerable trouble, in all his research, to weed out any reports or information he considered might be the result of unreliable sources, hoaxes, simple misinterpretation, or 'abnormal psychology'. Nevertheless, it would be fair to say that with our even more cynical and sceptical hindsight, one criticism we could level at Gould's work is that he was so intent on the gathering, ordering and analyzing of the information and reports of the mysteries he studied, he was occasionally inclined to under-estimate the likelihood of what Mike Dash terms the 'psycho-social' explanations. Dr Dash defines the term as ' "theories suggesting that many unexplained phenomena have their origins in the human brain, mediated by the culture in which the percipient was brought up" In other words, various psychological and physiological processes can cause people to think they have seen UFOs, ghosts etc., and their descriptions of what they saw will be conditioned by expectation based on what they already know of the subject'.[256]

There is also evidence that in some cases Gould would give too much weight to written sources simply because of the social status of the reporter. Is the evidence of an aristocrat, or perhaps a local dignitary who is a full-time professional in the community, necessarily more trustworthy than that of a labourer?

It must also be said however that Gould himself was very much aware of his own fallibility, and was never content with his publications. Just as with his book *The Marine Chronometer*, each time one of these books was published he began annotations in his own copy (most of these personal copies, happily, have survived) in the hope that, one day, each might have an improved version. As it turned out though, only two would go to second editions.

Wide scope

The scope of the subjects in the books was extraordinarily wide: from Astronomy to Perpetual Motion, from Mathematics to Crypto-zoology (the existence of unknown species of animal). The subjects he discussed were so different from ordinary life, in some cases so unworldly, that perhaps these too served to protect Gould from his recent hurtful memories; even in later years he described the mental scars of the court case as 'only skimmed over'.

To introduce the various strange subjects Gould considered in the first two of these books, *Oddities* (1928), *Enigmas* (1929), and to assess the strength of his arguments in their analysis, Appendix 4 summarizes their contents and takes a retrospective look at some of the topics in the light of current knowledge. Those wishing to have a closer look to evaluate his research and writing are referred to that Appendix.

Oddities

The first, and perhaps the best known, of Gould's 'mystery books' was *Oddities*, released on 30 August 1928, and published by Philip Allen & Co. (Allen and J.G. Lockhart). Gould had personally introduced John Gilbert Lockhart to the Sette of Odd Volumes as a member the previous year. Significantly, Lockhart, who was an ex-naval man himself and had already written on mysteries of the sea, including coverage of the *Mary Celeste* mystery,[257] took the Sette name of *Investigator*. As already noted, Lockhart had remained Rupert's close friend during and after the court case and it probably seemed just that Lockhart should assume the title left vacant by the resignation of Rupert's ex-colleague Lt. Cdr A.C. Bell just weeks before. Philip Allen then also joined the Sette in 1928, and *Oddities* was dedicated by all three, author and publishers, to the Sette itself.

Oddities was Gould's saviour at this time. He absolutely threw himself into its writing, taking under one month to complete the 75,000 words. The material for it had of course been collected over many years, Gould had been storing away facts and figures since he was 18 years old, but to produce a 335-page book with Preface, twenty-seven of his own carefully drawn illustrations and an index, in less than 30 days, is rapid work by any standard.

The dust jacket for the book is undoubtedly intended by Gould as a coded statement of his own feelings and situation at this difficult time, and is worth a closer look. It was an adapted drawing first done by him in

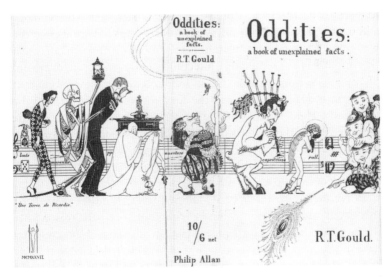

46. The dust jacket for *Oddities*. Full of symbolism, the stately funeral procession was an emotional and dramatic summary of how Gould saw his life at the time. (© Sarah Stacey and Simon Stacey, 2005)

1921 and titled *La Tierce de Picardie*, evidently drawn for the Sette of
Odd Volumes and possibly for inclusion in the menu on the night of
the inauguration of the musician Frederick Keel ('Singer') as President
on 25 October that year. At face value it represents, in the form of a
pictorial score, the musical expression of the title's name, denoting the
closure of a minor movement by a major third.

A sombre, burlesque funeral procession, representing the melancholy minor key, ends suddenly with the cheerful major group of three
happy children. But Lt. Cdr. Gould's presence in the procession, face in
hands, with death at his arm and time running out ('morendo': dying),
is hard to miss. The score begins with the ominous and foreboding marks
of 'Black Jack' and the 'Curse of Scotland'; the figure of Paganini (whose
music and life always fascinated Gould) often believed to have been in
league with the Devil, follows the Commander. Then comes the little
coffin with Gould's cap surmounted by a tiny figure of Captain Cook
(perhaps representing the death of Gould's promising, fledgling career
as a navigator). The chief mute (perhaps a reference to Rupert's inability
to speak during his first illness), proud but silent, advances majestically.
Then comes the Satyr, whose lustful caperings are governed by the
direction 'capricioso'. One possible interpretation is that this is a
reference to Val Hirst, who Gould noted had a strong libido (see note 245)
and whose very close friendship with him ended bitterly during the
separation. The procession hesitates at the captive Amor, which is open
to a number of interpretations, but could simply refer to his distress at
his separation from his wife. This is followed by OV, the dedication to
the Sette of Odd Volumes and the final fortissimo major third, the
uppermost child holding aloft the 'accidental' which converts the
minor key to major, a triumphant and happy ending to a sad tale. As
originally drawn in 1921 there was no Satyr, and this was added when the
drawing was redone, and titled '*Une*' [as opposed to *La*] *Tierce de Picardie*, for
the book cover in 1927.

In the Preface to the book, Gould modestly dismisses the subject
matter as possibly superficial, but he points out, with good reason, that
the facts he puts forward are indeed facts. While some of the mysteries
have subsequently been solved, others remain hotly debated to this day
and, although an in-depth assessment of each is impracticable, interested
readers are referred to Appendix 4 for a summary and commentary.

The subjects in *Oddities* cover a very wide range of topics: from
'The Devil's Hoofmarks' which were apparently left in the snow in

nineteenth century Devon, to the perpetual motion machines of the eccentric eighteenth century pseudo-scientist known as Orffyreus; from the Frankenstein-like creatures said to have been created by Andrew Crosse in the 1830s, to the mysterious number-theories expounded by the great French mathematicians Mersenne and Fermat. Fittingly, the last chapter (on the soothsayer Nostradamus), and the final part of the book, ends with a remark that could be applied to the whole work:

'However, I suppose that there will always be many who . . . regard a statement as incredible because it is well-attested—in the manner of the countryman who, confronted with his first giraffe, remarked "I don't believe there's no sich creature". It is probably a waste of time to commend to their attention the essential cowardice, as well as virtue, of "philosophic doubt" in the presence of unexplained facts'.

A notice in *The Times* newspaper on 31 August 1928 briefly covered the book's content, and *Oddities* was then given the best part of a column when reviewed in *The Times Literary Supplement*, being praised for the great variety and interest of its subjects. The publisher's advert for the book in *The Times* on 5 October, puts the price at 12/6 (Twelve shillings and sixpence), and trumpets a few sound bites from other reviews, *The Daily Telegraph* declaring it 'A fascinating book', *The Sunday Times* considering it 'delightfully unexpected' and the *Illustrated London News* going so far as to state: 'Vastly entertaining. None cultivating the curious could rest without it on his shelves'. Sales of the book were immediately successful and soon afterwards the publishers were happy to instruct the author to begin work on a second book of the kind, to be titled *Enigmas*.

Life at Downside

With *Oddities* complete, and a small income, with prospects for further business in future, now established, there began in Gould's life a ten-year period of stability and good health. There were periods of depression on occasions—this illness would dog Gould all his life—but the 1930s were generally to be the most settled, organized and happy years of his life.

Downside itself was far larger than Dodo needed but, as has been observed, she had never much liked the house in Southsea and longed for something grander. Given her family background, this first home of the Goulds was relatively modest and, by the Hilton's and Skinner's standards, humble. Downside was probably therefore Dodo's way of returning to some of the grandeur she had known in childhood. So there was plenty of room for Rupert and the arrangements were ideal.

Not only did Dodo and Rupert have large bedrooms for themselves, there were separate bedrooms for Cecil and Jocelyne when they visited, as well as rooms for the live-in staff. A large staircase in the centre of the house led upstairs from the main hall, at the top of which was a huge oak bracket clock (weighing over 100 lbs according to Rupert) by Dent (the manufacturer of the 'great clock at Westminster', known as Big Ben) and which could be heard throughout the house when it chimed the quarters.

Dodo and her staff, the chauffeur 'Partner' (Cecil Charles Partner), a gardener, a maid and a cook (the latter two living in) were always around for company. But Gould had his own room and the attic workshop for his writing and practical work when he needed solitude. Cecil described this haven of his father's: 'Dodo had made over to him a large room at the top of her house. It was called the "work room," not the "workshop." You went up a flight of narrow, creaking stairs from the first floor. Facing the staircase were two or three more steps which led into what was called the "Passage room". It was quite long, but the roofs sloped to such an extent that it was virtually useless. At the end was the door of the work room, which opened at right angles to it, towards the right . . . Around such of the walls as were vertical were my father's books. There were about two thousand of these . . . many were horological, and there were sections on other scientific subjects. There was a shelf or two on art and music . . . but though my father appeared to have read everything, there was little literature in any language'. At one gable-end of the room, in front of the window which looked out over the gardens at the back of the house, was Gould's workbench. 'If he were cleaning one of the Harrison chronometers or the original Orrery . . . we [the children] were allowed, and indeed encouraged, to watch him, though he must have known that neither of us had inherited his interest in mechanics'.

Once a month Rupert had weekend custody of the children at Downside and, with their grandmother there too, the children looked forward to the stay as much as Rupert did. The only unpleasant side to these weekends was the collection of the children from Muriel and Miss Gurney who were then living in a house on The Green (No. 23) at Kew. The collection was usually undertaken by Dodo, but if it were on a Sunday, when Partner was not working, Rupert drove instead. Cecil recalled that they would look out for the arrival of the car 'Occasionally

Miss Gurney came out with us and, from the pavement, hissed a message through the windows of the car. On these occasions Dodo replied as briefly as possible, but civilly. My father either nodded without saying anything, and without even looking at Miss Gurney, or he rapped out a few monosyllables. If this charade occurred on the return journey, we had the benefit of hearing Miss Gurney's reaction to it, in the form of an explosion on the subject of my father's "rudeness." It evidently never occurred to her that he had very good reason to detest her'.[258]

As for social life at Ashtead, Rupert soon got to know a number of the local residents and had friends to play billiards with at Downside when so inclined. Annual holidays were taken regularly in late Summer with Dodo, usually at The Grand Hotel, Sheringham on the north coast of Norfolk, with its bracing sea air, ideal for another of Gould's passions: kite flying. Money was always short (Dodo was, understandably, still not prepared to fund him) and the succession of writing commissions he undertook during the 1930s only just kept the wolf from the door.

The Sette of Odd Volumes

Indeed, there was a time when it looked likely he would have to resign from the Sette of Odd Volumes but, refusing to lose such a valuable brother, the 'Odd Council' made him an Emeritus Member, so he continued to attend dinners. During this period, producing illustrations for menus and contributing to the Sette's activities were the mainstay of his social life. From the summer of 1927 and for the following 10 years or so, Gould attended virtually every one of the Sette's dinners. He took the historian of science, R.T. Gunther (1869–1940) and P.A. Phillimore as guests to the dinner on 28 February 1928, when he spoke on 'Typewriters' (no doubt a shortened 'dry run' for the paper given at the Royal Society of Arts on 21 March). On 24 April 1928 he attended the huge Jubilee dinner (50 years) of the Sette (at the Savoy as usual), with Dodo and two guests, Frank Bowen and Capt. C. J. Fearfield MC, the photograph of the dinner happening to have the Gould party right under the camera, with the back of Rupert's head in the lower centre of the picture. In early November 1928 Gould was lecturing in Leeds on a great hero of his, Captain James Cook, and on the 27th he gave the same paper to the Odd Volumes, taking W.G. Perrin, the Admiralty librarian as a guest on that evening. He would soon be under contract to write a biography of Cook, though what finally appeared was not as ambitious as he had planned.

A few minor publications occupied Gould at this time while he made preparations for his next book. A long and detailed letter on the subject of the marine chronometer was written by him for *Lloyd's List and Shipping Gazette* in the Spring of 1928. It was in answer to a correspondent who had most unwisely asserted that it was in fact Christiaan Huygens who had invented the chronometer, not John Harrison. The *Mariner's Mirror* of October 1928 contained two articles from Gould's pen, 'Some Unpublished Accounts of Cook's Death', including a map drawn in illustration for it, and 'Notes on Cook's Last Voyage'.

Enigmas, Another Book of Unexplained Facts (1929)

Within a year of *Oddities'* release, Gould had produced its sister publication, though he admitted that, in case a sequel was not required, he had included most of his best material in *Oddities*. Nevertheless, *Enigmas* has some intriguing stories to tell and is generally regarded as a fitting pair to its predecessor. The book is slightly smaller than *Oddities* and contains nine mysteries for discussion, summaries and commentary on which are in Appendix 4. This time the book was dedicated to Dodo, perhaps settling a slightly difficult diplomatic tension. It is possible Dodo saw the dedication of the earlier book to the Sette (who, it could be argued had contributed to Gould's predicament) as a little surprising when she had supported him throughout his most difficult trials.

As with *Oddities*, *Enigmas* was given a dust jacket illustrated by Gould, this time featuring the great Sphinx and the statue of Memnon. The book, priced at the usual 12/6, was announced in Allen's advertisement in *The Times* for 1 November 1929, and was given nearly a column of approving review in the *Times Literary Supplement*. Just as with *Oddities*, a second edition of *Enigmas* was prepared by Gould during WWII, appearing in 1946, though unlike *Oddities* (the chapter subjects of which remained the same), the second edition of *Enigmas* had two chapters replaced with others, the earlier ones being deemed of insufficient interest for the general reader.

By the time *Enigmas* was being reviewed in the autumn of 1929, Gould was well under way with his next book, one of two on a subject for which he would be especially well known in the years to come.

❦ 12 ❦

The Case for the Sea Serpent 1930

Of the many and varied scientific mysteries for which Cdr. R.T. Gould is known, the one subject for which he is, even today, most closely associated is the question of the existence of sea serpents. The term refers to an as-yet unidentified species of sea creature, by some accounts resembling the prehistoric plesiosaur, sighted occasionally, over the years, by mariners and sea-going passengers, but never yet captured nor clearly photographed.

In this most controversial of Gould's scientific mysteries, he stepped down from his usual impartial look at the evidence and 'nailed his colours firmly to the mast'. He was convinced some species of such creature did, and therefore perhaps still does, exist in the oceans of the world, and at least in one case, in a certain freshwater Scottish Loch.

In an impassioned note in the BBC publication *The Listener*, in 1937 (after two radio broadcasts on the subject) he speaks of the overwhelming evidence which should leave no doubt in any impartial and unprejudiced mind that something akin to sea serpents really do exist.

Given his naval schooling, the lore of sea monsters must have been introduced, in one form or another, at an early age, and in his usual method-ical way Rupert was collecting notes on unexplained sightings from his teens. Though he never claimed to have sighted one himself, he had many first-hand accounts from those who did, including a colleague of his in the Hydrographic Department itself, one Captain F.E.B. Haselfoot.

As told in Chapter 9, Gould's first paper on the subject was given to the Sette of Odd Volumes on 25 March 1925, in which he cites 12 cases of posi-tive and well-documented sightings. Having thus prepared the ground for further research, and 'staked his claim' to the subject, he began plan-ning a larger account in early 1929 and was ready to embark on writing immediately after the typescript for Enigmas was finished that summer.

The Case for the Sea Serpent

Given that so much has been written on the subject since Gould's day, and given the controversial nature of the subject, on which the author of this biography is wholly unqualified to express an opinion, it is best simply to put on record a short summary of the book's contents and Gould's conclusions, along with a few comments by others, better placed to judge the quality of Gould's work on the subject.

This was not the first time the subject had been discussed in book form. One of Gould's largest sources of material was A.C. Oudemans' seminal work *The Great Sea-Serpent*,[259] a work described by Gould as 'monumental . . . a mine of curious facts and equally curious English', and one of which he was also highly critical.

Stretching to 291 pages and 13 Chapters, including seven plates and 30 drawings (19 of them his own) *The Case for the Sea Serpent* is a thorough and, by all accounts, an even-handed discussion of the evidence available to Gould, covering sightings from the eighteenth century, many from the nineteenth century and several 'post Oudemans' from the twentieth century. In this work, Gould not only reported on cases that had been published before, he went back to the source of those cases, checking the facts for accuracy and often digging up new information which enabled the case to be seen in a wholly different light.

The Belgian zoologist Bernard Heuvelmans, one of the most notable post-Gould authors on the subject, praises Gould's work in his comprehensive work *In the Wake of the Sea Serpents*.[260] In a brief biographical note on Gould he states

Naturally enough the riddle of the sea-serpent attracted him greatly, and he attacked it with learning and care as no one had seriously done since Oudemans. It is no credit to zoology that the next in the list should be an amateur as regards that science, though he was an excellent mathematician, and The Case for the Sea-Serpent (1930) is a model of scientific rigour . . . Gould aimed at quality. He offered only a selection of reports, but they were all scrupulously authenticated and carefully checked by a man with an intimate knowledge of seamanship, and who had access to official papers and specialised libraries. Log-books, meteorological records, naval archives were all consulted to make sure the witness was telling the truth. We can therefore be virtually certain that the couple of dozen sightings . . . he cites in the book are genuine. Thus he greatly strengthens the case for the sea-serpent, though from a zoological point of view he has little new to add.

In the Introduction, Gould very sensibly tackles head-on the probability that many readers would already be sceptical, but urges them to keep an open mind. 'I submit that when the case for the sea serpent's existence is examined in detail one can scarcely fail to be struck with the consistent and weighty character of the evidence, and the almost puerile nature of many of the numerous (and inconsistent) attempts to discredit and belittle it by applying some naturalistic explanation'.

Noting however that some contemporary writers *are* attempting to keep an open mind, Gould quotes E.C. Boulenger, a friend and, as 'Brother Shark', a fellow member of the Sette of Odd Volumes who was Director of the Zoological Society's Aquarium in Regent's Park, London. Boulenger pointed out in 1926 that, as there were still many hundreds of square miles of unexplored ocean (to which Gould adds that this is itself enormously under-stated), we should at least give the sea serpent the benefit of the doubt. It would not be long though before Brother Shark was expressing rather different views on the whole question and would prove not such a close friend as Gould supposed.

Mackintosh Bell

Of the dozen or so cases cited by Gould (mostly sightings in the Atlantic, but one or two from across the world), the majority are entirely new and unpublished. Mike Dash has pointed out the importance of one in particular, the Mackintosh Bell sighting, as potentially the most important. He notes: 'This case is unique in providing reasonably credible evidence of the overall appearance of a sea serpent'. The sightings of the animal (in the Orkneys) occurred repeatedly over a period of several years, the witness claiming to have been able to look down on it swimming beneath him through clear water, providing an exceedingly good view.

One of the arguments most frequently raised for the non-existence of *Megophias Megophias* (Oudemans' Latin classification name) was the absence of any bodies of dead serpents. A whole chapter in the book discusses this question, pointing out that there have in fact been several cases of unidentified carcasses washed up on shore in various places across the world (though none positively 'sea-serpentine'). However, Gould points out that one would not generally expect to encounter such things, as the bodies of marine animals do not normally float when dead, save for a short period during decomposition, when buoyed by gas.

In the final Chapter, 'Theories versus Facts', Gould summarizes the debate by producing a typically ordered classification for the various theories ascribed to answer these mysteries. Under 'General theories' he considers the possibility of 'Deliberate deception' and 'Collective hallucination'. Then, assuming something was actually seen, a category considers inanimate objects such as floating seaweed or tree trunks. Then, 'Living oceanic creatures of known species', either in groups or singly, are listed as possible contenders, followed by 'Other living creatures of known species', 'Supposed gigantic examples of known species' and 'Supposed survivals of 'extinct' reptiles'.

In conclusion, Gould states his belief that the many and varied sightings he has quoted are not deceptions but are most probably explained by the existence of three different, shy and very rare, creatures, not yet scientifically described: a long-necked seal, a gigantic turtle-like creature and (the explanation for the majority of the reports), 'a creature resembling in outline and structure the Plesiosaurus of Mesozoic times. I do not suggest that the last named is actually a Plesiosaurus, but that it is either one of its descendants or has evolved along similar lines'.

Writing on *The Case for the Sea Serpent* had begun in the autumn of 1929, but the continuation of H3's restoration then intervened, and it was only in early 1930, with H3 still not finished, that Gould got down to it. As usual, he was leaving things until the last minute as he was contracted with Philip Allen's to complete the typescript by the end of May. On 24 June 1930, Gould was attending the monthly OV dinner as usual. That occasion was Ladies Night however—Dodo was his guest—and Vyvyan Holland, David Low, Ralph Straus and he were billed 'to amuse the company'. We are not told how, but no doubt Sea Serpents got a mention there somewhere.

Proofs for *The Case for the Sea Serpent* were checked while on Holiday at Sheringham in September, the final proofs being accepted by him on 2 October 1930. The book was sold with a dust jacket also illustrated by Gould, rather unwisely depicting a huge and apparently savage sea serpent attacking a modern battleship:[261] hardly likely to induce an open mind in the serious reader! The book's dedication would have been no more serious, had it not been for a tragedy that befell one of Gould's close friends at the time.

Until a few months before publication, the book's dedication page read: 'This book is dedicated to my friend BILL, who will never read it (being a grey Russian cat) but who is very fond of fish'. A proof of this page survives in Gould's own copy of the book (Gould was, needless to say, actually very fond of Bill; one accepts that pet-owners often fall into one of two categories, 'cat people' and 'dog people', Rupert was decidedly in the former group). However, in July, Armorel Heron-Allen, daughter of Gould's fellow Sette brother and friend, Edward Heron-Allen, was killed in a car crash[262] and, as a token of his condolences to the devastated father, Gould dedicated the book to her memory 'Armorel Daphne Heron-Allen 15 June 1908 to 3 July 1930', though he had in fact only met her once.

Reviews

The Times leader for 30 October dedicated a whole column to Gould's work. Headed 'The Amende Honorable', the paper waxed lyrical: 'To be big and splendid and yet to be ignored, is a poor life. Lovers of animals, whether they subscribe to the RSPCA or not, will hear with joy that the Sea Serpent has had a kind book written about him at last. It comes out today and, as an additional compliment, it is by a naval officer, Lieutenant Commander Gould'. The piece congratulates Gould on covering a subject well known to naval officers and seamen alike but which few of them care to discuss for fear of ridicule, ending: 'If men do not see him very often it is perhaps because in his serpentine wisdom, not knowing what medicinal oil he may not contain, the sea serpent stays below, and prefers to be rare and semi-fabulous instead of being extinct'. The following day's *Times* included the title in its list of new books, Philip Allen putting an advert in the same issue, headed 'A Challenge to Public Opinion', with the book marked at the usual 12/6.

There was no second edition, but in 1935 a German edition appeared, including text from *The Case for the Sea Serpent, The Loch Ness Monster and Others* (see Chapter 16) and some relevant new material added by the joint author, Georg-Gunther Freiherrn von Forstner, himself a sea serpent witness.

The Case for the Sea Serpent was reviewed, *inter alia*, in *The Times Literary Supplement* (TLS) on 8 January 1931, the reviewer describing Gould as 'an extremely readable specialist on odd and exceptional things'. The review prompted a letter in the TLS[263] from a distinguished geologist, Dr F.A. Bather. He refused to accept the Plesiosaur theory on the grounds that no fossil remains of Plesiosaurs had been found after Mesozoic and Secondary eras, and that the evidence was that they were definitely extinct. Gould replied[264] that absence of fossil remains did not guarantee extinction, giving the examples of the Chimaera (a species of fish), the long-necked river tortoise, and the iguana.

Gould also stated he was not saying these were Plesiosaurs, but descendants, or something evolved along similar lines. In a further reply[265] Bather answered that in fact there is fossil evidence of all the three species Gould cites, and therefore his case stood that the sea serpent could not have evolved from a Plesiosaur. A final pair of letters appeared,[266] Gould citing a well-known zoologist as stating that absence of fossil evidence is not sufficient reason to assume a species is extinct, with Bather replying that he could not agree and, while wishing to point out that he is not saying there is no such thing as a sea serpent, he doubts it is even anything like a plesiosaur. The Editor then closed the correspondence. There were however other sceptics who would take the opportunity to criticize Gould's work heavily in the years to come.

Man and the Natural World

Before leaving a discussion of Sea Serpents, this would be an appropriate place to mention another subject Gould felt passionately about: man's poor treatment of his fellow animals. In one of his characteristic parting remarks in *The Case for the Sea Serpent*, Gould hoped that one day one of these creatures may be photographed and studied, but *not* captured. He reminds us that of the several enemies such an animal may have, by far the worst would be Man: 'The list of creatures which Man has swept out of existence for no reason but his own greed and selfishness—the great auk, the dodo, Stellar's sea-cow, the passenger pigeon and the like—is a pitiable one. Even his brother man has not escaped—witness the fate of the Tasmanian Bushmen, and the steady degeneration of savage communities under the influence of "civilised" man's inseparable comrades—drink and disease'.

An inveterate misanthropist, over the years Gould took every opportunity he could to ram the message home in his publications about Man's dreadful behaviour in the natural world. He even spoke to the Sette of Odd Volumes on it, giving a paper entitled 'The Dodo and Other Creatures', at the Sette's 'Extinction Night', on 25 February 1930.

On the question of killing animals, it's true that Gould joined his father in shooting rabbits and fishing when they were in Scotland, but this was for the table, and he had never been squeamish about killing for food (in fairness, neither should anyone be, who eats meat). Blood sports like foxhunting on the other hand—'killing animals for fun'—was an entirely different matter and he was utterly opposed to it. On several other occasions Gould spoke out eloquently about man's cruelty to his fellow animals.

As an aside in his chapter 'The Auroras and other Doubtful Islands' in *Oddities*, he commented on the result of the submergence of Garefowls Rock, an islet off the coast of Iceland. The inaccessible Garefowls Rock had harboured the last remaining colony of the famous sea bird, the Great Auk, which, on the disappearance of the island, was obliged to adopt another refuge. This was the islet of Eldey, which was much nearer the Icelandic coast and more easily accessed by man. Speaking passionately of the now-extinct Great Auk, Gould rams home his misanthropic message sarcastically: 'their numbers were rapidly depleted by the hardy Icelanders, who dared not only the perils of a six-mile voyage, but also the grave risk of getting quite a sharp nip in the slack of their trousers before they could safely knock their formidable quarry on the head. Rabbit-shooting itself could scarcely offer more thrills and dangers. It was on Eldey, in 1844, that the last known pair of Great Auks were murdered by two heroes named Jon Brandsson and Siguror Islefsson, both natives of Iceland. It is permissable to hope that by now they are experiencing a much hotter climate'.

In a similar vein, when describing the discovery, by a 'sealer', of the group of islands in the Antarctic (The South Shetlands), Gould took the opportunity to remind the reader what a cruel practice 'sealing' was.[267] The 'sealer' itself was the ship, the crew of which made their living by 'hunting' (really just 'finding') and killing seals, selling the pelts for their fur.

When the South Shetlands were discovered, vast numbers of seals were found: 'To the commercial eye, the spectacle was . . . much as if the shore had been carpeted with guineas'. The underlying story is a hideously cruel one from a modern perspective and, though only incidental to the main subject of the article (the sealers' charting of the Antarctic), it was one Gould was determined not to underplay: 'the seal were so tame that they would come placidly ashore right among the men who were slaying and skinning their companions. In the Antarctic Summer of 1820–21 there was a wild rush of sealers to the group from all parts of the world . . . An indiscriminate massacre followed. As fast as men could work, the seal—bulls, females and pups alike—were clubbed to death (not always even to death) and quickly flayed, the pelts being stowed in salt until the holds would take no more, while on the gory beaches a host of sea-birds pecked and tore at the naked carcasses, many still feebly moving. So it has always been . . . so, I suppose, it will always be. Of all animals, man is the cruellest and the greediest'.

At the beginning of 1931, with *The Case for the Sea Serpent* now launched, and no more commissions for 'mystery books' currently on the stocks, Gould turned his attention to Horology again. Finally he had time to pick up again on the extraordinary saga of the restoration of the Harrison timekeepers.

13

The RAS Regulator 1927–1929

To pick up on the story of Gould's work on the restoration of the timekeepers by John Harrison, a brief summary of what had been done by Gould on the timekeepers so far, might be useful.

A short recap

H1 had had a preliminary cleaning at Gould's home in Kensington Park Road in the summer and autumn of 1920 and H4 was overhauled, between July 1921 and early 1922, during which time the family had moved to Epsom.

After publication of *The Marine Chronometer* in April 1923, from October that year Gould took on both H2 and H3 for restoration, H2 being finally completed, after many interruptions and bouts of illness, in September 1924. In 1925 it was exhibited at The British Empire Exhibition at Wembley, after which it was sent on loan to the Science Museum. At the time H2 left Epsom, most of H3, completely in pieces, was taken back to Greenwich, with the intention of sending the parts to William Buck of Westbourne Grove for polishing.

With H2 gone and H3's parts in London, Gould sought to start on yet another Harrison masterpiece, the regulator of the Royal Astronomical Society (RAS). However, although permission was granted on 20 November 1924, even Gould's enthusiasm for horology was overcome this time. Another major nervous breakdown was imminent, and the RAS regulator, and H3, would have to wait. The RAS clock was not taken on and, after Gould's period in hospital in July 1925, the other parts of H3 were returned to Greenwich, in boxes.

In November 1925, Gould arranged for H3's plates to be sent from Greenwich to Buck in Westbourne Grove for polishing, the dials going for silvering there at the same time, at a cost of about £5 to the Observatory. Then, in July 1926, H3 (in pieces) was brought back to Gould, now at Leatherhead, though nothing further happened to the timekeeper at this period. As Gould tells us 'I was too busy and too worried to do much'.

The restoration of the RAS regulator

Gould chose to be busy, no doubt seeking distractions from the worry of the dysfunctional family arrangements at Glenside. The distractions he chose were, of course, more horological projects. The Fitzwilliam watch catalogue was a major task on-going at the time, but in addition to this, Gould decided, in spite of having H3 in pieces on the bench, to take up the matter of the Royal Astronomical Society's clock again.

On 24 January 1927, following a visit to thoroughly inspect the clock at Burlington House, he submitted a detailed report to the Council of the Society, indicating that the clock was very dusty and dirty, the remontoir was defective and that the winding ratchet was partially seized.[268]

His recommendations included cleaning and lacquering the movement and re-silvering the dials. He also proposed replacing the very small (apparently non-original) remontoir fly with a much larger one, 'in keeping with Harrison's intentions', and replacing the existing wooden pendulum rod with a gridiron. Convinced the pendulum rod must be later, Gould described it as analogous to 'a Rolls Royce motor car chassis with a body made of packing-cases'.

A further recommendation was that 'the present case should be discarded, for two reasons'. The reasons given were that the case was not dust- or damp-proof, and the clock would thus need cleaning very regularly, and secondly that the case obstructed a good view of the 'unique mechanism'. He explained: 'Its chief interest, apart from its association with the man who was probably the greatest, and certainly the most original horologist that this country has ever produced, resides in its escapement, its remontoir, its huge pendulum arc, its innumerable contrivances for converting sliding into rolling friction, its wooden pallets and pinion rollers, its wonderfully cut teeth and the admirable square-ness and accuracy of its layout and construction'.

Enclosing a rough sketch, he proposed the movement and pendulum be mounted on an iron plate on the wall. It was to be surrounded by an airtight case that was dust-and damp-proof and one which provided a full view of the movement, Gould noting 'The present case fulfils none of these conditions'. Citing the larger marine timekeepers, he supports the idea of a new case by pointing out that, like them, the regulator runs without oil and should, with a dust-proof case, be able to run continuously for many years without needing attention other than weekly winding.

From the extant historical evidence we can be certain that by the early nineteenth century this clock had lost its original case and that the one it occupied in 1927 was a later replacement. But surprisingly, Gould appears not to realize this, simply stating: 'The present case could, of course, be stood alongside this to show how the clock was originally installed'.

On 16 March 1927, the Secretary to the Society notified Gould of the Council's decision that they would be very pleased for him to go ahead, as proposed. The letter also informed him that as the clock had been overhauled not many years previously (1909) by E.T. Cottingham FRAS, (1869–1940), the Astronomer Royal, Frank Dyson had written to him to let him know, out of courtesy, what had been proposed, and Cottingham's reply was attached.

Cottingham's letter was a rather strange and grammatically rather contorted note (e.g. 'Commander Gould, like myself is an aesthetic on these "Gems of Horology," and I think he goes beyond the trend of your letter . . .'), and it seems likely from its tone that he would have preferred to have been asked to look at the Harrison clock himself.[269] But, as it was Commander Gould who was being proposed, he could only agree. He does however suggest that Gould should not be allowed to make any alterations to the original clock, and proposes an automatic device for stopping the clock just before it runs down, to keep the remontoir set up correctly. He also suggests that a new case should only be considered if it be fully airtight and that an autowinding system be fitted within it.

As has been noted already, at various times in Gould's horological career (and often since his death), there have been professional horologists who have, to a greater or lesser degree of subtlety, implied that he might not have been the ideal man to carry out the work he did. Cottingham's letter, although outwardly supportive of him, was suggestive of this too. Gould was evidently irritated, remarking in his reply of 20 March to the RAS Secretary that 'I am not quite certain that I have fully understood all the expressions which he uses, particularly in regard to myself, but I gather that he does not object to my undertaking the repair of the clock'. He goes on to criticize Cottingham's suggestions while stating that he has the highest opinion of him as a craftsman, but then pointing out that he personally has more experience with Harrison clocks than Cottingham.[270] Final approval from the RAS Council was then sent 2 days later for the work to begin at Glenside, in Leatherhead.

The Council of the Society could have had no idea of the emotional scenes being played out in the house at the time, and (as told in Chapter 10) things were soon to get much worse, with Muriel and Miss Gurney eventually storming out for the final time, with Jocelyne in tow, on 21 May 1927. But at the end of April when the RAS regulator arrived, they were all still at Glenside and, though clearly distracted and upset, Gould could only continue with his horological muse. The RAS regulator was set up in the lean-to at Leatherhead on 2 May 1927, so he could see how it ran and carry out experiments on the escapement.

The RAS Regulator Notebook (No. 16)

Entries in the notebook he was using at this time to record his work on the RAS regulator, give an indication of the state of his mind. It is a small, lined exercise book, but without an obvious front or back on the covers, only a 'title' page inside with a printed note to indicate the 'front'. The entries begin at the 'back' of the book, continuing, off and on, for half a dozen pages or so. Then, on the 30 May, Gould makes an entry at 00.30 a.m. but, apparently getting confused, the book is then turned over and, 'the right way up', entries begin again, the same night, on the first page of the 'front' of the book. The entries then continue at the front of the book from the 1 June, but on 3 June an entry appears at the front and also continues the same night at the back of the book, picking up from 30 May entry. The next day Gould starts at the front of the book again for one entry, but on the 5 June he is back to using the back of the book and continues at the back until the move from Glenside on 28 July. After the move, a couple more entries, for the 13 and 14 August, appear at the back and then the entry for 14th continues at the front. The record then stays in the 'correct' sequence and Gould now works through the book normally, except when reaching the back pages, when he *writes over* and *past* the earlier (upside down) entries!

Looking at the work he recorded (the Glossary in Appendix 6 may be of help for the interested non-horologists here), on 30 May, he noticed the clock making a 'scraping' noise which he didn't understand, but supposed must be the pallets of the escapement disengaging. He tried *oiling* both pallets (nothing is supposed to be oiled in these clocks) to see if the noise disappeared, but it did not. Then he reduced the counter-weights of the pallets, which he found greatly reduced the scraping noise and he also reduced the experimental going weight.

In early June an experimental 'kite-shaped' 'Gould pendulum' made out of thread with a horizontal 'spreader' near the top, was fitted and the clock ran for a few days successfully while he took measurements and made sketches for the new proposed case. Personally it was of course a very distressing time and the notebook entry for 6 June ends pathetically: 'It being nearly midnight, I turn in. A full Whitsun, although scarcely a happy one—yet, I don't know . . . what is the use of repining?'

On 8 June Gould submitted a formal report to the RAS of his initial findings on 'the amount of replacements, repair and recondi-tioning . . . to restore it to as perfect a state as when it first left its makers hands . . .' Amongst other work, he notes that he has already been able to reduce the driving weight from 20lbs to 12lbs, noting 'the remontoir train was . . . slowly but surely hammering itself to pieces'. The proposal now included new pallets from *lignum vitae*. He stated 'the original pallets appear to be mahogany, but *lignum vitae* is preferable for many reasons'. He does not give those reasons, but it should be said that, at least in the view of this author, *lignum vitae* was not, by any means, a preferable alternative to a hardwood like mahogany or oak, which would have been tougher, less likely to chip, and less likely to slip off the escape wheel teeth. But *lignum vitae* was what Gould used.

He also proposed a new 'fork arm' for discharging the remontoir, to cost £2; new driving weight, also £2; cleaning and polishing of the brass plates and re-silvering of the dials, to be done by Buck, as before at £3. In addition, he estimated for other parts to be made by him, and new stopping gear, based on a design by Ferdinand Berthoud, seen by him in Berthoud's marine clock No.2, but based on H2 & H3, at a cost of £2. Finally Gould proposed the new gridiron pendulum, be made by 'a personal friend', the instrument and clockmaker J.H. Agar Baugh of Hammersmith, who he felt sure would make it a matter of personal interest. Ideally, he stated, the brass and steel to be used should have their coefficients of expansion tested at the NPL (National Physical Laboratory), to ensure the pendulum compensated properly, and if this was done the finished pendulum would cost £12, but if not, just £6.

For manufacturing the new air-tight glazed case, Gould also recommended Agar Baugh, and quoted £25 for a case with a small iron mounting set into a mahogany back, or £40 for a case with a solid iron back throughout. He invited any member of the Council who pleased, to go and see the clock running at Leatherhead.[271]

The Council responded via the Assistant Secretary three days later that they would accept the lower estimates for the pendulum and case and approved the cost of repairs, making a total of £40 in all. Gould got straight on the telephone to Agar Baugh who reported by post on 14th that he was having the coefficients of the rods measured anyway, in a technical school, 'for my credit's sake', and asking about Harrison's form of pendulum bob. Full drawings of the front and side of the clock movement were sent by Gould to him on 14 July to get the case construction started.

The entry in the notebook for 22 July 1927, is headed by Gould 'Last days at Glenside' and notes of the RAS regulator 'he will probably be stripped at weekend for cleaning—I have made travelling case for movement'. The final day came on 28 July 1927, Gould marking the page: 'Good-bye to Glenside'.

It must also have marked the final nail in the Gould's marriage, as Rupert made his way over to his mother's house in Ashtead with the remaining contents of the house including, one assumes, the RAS movement and H3 in boxes.

The RAS regulator continued

The notebook reveals that in August 1927 the stripping of the RAS movement began, Gould recording progress, with sketches, notes of inscriptions, and instructions, in the usual way.

But it is not wholly clear *where* this work took place, as it seems he was initially not in the attic workroom at Downside that Dodo eventually allocated to him. The entry for the 14th demonstrates the rather ad hoc arrangements he now had: 'Owing, I suppose, to a leak in workshop roof, found box wet. Dried all parts in it, stood box on end to dry'. Perhaps he was temporarily in an outbuilding, but a remark in the notebook a few days later seems to suggest the workshop wasn't even in Ashtead, let alone at Downside.

On 18 August 1927 the day had been spent completely dismantling the movement—rather easier to take apart than any of the marine timekeepers—the whole job taking just 2 h 40 min. Gould recorded packing the dials separately for Buck and putting 'all H mechanism, except the 5 train wheels' in an attache case. He then notes: 'Took this and dials to Ashtead'. But nowhere else is there a suggestion that the work wasn't all done at Downside, so perhaps the reference means he was taking those parts from the house into the village (the centre of Ashtead is half a mile down the road) for some reason. One does also wonder where the 700+ parts of H3 were being kept at this time, as the references in the notebook are ambiguous, but at least by the end of 1927 everything was definitely at Downside.

The court case intervenes

There is now a 5½ month hiatus in the notebook, and the evening and weekend work on the RAS clock comes to a halt as the judicial separation and its implications are all dealt with by the respondent. A brief horological task turned up in September 1927, when the Director of the Science Museum, Sir Henry Lyons, wrote to tell Gould that H2 had developed problems with the balance cross-wires, one of H2's particular 'Achilles heels'. With the permission of his boss, Admiral Douglas, Gould attended the Museum in early October 1927, fixing the wires and producing a beautifully illustrated 14-page note for future repairers (now part of notebook 14) on carrying out this specific but very tricky task successfully. The Hydrographer duly sent copies to the Science Museum and to the Astronomer Royal.

As described in Chapter 10, the court case was heard in November and by the end of December Gould was in dire straits, having lost his job, his home, his wife, and custody of his children.

RAS Concern

Naturally enough, knowing of the judicial separation and Gould's move from Leatherhead, the Royal Astronomical Society were concerned about their Harrison artefact, and on 5 January 1928 Gould invited Theodore Phillips (1868–1942), the Society's President, to Downside to see how progress was. The same day Gould wrote Phillips a letter outlining the present state of affairs as a summary of their meeting.[272] Work had not really progressed at all since the previous August. The polishing of the movement and the dial re-silvering still had to be done by Buck, the pendulum and glazed case still had to be made by Agar Baugh (though he had, by this time, supplied Gould with a wooden mock-up of the iron movement bracket to be sure it fitted), and the parts as specified, still needed to be made. Gould stated that 'Mr Hopwood' (sic: should read Hopgood) of Blackheath would be making all of these except the lignum vitae pallets, which he'd do himself. He estimates 4 months for the work to be completed—of course Gould now has much more free time to dedicate to this task—but in fact it would be another year before the clock's return to Burlington House.

Agar Baugh

Correspondence with Agar Baugh at this time reveals Gould and he becoming good friends (prefixes were soon dropped and 'Dear Gould' and 'Dear Agar Baugh' adopted). Professionally a talented metallurgist and scientific instrument maker, Welsh-born James Harold Agar Baugh (1872–1935) was well travelled, and fluent in French and German. Familiar with the main observatories and scientific institutions on the continent, he numbered among his acquaintances and friends the chronometer maker Paul Ditisheim and the distinguished chemist and metallurgist Charles Edouard Guillaume. Agar Baugh now specialized in the use of invar, Guillaume's new (1898) nickel–steel alloy which had an exceedingly low coefficient of thermal expansion (i.e. it didn't expand or contract with temperature changes). Making and supplying invar pendulums for regulators was a specialization of Agar Baugh's and he was a natural choice for making a special pendulum for the RAS clock.[273]

Agar Baugh followed Gould's several activities with interest, their letters ranging across many other subjects. For example, in a letter of 21 July[274] he tells Gould he has found a fine chronometer by Arnold[275] and adds: 'Clock by Harrison. I saw one yesterday in private hands. It has wooden wheels and is, I should think, quite genuine. Pendulum might not be original'.[276] In the same letter he added: 'Sale of Antique Watches at Christies. I shall have something very interesting to tell you about a sale which took place there this week'.[277]

Hopgood

R.J. Hopgood, on the other hand, was the archetypal, deferential tradesman. Originally from the watch-making and instrument-making centre of Clerkenwell in East London, Hopgood was now based just down the road from Greenwich Observatory, at Levett's Yard, Dornberg Road in Blackheath. Dornberg Road is on the borders with Charlton (Hopgood's home was a couple of blocks away at 40 Mayhill Road in Charlton), though most of Dornberg Road, including Levett's Yard, was swept away in the late 1960s when the A2M motorway connecting the Blackwall Tunnel approach road to the old A2, was begun.

Hopgood described his firm as 'Contractors to His Majesty's Government' and carried out occasional adjustments and repairs to instruments at the Observatory as well as other Government departments. Gould was almost certainly introduced to him through the Observatory, and the RAS project was Hopgood's first main contribution to the Harrison restorations, in the summer of 1928.

In all the correspondence between him and Gould, Hopgood only ever addressed his letters 'Dear Sir' and 'Yours Obediently', and was only ever addressed by Gould as 'Mr Hopgood'. Equally, the contents of their letters only ever discussed the metalwork for which he was being paid. Gould never played the 'class' card though; he invariably got on very well with such tradesmen and over the years Hopgood and he had the most friendly of relations, with Hopgood carrying out a great deal of high quality work for him during the later restorations of the Harrison timekeepers.

Horology at Downside

The Summer and Autumn of 1928 saw work on the RAS clock gather momentum, and Gould's first major restoration job in the workroom at Downside got underway. As described in the previous chapter, this was a large room at the top of the house and doubled as Rupert's 'den' and place for writing. Cecil remembered the room having a table with a vice attached, serving as a bench, and drawers of tools.

It seems Gould's clockmaking activities were limited to work with hand tools. Apart from dismantling and re-assembly, the only creative work possible was that which involved sawing, cutting and filing. Any fine drilling or turning had to be out-sourced, usually to Hopgood in Blackheath. Dodo would have kept a 'weather eye' on activities in her house too, and would probably have objected if large and noisy machine tools began to appear (her bedroom was directly beneath the workroom at the rear of the house).

Experimental work to get the RAS clock running reliably was only completed on 11 August 1928, and Buck was polishing parts and Hopgood making the various pieces during September. On 7 September Agar Baugh wrote concerning the pendulum: 'As the idea is to make one similar to those actually made by Harrison, I find I should like to examine again the one we saw in the office near the Strand'. This most interesting reference is to a clock which Agar Baugh and Gould had seen in the possession of a Mr Crouch of 17 Surrey Street, Strand, but exactly what this clock was, is still uncertain.

Agar Baugh's foreign contacts were not just of horological use. Writing later in the month, he arranges for a consignment of four thousand bulbs from Holland, some of which he has allocated for Dodo and the garden at Downside. He also asks about the publication of *Oddities* and for more information on Harrison-type gridiron pendulums. Gould replied that the only genuine example is that at the Clockmakers Company collection in the Guildhall and suggests he writes to J.L. Douthwaite, the Librarian and Curator at the Guildhall Library. Of *Oddities*, Gould remarked: 'So far, I have had many reviews and not a single bad one'.

The RAS Pendulum

On 28 September Agar Baugh wrote again asking about the Guildhall clock's pendulum and Gould answered the next day from his bed, 'with bad feet, which I cannot walk on' and giving him much help and advice about Harrison and the pendulum construction. Agar Baugh took a great deal of trouble over sourcing brass and steel of a similar kind to the Guildhall clock's and managed to find some very similar, which was only slightly larger in diameter. He duly visited the Guildhall in October, the Company arranging for the clockmaker W.J. Barnsdale to be present to oversee the study, Agar Baugh noting 'He proved very aimable' (sic).[278] Cleaning work on the clock's parts was all done by Gould during November 1928, his list (dated 26th of that month) of parts done totalling 521 in all. Correspondence with Hopgood continued right through November on the design and making of the special stopping gear for the clock, and drilling of the lignum pallets Gould had made. December saw Gould submitting final bills to the RAS for Buck's and Hopgood's work, unfortunately both much exceeding the original estimates,[279] for which Gould apologized profusely. He said that he would happily have paid the difference if he had been able, and that the work has at least been done to a very high standard.

In mid-December Gould had an accident with the clock, dropping the great wheel assembly, with the roller bearing attached, breaking two of the brass roller pins. The bearing was posted to Hopgood who replaced the pins as soon as he received them and at no charge. A few days later the other bearing was also accidentally damaged while Gould was getting it in, and that too was sent to Hopgood for repairs. Gould noted in irritation 'It is a curious fact that I have had the whole clock to pieces and put it together again, four times with out having any trouble of this kind at all. Now when everything is clean and it is going together for the last time I find things going wrong like this'.

The escapement

On Christmas night, 1928, at 11.27 p.m., Gould started the clock for the first time since the overhaul. The suspension was somewhat distorted and the pendulum wobbled badly, so he made up a new suspension the next day from a piece of spring steel 'given me long ago by A.W. Curzon' (Curzon was a highly respected Clerkenwell watch maker and teacher). Between Christmas and New Year, a series of experiments were carried out on the escapement. The clock's running was proving unreliable and Gould had noted that the impulse delivered by the two pallets was very different and wondered if this could be quantified, and in some way remedied. He concluded that it could not, but that it should not in fact matter.

Gould's work was briefly interrupted at this point by an urgent job for *The Times*, his first, he notes, as 'A Naval Correspondent'. The piece concerned Sir Hubert Wilkins' aerial survey of Graham Land, the first ever made over the Antarctic. It was published on 2 January 1929, Gould describing the survey as: 'a pioneer flight which has set a completely new standard for future exploration of the Antarctic'.

Several small repair jobs on the regulator now took the project over into the new year but, with further adjustments to the escapement and remontoir, Gould at last had the clock running reliably on 3 January 1929. It was then necessary to carry out tests to determine the size and weight of pendulum bob needed. The clock was left running on 7 January when Gould travelled to Edinburgh for a couple of days, and was still going well on his return.

Further experiments were now carried out on the remontoir torque, as he had noted that the power delivery from the remontoir appeared to be rather uneven. It was at this time that Gould invented and made his intricate and very sensitive little torque-meter (see p. 242), which would prove so useful in later work. On 15 January 1929 Gould records in the notebook 'I suspect my work is almost finished . . . I shall clear up decks and begin my new book "The Sea Serpent" '. In fact there was also quite a bit still to do on the completion of *Enigmas*. On 18 January some of the final adjustments had to be carried out by candlelight 'the electric light having failed, as is its custom at Ashtead'.

Near disaster

On 1 February 1929, the movement was packed and transported to Agar Baugh's premises in Hammersmith to have it fitted to the case and tested with a temporary pendulum. In the note book Gould records that the clock came very close to disaster at this time: 'About three weeks after the clock had left my workshop, the latter was partially wrecked by fire (18.2.1929) [this was caused by a defective flue] The floor collapsed within a foot of the spot where the clock had been standing. It has been there for nearly two years & was not insured'.

Agar Baugh reported to Gould on 2 February that the clock was now running, at 15 degrees total arc, but going rather slow and needing a new pendulum bob. Nevertheless, the clock was set up on the wall of the Grove-Hills library in the Society's rooms on 6 February 1929. Gould formally wrote to the Secretary of the Society the next day to report that the clock was now running satisfactorily, though only with a temporary pendulum bob, made from that of the old pendulum. The new pendulum bob was fitted later that month and the RAS Monthly notices of February 1929 contained a detailed account of the work carried out.

Visits from Gould in the summer of 1929 and, occasionally, over the next few years, ensured the clock was running reliably and on time. Apart from a repair in 1937 and a 5-year sojourn in Oxford during WW2, it would be nearly half a century before the clock needed any further attention.

The Affair of the Queen's Watch

It was at this period, in the late 1920s, that Rupert Gould and his friend Paul Chamberlain became embroiled in something of a horological scandal, entirely hushed up at the time, and a sad tale of hubris, self-interest and deceit (none of this Gould's or Chamberlain's, happily). However, as some of the story does not directly involve Gould, the details are contained in Appendix 5 for those who are interested.

Supplementing his income

Always trying to supplement his tiny income while living at Downside, Gould continued with the writing of articles and acting as a consultant. He was commissioned by *The Practical Watch and Clock Maker* to write a three page article on 'The Legacy of the Old Masters', looking at the contribution of the great English and French makers, for the magazine in March 1929. The PW and CM[280] was a relatively short-lived but lively and interesting rival to *The Horological Journal*. It was edited by the dynamic Arthur Tremayne (1879–1954), who then took over the editorship of the HJ for some of its best years.

Gould was often asked to help colleagues and friends with publications of their own, and around this time he helped with two such projects. In 1930 Jules Sottas published an article in the *Mariner's Mirror*, on the French Corvette L'Aurore which, in 1766, had taken Pierre Le Roy's timekeeper on a trial to Newfoundland. Gould advised on technical and historical points and provided notes on a special navigational instrument described therein.[281]

Another task he undertook at this time must have had considerable resonances for him, as it involved a close reading of a translation of a German account of life in a U Boat during WW1. *U-Boat Stories* was written by Claus Bergen who, in 1917, underwent a long voyage as war artist with the crew of U53, in wartime conditions, to draw and paint his experiences. The book was illustrated by Bergen and was described by him as a series of stories told by individual seamen. All the accounts however seem remarkably alike in their extraordinarily jingoistic style and probably underwent considerable editing by Bergen. The stories all describe the harsh existence when submerged, but the delight when sinking enemy ships and generally giving 'Tommy' a good seeing to. It must have made odd and rather depressing reading for Gould.[282]

14

H3 is Completed 1929–1931

At the time of the completion of the Royal Astronomical Society (RAS) regulator's restoration in February 1929, Rupert was busy finishing *Enigmas*, his second book on scientific mysteries, and had already got started on his third, *The Case for the Sea Serpent*. But, now that the RAS clock was completed, there was another long-outstanding horological task also beckoning his attention; the completion of H3.

Following the court action and Gould's dismissal from the Admiralty in December 1927, a predictable consequence had been that his ex-boss, the Hydrographer, Admiral Douglas, demanded the immediate return of H3, which was then completely in pieces at Downside. Douglas informed the Astronomer Royal on 19 December 1927 that a letter had been sent to Gould 'telling him to forward forthwith to the Admiralty Harrison No.3, with any remarks he may have to offer'[283]

Gould decided not to do this, but the following day personally delivered a formal note directly to Douglas (but intended as a summary for the Admiralty Board too). The note read:

The machine is at present entirely in pieces, of which there are over 400 [actually many more]. Some of these have undergone, and others are undergoing, various cleaning processes. The experimental work in connection with getting the machine going again (it has not gone since 1775) is completed, and the new pallets, which are absolutely essential to this end, have been designed, and I am now cutting them.

Until the cleaning is completed, it would be impracticable and useless to re-assemble the machine. I can undertake to have it together and going, and in all respects to the Astronomer Royal's satisfaction, in 6 months time (i.e. by the next Visitation of the Observatory)—although I should prefer to have twelve months time to do this in. In view of the time which I have already spent on the machine, it may be pointed out that my work has been interrupted by illness, and that the machine is exceedingly complicated, and took, originally, no less than seventeen years to construct.

I cannot undertake any responsibility for returning the machine in its present state, and I gravely doubt whether any professional horologist would undertake to reassemble it—if any did so, he would probably charge £75 to £100 for this, and a considerably greater sum for restoring it to going order.

I may point out that I volunteered, some years ago, to clean and repair all four of the Harrison timekeepers preserved as national property in the Royal Observatory; these being at that date, and for long previously, in a dirty and damaged condition. I undertook this work in a private capacity, without any remuneration (in fact, at some personal expense) and in my own time. The result, so far, of my seven years work has been that three of the four machines have been completely cleaned and rendered incapable of deterioration, and that two of them (the second and fourth timekeepers) are in going order. The second machine has been going for the past three years, and after been [sic] exhibited at Wembley, is now on exhibition, going, at the Science Museum, South Kensington. This work, which has, so far, met with the complete approval of the Astronomer Royal, has not been done in my Departmental capacity, but purely as a result of arrangement between Sir Frank Dyson and myself.

As I feel that I cannot fairly be held responsible, as a craftsman in such matters, for returning the third timekeeper in its present state, I have the honour to request that I may be permitted to return it cleaned, re-assembled and going. Failing this, I ask that I may be discharged from my present responsibilities in the matter by an order of the Board[284]

Douglas copied this to Dyson asking him to reply, and commenting 'The remark in his submission to me about an order from the Board of Admiralty is of course a fatuous one'. Dyson then wrote to Gould, stating that he had 'carefully considered' the statement, but that 'the machine should be returned at once, in its present condition'.[285]

And so in January 1928, H3—unfinished and in pieces—had been sent back to the Observatory for a second time.

The Completion of H3's restoration

One year later, having set up the RAS clock in the Society's library in early 1929, Rupert's attention turned once again to H3. What a saga this work was turning out to be! Just as John Harrison had had the greatest problems with this machine, so it was with Rupert: H3 seemed fated to be embroiled in the most complicated and contorted series of events during its lengthy restoration.

He doubtless also felt partly responsible for the fact that H3 was in a vulnerable state, still in hundreds of pieces in boxes at the Royal Observatory. So, with *Enigmas* almost completed, and the writing of the next book, *The Case for the Sea Serpent*, underway, Gould wanted the completion of this outstanding job started again too, enabling him to do practical work or writing, as the mood took him.

But there really were difficulties now, as the Hydrographer, Admiral Douglas himself, had been the cause of H3's return to Greenwich and he would not be easily persuaded to allow it back to Gould. In notebook 3, Gould tells us what happened then:

In the Spring of 1929 I began to draw up a statement of the whole matter which I intended to lay before the Board of Admiralty. This becoming known to the Officer (*The late Vice-Admiral Sir Percy Douglas R.N.) who had caused the stoppage of my work on No.3, I was invited to resume this . . .[286]

In fact, Gould had been canvassing support from other scientists and academics, and pressure was applied from this quarter too. On 8 February he attended the Royal Astronomical Society's Club dinner as one of the Club's guests alongside Herbert H. Turner (1861–1930). Turner was Savilian Professor of Astronomy at Oxford University and had been chief assistant at the Observatory between 1884 and 1893.

It seems Gould took the opportunity at the dinner to have a word, as Turner then wrote to the Astronomer Royal, on 6 April 1929:

I wonder what you feel about Gould continuing his work on Harrison 3? I am fully aware of the trouble at The Admiralty, but I think that the punishment has been a little severe, and I am inclined to help Gould if possible. I asked Loomis whether he was in want of an expert for his work, but he has replied saying that he has no need at present; at the same time he has expressed interest and sympathy, and sent me a sum of money which might pay for some work which I might think it desirable for Gould to do. This was quite spontaneous on his part. Thinking the matter over, I remember that when I saw Gould at the annual meeting (at the RAS) he expressed a wish to finish the work on Harrison No.3 if your permission could be obtained; and it seems to me that this benefaction from Loomis might meet the case by avoiding any question of using Government funds to compensate Gould.

Alfred L Loomis

Gould's benefactor was in fact the charismatic and powerful American tycoon and scientist, Alfred Lee Loomis, of Tuxedo Park, NY, the man who would 'turn his laboratory . . . into the meeting place for the most visionary minds of the twentieth century . . .' and who, *inter alia*, would personally bankroll much of the development of radar and radar detection during the Second World War.[287]

Loomis had visited the United Kingdom some years before, buying precision instruments and clocks, and probably met Gould when on one of those visits. It is interesting to speculate whether Gould would have gone to the United States had Loomis decided to offer him work at the Tuxedo Park laboratories. It is equally fascinating to imagine what might have been the result.

In their financial circumstances, the two could not have been more different: Loomis was exceedingly wealthy, while Rupert Gould hadn't a penny to his name. In other ways though, they were bound to be sympathetic to one another. Both had brilliant minds and many similar, technological interests. Both had lost a brother-in-law during the First World War and marital problems were something they both understood: Loomis would have clearly remembered the distress of his parents' bitter divorce, which scandalized New York society when he was a young man.

John Arnold's first marine timekeeper

The amount of money provided by Loomis is not known, but Gould was evidently very grateful. At the time Gould owned what is believed to have been the very first prototype marine timekeeper by John Arnold. It is a hugely important artefact in horological history, yet Gould chose to present it to Loomis after receiving Loomis's help. Sadly, the timekeeper is now missing. Gould tells us, in an annotation in The Marine Chronometer,[288]: 'I gave this timekeeper, in return for much kindness, to Mr. A.L. Loomis of Tuxedo Park'. Unfortunately, when Loomis moved his laboratory from Tuxedo Park during the Second World War, the timekeeper disappeared, and has not been seen since.[289]

With Loomis' money on the table, Dyson felt able to reply positively to Professor Turner's request for Gould to start H3 again, but said he needed Admiralty approval. Writing to Douglas with the proposal, he remarked 'If such approval were likely to be interpreted as an admission of unduly severe treatment in requesting his resignation, obviously it could not be given . . .'. But he pointed out that Gould was probably the only person who could put H3 back together again. Douglas replied curtly that 'on the face of it I see no objection to your proposal that the officer you mention should do the work you suggest . . .'

So, Dyson wrote to Gould on 20 April 1929 asking for the approximate cost and time for the reconstruction. Naturally Gould was delighted and replied immediately with estimated costs of 'about £15', including the cost of making of one of the caged roller bearings which was missing. He estimated the time at 6–9 months and recommended that the clock be insured for £1,000, on the rather curious basis that 'for that sum, should any misfortune befall it, an exact duplicate could be constructed from the scale-drawings which the Observatory possesses'. This is a reference to the Bradley drawings, which are illustrative but are nowhere near detailed enough for the creation of 'an exact duplicate'.

To cover Gould's costs of insurance and incidentals, Dyson added £5 to the estimate and suggested a date for completion of 23 April 1930 in his letter of acceptance to Gould on 15 June. The timekeeper was delivered to Ashtead on 8 August 1929, and so began the next saga in H3's adventurous life.

No work took place on H3 until October, Gould's holidays and writing the final chapters of *Enigmas* occupying all his time. On 3 October he wrote to William Bowyer, the Astronomer Royal's Assistant at Greenwich, discussing the plans for H3's showcase. He reported having terrible trouble getting the hardened lacquer off H3's plates and asked for an advance of £3 towards his costs, saying: 'I like to pay my way with tradesmen if I can and as you know, I am not overburdened with money nowadays'. The first notebook entry for this period begins on 9 October with a 'Note on the history of my dealings with Harrison's No.3 Timekeeper', summarising the difficulties he had had with it so far (quoted in part earlier). The second entry, on 10 October 1930, begins with a note that he had visited Windsor Castle that day to find the Mudge watch, a reference to the story told in Appendix 5.

In sorting out all H3's parts, he recorded getting 'out rollers of rem-wheel pinion. In so doing, split a roller—the first time I have done this in 10 years experience of Harrison machines', though on getting the balance arbors out two days later he decided to leave the job to Hopgood, who had better resources, as he had already damaged one of them badly. The 15th sees him drawing up a list of parts, concluding that H3 consists of over 750 individual pieces. On 16th a long list of 16 repairing and cleaning jobs went off to Hopgood, all carefully described and illustrated, the parts to be handed over to Hopgood by Mr Wakeman, one of the Admiralty's and Observatory's technical staff.

The metalwork to be done by Hopgood included: repairing the balance stopping gear; repairing the tensioning arm of the saddle-piece (part of the elaborate 'isochronising' device); making a new cam to go under the saddle piece (original missing); new remontoir springs (the old ones were badly distorted); repairing the fork arm of the remontoir detent; the attachments for the cross-over wires for the balances; several lignum vitae rollers from the pinions and one of the balance pivots, made of a high tin bronze. In his covering letter to Hopgood, Gould added: 'I imagine that this work will take some considerable time. There is no great hurry, as the cleaning of the parts will take me at least a month to six weeks. Provisionally, I am hoping to get the machine, with your cooperation, together and going by Christmas'.

A note on 14 December brings Dyson up to date with progress, and asks for an extension of time (and the insurance) but assures him 'there is no question of its being back with you long before the next Visitation' (the annual visit of officers from the Royal Society to the Observatory). A note from Hopgood on 9 January 1930 confirms the completion of virtually the last of the jobs on H3, the blueing of the balance spring, carried out by the mainspring makers Cottons in Clerkenwell. Gould replied saying that he was generally delighted with the way the work has all been done and sent a letter to Bowyer at the Observatory on 23 January enclosing Hopgood's bill of £19/18/- (nineteen pounds, eighteen shillings) for work on H3.

At the same time, he asked Bowyer for the showcase to be sent over to Ashtead as soon as it was ready. This arrived at Downside on 4 March 1930, though it seems little had been done on H3 in recent weeks. On 9 May he sent a letter to Dyson to bring him up to date again with progress, explaining that the parts were all taking much longer to clean than anticipated. He explained: 'I have come to the conclusion that when Arnold and Dent returned the machine to the Observatory in 1840 they vaselined it, and that the vaseline used was not acid-free. The parts . . . have a most peculiar patina on them, very difficult to remove . . .'. He also pointed out that he had other matters to deal with: 'including a new book which I am under contract to finish by the end of this month', a reference to *The Case for the Sea Serpent*.

But still he reckoned that H3 would be ready by the middle or end of July, a highly over-optimistic estimate as it turned out. Dyson must have suspected this, and in spite of a further letter from Gould on 6 June still promising the end of July, Dyson extended the cover till the end of December 1930.

In August, the Director of the Science Museum approached Dyson saying he had heard that Gould was now cleaning H3 and asking if they might display it alongside H2 once it was completed. Dyson replied positively, subject to Admiralty consent, suggesting H1 might join them, as 'the Science Museum seems a most suitable place for them . . .' It would be 4 years before the National Maritime Museum at Greenwich was even approved by Act of Parliament, and 7 years before its doors opened.

H3 goes on holiday (in part)

September 1930 saw Gould and Dodo off on holiday to Sheringham, as usual. But this time, aware of the long delays in getting H3 completed, Gould took with him several packets of parts to polish (Dr Jackson of the Observatory staff had approved this unusual request). Typically for Gould, this work was usually done during odd hours late at night or early in the morning. It must have been one of the stranger activities engaged in, in a room at the Grand Hotel. Work started the same evening they arrived, and from 10 pm on 2 September, Gould could be found polishing the compensation curb, the fly, the escape wheel stopping detent, and the remontoir detent. Up early the next day, he then tackled the arbors of the eight anti-friction bearers and their centreing springs, all polished before breakfast, between 7.00 a.m. and 8.50 a.m. A few hours every day were spent this way, the work on the 5 September stopping abruptly, 'proofs having arrived' (for the book *The Case for the Sea Serpent*).

Back at Ashtead later in the month, work pressed on with cleaning H3 parts again. Gould recorded in the notebook on 2 October that all the train wheels had now been polished (and that he had also passed the final proofs of *Sea Serpent* that evening). There was however still much to do and work continued late into the evenings during this last stage in the restoration.

A letter to Bowyer at the Observatory on 8 October told of a problem with the power delivery through the train, which Gould had great difficulty identifying, so he decided to re-polish *everything* again, in case it helped. He remarked wistfully: 'Incidentally, if you want a recipe for a headache, take an escape wheel of 120 teeth, so thin that you can bend it in your fingers, and polish every tooth separately, front, back and sides. This takes about two hours, but you get the headache much sooner'. Apparently the second cleaning was very worthwhile as the whole machine then looked markedly improved in appearance. Gould hoped to have H3 up and running by the end of the month but the work was suspended for a fortnight owing to his having accidentally gashed his right hand.

Some more jobs now went to Hopgood, with the usual drawings and instructions, including the making of the winding crank and the drilling of the new pallets he had made. The 9 November sees Gould fitting the balances: '(Armistice Sunday) . . . I got the balances in successfully, but it was an APPALLING job . . .'. Gould notes later (in his 1935 lecture) that to adjust and reshape H3's pallets it was necessary to get them in and out of the machine many, many times. Each time meant dismantling the movement which, in turn, took *4 hrs*, with another *4 hrs* to reassemble it. One begins to understand why the whole process took so long. After 9 November work was suspended, Gould noting in the book 'I have a show card to do for the Sea Serpent', presumably a drawing for the publicity material for the new book.

Not surprisingly, at this point Dyson was becoming anxious that H3 was not going to be ready even by the end of the year. After hearing from Bowyer about a conversation with Gould in which he expressed his doubts about the time needed for final adjustments, Dyson wrote a very firm letter to Gould on 24 November saying that H3 must be returned, finished or not, before the insurance expires.

Gould replied on 3 December with a long and serious letter, consisting of one and a half pages of close type, earnestly requesting him to allow the work to be finished. He pleaded:

I will not go into the past history of the machine from 1924 to 1929. But, since I last received it from Greenwich, I can fairly say that I have spared neither time nor trouble on a piece of work which, in its difficulty, has vastly exceeded anything which I ever undertook in my life, and which has made demands upon every faculty and resource which I happen to possess. Quite sincerely, I would sooner undertake to do No.1 or No.2 again, using my right hand only, than to repeat the work which I have put into No.3 . . .

He went on to list the many contrivances he made especially for this work, the trouble he had taken to keep costs down, and he admitted there is still some work that he needed Hopgood to do, which will need paying for. But he said he would be happy to show any official the present state of the work and demonstrate how close it is to completion. A curt reply from Dyson cites Gould's own letters of earlier date, stating that virtually no further expense will be necessary, and announcing that he must adhere to his decision to ask for H3's return before the end of the year.

Gould says in his notebook 'I shall take time and advice before replying', but under these circumstances there would only be one sensible course: Gould must try and persuade Dyson personally. And this is of course what he did, the notebook recording 'I intend to call at the Observatory tomorrow morning'. The next day's entry reads 'Visited Observatory this morning. All is well. Time is extended to March 31 if necessary. Insurance will be renewed and Hopgood's bill paid up to £5. I could ask no better terms'.

Several more jobs revealed themselves as the complex and very time consuming work of reassembly and adjusting progressed and kept Gould up late into the evenings during this last period. Over New Year's Eve into 1931, in spite of the urgent work on H3, the notebook records Gould taking a break and carrying out a few repairs to a clock (the 30-hr longcase clock he had made the 'winking pendulum' for back in 1923) which he had bought from a neighbour that year, and given to Dodo. Work on H3 appears to have resumed immediately after the New Year celebrations were over. More work was required on the saddle piece and its controlling spring, something Gould hoped was the last job he would have to ask Hopgood to do.

By this date it would seem Rupert felt he was *persona grata* again with the BHI, as a lecture on 11 February 1931 at the Institute, to be given by S.J. Smith on 'the Free Balance', was announced with the remark: 'an additional incentive to be present is that R.T. Gould has kindly consented to preside at the meeting'. Indeed, the following month he was invited to re-join the BHI council and would remain a member for the rest of his life.

H3 is finally running

Finally, at 1.15 p.m. on Tuesday 17 February 1931, Gould records the timekeeper running under its own steam: 'for the first time, so far as records go, since May 25th 1767. Well, I have something to show for my seven years'.

But, in spite of H3's initial running it did not prove reliable and further jobs were needed of Hopgood after all, including the making of new remontoir springs. These too had to be remade as the first pair were insufficiently strong, the replacements sent by Hopgood on 20 February 1931. Hopgood's actual last job on this project was the adjustments to the brass pillars upon which H3 was to stand on its baseboard within the showcase. A letter from Gould to Bowyer on 10 March reports that he was just bringing it to time and that 'It looks like new and I feel confident you will be pleased with it'.

Things seemed to be finally sorted out but, on 14 March, Gould was showing H3 to two visitors to Downside, the horological collector and dealer F.H. Green (who had just published his book 'Old English Clocks')[290] and his son, when H3 stopped. Stretched balance connecting wires (made of copper) appeared to be the problem and this was soon corrected by fitting phosphor bronze instead.

One last, rather controversial, task remained. Gould inscribed the back of the dial:

'To the immortal memory of John Harrison. This is the dial of Harrison's third marine timekeeper, begun in 1741, it was completed in 1757 and became in 1765 the property of the nation. It was deposited at the Royal Observatory, Greenwich, in 1766 and was tested there from October 1766 to May 1767. Having been much damaged by neglect, it was cleaned, but not repaired, by Messrs Arnold & Dent in 1836–1840. It was cleaned and repaired by me, with the assistance of W. Buck & R.J. Hopgood, in 1924–1931, not without dust and heat. It was definitely restarted on March 8th 1931, not having gone, so far as records show, since May 25th 1767. τετελεσται Rupert T Gould, Lt Commander R.N.Retd, 16 III 31'.

The Greek expression *tetelestai* literally translates as 'It is finished'.[291] That evening 'Humphrey & I' (Gould's neighbour Humphrey Eggar) put the machine into its glass case and determined that it wound well through the glass hole in the front. At this point in the note book Gould inserts one of his typical epilogue-type notes: 'My work is finished . . . I have spent long and sad hours—long and sad years—with the machine. I shall miss my companion, even though I shall only have known him, as an entity, for a few days. It was worth doing, & if need be I would do it again willingly'. On the morning of 23 March 1931, helped by the chauffeur, Partner, Gould got H3 out of its case and onto the carrying frame he had made for its transport downstairs, to await the arrival of the Admiralty lorry with Bowyer.

En route to Greenwich, and still in 'record' mode, with the notebook on his lap, Gould recorded the direction the lorry took 'short cut at

Morden, 1½ miles of awful potholes, then went right up to Camberwell via Tooting! He must have gone at least 4 miles further than the direct . . . route'.

On arrival at Greenwich the machine was soon going and Gould noted that ' *"Times"* photographer (a nice young fellow named O'Gorman) took 4 photos'. The next day he was at the Nautical Almanac office, where he wrote an article for *The Times*, and saw the photos ('good') taken the day before. The article appeared in *The Times* of 13 April, signed 'From a Correspondent' and telling the story of Harrison, the Longitude and the timekeepers and giving a brief summary of the work on H3.

The article prompted a letter to *The Times*, of the 'chiming in' variety, from one Maurice Brockwell, providing further information about Harrison, but phrased as a kind of corrective—guaranteed to irritate Gould—and he responded on the 17 April, pointing out 'one or two slight inaccuracies in Mr Brockwell's letter'. The *Times* article was then published in *The Horological Journal* in the May issue, also prompting a letter, the next month, this time questioning the statement that H3 requires no oil whatever. Gould was asked to reply in the same issue and confirmed that it does indeed need no oil on any pivots.

H3 had been a monumental task and its completion had consequences. Just as, back in 1924, Gould had sunk into a severe depression after the completion of writing *The Marine Chronometer*, now the stress of recent overwork on writing and restoring, and the continuing burden of commitments, led to another minor breakdown.

Writing commitments

A clue as to why he took on so much was given in a letter to Miss Drake of the Scott Polar Research Institute back in January 1930.[292] He explained that the 'difficulties consequent upon my having to leave the Admiralty . . .', hadn't got any better, and that as well as working on H3, he was having to spend much of his time writing for money. It seems his fear of having no income unfortunately caused him to take on too many literary commitments and now he was unable to deliver. Miss Drake had written to chase him up on a translation of the Russian Capt. F.G. von Bellingshausen's account of his voyages in the Antarctic regions, Gould having apparently accepted the manuscript some time before.

But at this time there wasn't the slightest hope of his honouring this commitment. In January 1930 he was in the middle of H3's restoration and was very behind with writing for *The Case for the Sea Serpent*, contracted for the end of May.

Argonaut Press

Nevertheless, he told Miss Drake, he had just signed a new (and very ambitious) contract with Argonaut Press to edit a complete edition of James Cook's journals of his three voyages of exploration, reprinted with full annotations and a comprehensive index, to be published alongside a full biography.

Not content with already having taken on all this, he told her he also had *three* other books (in addition to Sea Serpents) on the stocks: *9 Days Wonders*, *Mare's Nests* and *A Book of Cranks*, though as it turned out, none of these would ever see the light of day. This would be a recurring feature of Gould's literary career: starting several writing projects at once and only managing to complete a few. The Bellinghausen manuscript was not returned to Miss Drake because, as we have seen, the year 1930 was fully taken up with completing *The Case for the Sea Serpent*, and his work on H3.

Now, in June 1931, Miss Drake was desperate to have the manuscript returned, and on 20 July Gould was obliged to send a letter of apology to Miss Drake's boss, Frank Debenham, Director of the Scott Polar Research Institute, to explain his predicament.

Not surprisingly, his mental health had again been poor: 'for months past I have been in the trough of one of my periodical waves of lethargy and depression' and he was now in deep trouble with his publishers. For some time he had realized that he couldn't possibly honour his contract with Argonaut Press, and had been trying to pluck up the courage to tell them that he had to pull out. Unfortunately he had already accepted a £20 advance, and was saving up from his £9 a month pension to pay this back once he broke the news.

Meanwhile, he told Debenham, he couldn't go near Argonaut Press, where the Bellinghausen manuscript currently was, and asked if they would mind fetching it themselves. Debenham accepted the situation in a kindly letter, reminding Gould 'You must know by now the high opinion your friends have of your work when you really get down to it'. Thankfully, Argonaut Press also accepted the situation, once the advance was repaid.

Thomas Earnshaw's Commemoration

Towards the end of this period of depression, in July 1931, Gould was one of the special guests invited to the church of St. Giles-in-the-Fields, in central London, to witness the ceremonial unveiling of a plaque commemorating the life of the great chronometer maker Thomas Earnshaw (1749–1829) who was buried there. The driving force behind the project was Heinrich Otto, who had just been embroiled in the Queen's watch affair (see Appendix 5), the dust from which was just settling. The photograph of the event shows Otto, centre stage, accompanied by three descendants of Earnshaw and surrounded by a number of Horology's great names, including the Astronomer Royal Sir Frank Dyson, Gould and Frank Mercer, the owner of the celebrated firm of chronometer makers Thomas Mercer Ltd. It's an interesting snapshot of Gould among some of his horological colleagues at the time.

47. Earnshaw's commemoration, at St.Giles-in-the-Fields church, 28 July 1931. Front row (l-r):C. Cottleburns (Dents); C.E.Earnshaw (g-g-grandson); Hugh Rotherham (BHI); Sir Frank Dyson; E. Desbois (BHI); Mrs Otto; E.Green (BHI); Back row: H.R.Earnshaw (g-g-grandson); unknown; E. Houghton (with hat); Frank Mercer; Heinrich Otto (glasses and moustache); J. Savidge (BHI); C.Crow (BHI); Gould (standing behind with coat over arm); W. Beckman; C.Hay (Patek Philippe) and ? Earnshaw (g-g-g-grandson). (Courtesy of Charles Allix)

H3 joins H2 at the Science Museum

In October, Gould delivered a lecture at the British Horological Institute's Annual General Meeting on the subject of H3's restoration, giving graphic detail of the long and arduous work entailed. This was then published by the *Horological Journal* in parts between March and July 1932. H3 remained at the Observatory for some months after return from Ashtead, eventually going on loan to the Science Museum and being set up by Gould in Gallery XLV, alongside H2, on 2 June 1932.

On that date Gould also presented the Admiralty Chart Room with 'working drawings' of H3. These are probably no more than what is already in the notebooks and articles relating to that timekeeper. Notices about H3's public display at the museum were then distributed on 9 June to *The Times, Nature*, and '. . . the six principal News Agencies'.[293] In February 1933, the Science Museum wrote to Dyson asking for permission for filming of H2 and H3, a short film illustrating the history of Time Measurement, being made by Ideal Films Ltd., and Dyson was happy to give his approval. It is not known whether this film survives, but if it does it would provide the most wonderful souvenir of part of this extraordinary horological saga of Rupert Gould's.

❦ 15 ❦

H1 the Full Restoration 1931–1933

With H3 completed in the summer of 1931, there remained only one Harrison project as yet unfinished. H1, although clean, was unable to run, being by far the most deteriorated of all the machines. Gould simply had to finish the work destiny had decreed was his: the full restoration to working order of all Harrison's timekeepers.

So, on 4 June 1931, while facing up to giving Argonaut Press the bad news that he had done nothing on the Cook journals publication, Gould wrote to Dyson asking for permission to restore H1. He stated 'I am confident that I can make it go—and furthermore, that no ordinary observer would detect the fact that any part of it has had to be replaced. Actually, the replacements needed are small and inconspicuous—not amounting, in weight, to more than 2lb additional to its present 66lb.'

It must be said that although the missing parts were only 2 lbs in weight, they were critical parts, including both the pallets from the escapement, all of the balance springs and the connections to the compensation, so this was not going to be an easy job, even by comparison with the other restorations.

His estimates (no doubt on past performance Dyson would regard these as very much estimates!) were that the out-of-pocket expenses would not exceed £15 and that the work should not occupy more than 15 months at the outside. Dyson duly applied to the Secretary of the Admiralty for permission, outlining the history of Gould's work on the timekeepers so far. He pointed out 'This is an opportunity, which may never occur again, of completing the restoration to their original condition of the four historic instruments of Harrison. All new parts will be carefully marked but will be only replacements of necessary portions which are missing'. Again Dyson adds a little to the estimate and quotes £30 as the likely cost, including insurance, and suggests a value of £1,000 for the timekeeper if agreed. It is interesting that Dyson states the replacement parts would be marked, as Gould did not offer to do this, and all those which were eventually replaced were not identified in this way.

On the 22 June the Hydrographer, Admiral Douglas (in his last year of office), agreed and Dyson, writing to Gould the next day, ended his letter 'It will be a fine thing to have these famous instruments going once again thanks to your knowledge, skill and generous work'. Naturally Gould was delighted and, in his reply to Dyson, he hoped to be ready to begin work on the machine immediately on his return from holiday in Norfolk in the middle of September. He estimated that it should be possible to have H1 completed and running for the Observatory's next 'visitation' in June 1932.

In his letter to Dyson, Gould also remarked that he believed he had solved 'the mystery' of H1's fusee: 'how the two mainspring barrels both drive a fusee cut for a single chain', proposing the strange theory that as one chain unwraps off the fusee, the other wraps onto it, to equalize the slight variation in the springs. In fact he hadn't solved the 'mystery' at all, because he had misunderstood the mystery! Evidently Gould failed to notice when he had H1 apart in 1920, that the fusee is *not* cut for a single chain, but in fact is a 'double start' fusee, in which two chains follow each other round in a double helix. In this very clever design of Harrison's, the two mainsprings work together and, as they are disposed 180 degrees apart, pull off the fusee in opposite directions, equalizing and relieving the load on the fusee pivots.

Dyson insured the timekeeper for the period up to 31 December 1932 and had it delivered to Downside on 2 October 1931. True to form, in spite of Gould's promise to start immediately, no work was to take place for several months. On 28 January 1932 Gould wrote to Dyson with his preliminary report. In this he makes a proposal for fitting an improved winding mechanism, stating that 'most of the original winding gear is missing'.

H1's winding mechanism

Gould must have known this was a very misleading remark, and this whole aspect of his restoration of H1 in 1932 is both surprising and disappointing. Harrison designed the most delightful and elegant winding arrangement for H1. A cord simply wraps around a barrel on the front of the fusee, and it is only necessary to pull down on the cord to wind it. A simple, positive 'stop-work' prevents overwinding. The inner end of the cord is attached via a small brass segment of the wall of the barrel, jointed inside it. With the cord wrapped around the barrel this segment forms part of its wall, but on the last turn of the barrel, as the end of the cord winds off the barrel, the little brass segment hinges out, meets a solid block on the frame of the movement and comes to a positive stop. The little segment and the cord were the only bits missing from 'the original winding gear' and it would have taken 30 min at most to make the brass piece and attach a cord to it.

But Gould was determined that H1 should be converted to key wind. He cited an occurrence, recorded in the Observatory's archives, when in 1766, the brass segment accidentally unhooked itself, and used this single reference as the justification to claim that this original arrangement would be 'entirely unsuited for use by anyone who is not more or less of an expert.' He then recommended H1's conversion to key winding, spending some two and a half pages describing how it could be done. Accompanying this description he included several nicely executed sketches of the proposed arrangement.

As Hopgood was to be asked to make many of the other new parts needed, Gould told Dyson he would be inviting him to Downside to see the machine before disassembly. A revised timetable was provided, with delivery at December 1932. Dyson wrote in reply on 2 February agreeing to all Gould's proposal's including, alas, the new winding gear.

With such immediate approval given, and no books now on the stocks, one might have expected work on H1 to move ahead swiftly, but in fact nothing more was done to the timekeeper for over 7 months. Indeed, apart from giving a lecture to the BHI on 'The Original Waterbury Watch' on 13 April, and writing an article for the Society for Nautical Research on 'The Navigator's Equipment',[293] Gould appears to have taken a break from horology during this period. More time was allocated to other activities, including the beginnings of his service as a tennis umpire at Wimbledon, and spending more time with his children.

The Children

Cecil was by now a day-boy at Westminster School, as his late uncle Cecil had been (he would become a boarder from 1934). In spite of Muriel's mother paying one third of Cecil's fees, Muriel's financial situation was precarious to say the least, as Rupert's alimony would not have contributed much towards such an expense. Rupert tried to take a dutiful interest in their progress at school, and helped with their studies when he could, especially if the subject interested him. Cecil recalled his father, who had a love of English poetry, helping him with an end-of-term poetry reading, of Coleridge's *Youth and Age*. Going over the work several times and marking the text with marginal notes 'very quiet here', 'increase pace', 'climax' etc., he rehearsed his son so thoroughly, Cecil walked away with the school's Oration prize that year.

During the summer of 1932, Rupert had the children to stay at Downside for a fortnight. The weather was fine all week and the children evidently had a very happy time. Using the back pages of H1's notebook, Rupert decided to keep a brief diary of their stay. Cecil and Jocelyne arrived on 3 August, a day, coincidentally that Gould's friend Vyvyan Holland was there. That day Holland and he marked out the court on the lawns at Downside, and the children and they played tennis. Other days were filled with horse-riding for Jocelyne, a visit to the circus, a day out at Whipsnade Zoo and St Albans, and walks almost every day. Several times Gould took the children by bus to a local park which had a boating pond, to sail the model submarine 'K2'. Ostensibly this had been bought for Cecil, but there is no doubt who got the greatest fun from outings of this kind. These trips were invariably followed by tea or ice cream: for all three probably the highlight of the day.

Work starts on H1

It was only on 17 September, on Gould's return from Sheringham after his annual Norfolk holiday with Dodo, that he began H1, exactly a year after he said he would be able to start. However, that first day things progressed quickly and, over the following few hours, H1 was fully dismantled. In fact Gould worked through the whole night on it, producing over ten pages of notes, stripping the whole machine, the entry ending '11.20am 18.IX.32'. The next day, before sending the plates away for polishing, Gould noted that he must obtain all data concerned with missing parts needing restoration. After carefully straightening all the balance bearers, he proceeded to cut and open out several parts of the plates to provide 'better clearance' for the balance arbors.

Alterations and damage

This kind of alteration, of which there were quite a few during this restoration of H1, would of course be wholly unacceptable today, and a

1. Thomas Skinner MD (1825–1906). Rupert Gould was very proud of his maternal grandfather, and emulated his open-minded academic approach to his studies. (©Sarah Stacey and Simon Stacey, 2005)

2. David Beatty (1871–1936). Midshipman Gould served under him on HMS *Queen* for 18 months from January 1909. He idolized Beatty and this was the happiest and most inspired period of Gould's Navy career. (By courtesy of Felix Rosentiel's Widow & Son Ltd, London, on behalf of the Estate of Sir John Lavery, Collection ©National Portrait Gallery, London)

The Colourphone

TELLS · THE · TRUE · COLOUR · OF · EVERYTHING ·

When shopping of an evening or on a dull day ask your Draper for the "Colourphone"

3. *Valses des Fleurs*, 1910. Gould's interpretation of the scene of that name from Tchaikovsky's ballet, *The Nutcracker*. This was a commercial copy, taken without permission, and much to Gould's annoyance. (© Sarah Stacey and Simon Stacey, 2005)

4. H1. Created between 1730 and 1735, Harrison's first marine timekeeper astonished the scientific establishment when it was brought to London in 1736, and performed relatively well on a sea trial to Lisbon in 1737. (D6783, ©National Maritime Museum, London)

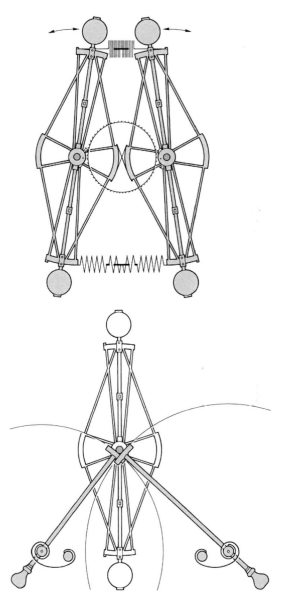

5. H1's balances. The springs connecting them at top and bottom provided what Harrison called his 'artificial gravity'. The wheel in the centre is the grasshopper escape wheel which drives them. The lower drawing shows the pairs of anti-friction rolls upon which each balance runs (David Penney (www.antiquewatchstore.com). Copyright)

6. H2. Built over two years, between 1737 and 1739, H2 was a refined version of H1, with a remontoir to ensure a uniform drive to the balances. A fault in the design of the balances caused Harrison to reject the clock and start work on H3. (D6784 ©National Maritime Museum, London)

7. H3. Harrison spent 19 years working on this immensely complex timekeeper. For many years he was convinced it would win him the longitude prize. It is the only timekeeper to retain its original carrying case. (D6785–5 ©National Maritime Museum, London)

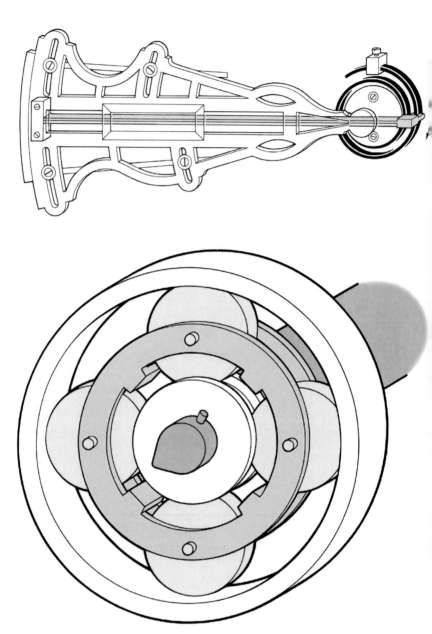

8. The bimetallic strip temperature compensation and the caged roller bearings in H3 are both highly important inventions and both still very much in use today in many technical applications. (David Penney (www.antiquewatchstore.com). Copyright)

9. John Harrison's 'RAS' regulator. Built approximately in parallel with H3, this was Harrison's 'state of the art' fixed pendulum clock. He claimed time-keeping as good as a second in 100 days from it. (D6786–1 ©National Maritime Museum, London)

10. H4 (left), 1759, 'the most important watch in the world', alongside Kendall's copy, K1, 1769. K1 enabled James Cook to navigate with great accuracy during his second and third voyages of discovery and was dubbed by him his 'trusty friend'. (B5165 ©National Maritime Museum, London)

11. The movements of H4 and K1. The oscillating balance, under the beautifully pierced and engraved upper part of the movement, was one of the main keys to the design's success. Harrison agreed that the copy K1 was even more beautifully made than his original. (B5166 ©National Maritime Museum, London)

12. H5. Visually simpler in design than H4, H5 was of very similar technical construction. The watch was tested, with great success, by King George III at his private observatory in 1772. (Clockmakers' Company)

13. Thomas Bradley's drawing of H2 c. 1835–40, partially dissected and shown in four views. The draughtsmanship and the finish of the drawings are excellent, but the technical details are not always 100% accurate. (RGO Archives, University of Cambridge ref: RGO6/58, fo.2r)

Y^e 365th Meeting of Y^e Sette of Odd Volumes
held at Y^e Imperial Restaurant (Oddenino's)
on Tuesday, y^e 23rd daie of November, 1920

Y^e HYDROGRAPHER *laboriously compileth, not without sustenance, hys odde*
DISSERTATIO DE INSULIS INSOLENTIBUS

Hys Oddshippe
Bro. Horace C. Beck, Visionary
in ye Chaire

14. Gould's self portrait for the O.V. Menu on 23 November 1920. The horo-logical bookseller Malcolm Gardner subsequently had the drawing published separately, simply titled: "Good Master Hydrographer, Chart me the Unknown seas". (©Sarah Stacey and Simon Stacey, 2005)

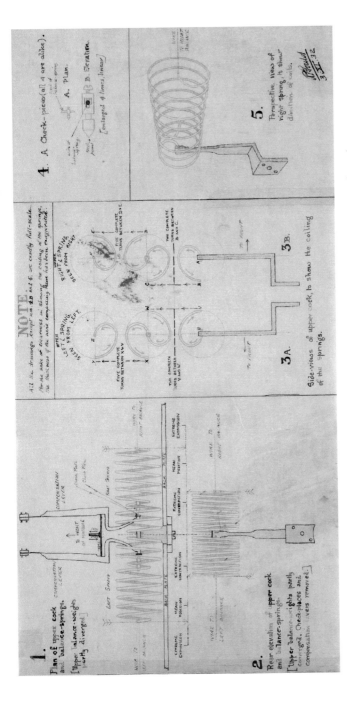

15. Some of Gould's construction drawings and instructions for Mercers. Using these plans the missing parts of H1's balance springs and compensation were remade in St. Albans in late 1932. (© Sarah Stacey & Simon Stacey, 2005)

16. Jocelyne and Cecil, with two of Rupert's drawings, outside the home they shared in Thorncombe in 1993 (Jonathan Betts).

full list of the accidental breakages, followed by the necessary repairs on H1 alone, would frankly fill several pages. But, as has been observed before, Gould's work on these machines has to be seen in the context of the time and consideration given to the alternative fate of the Harrison timekeepers had the work not been done at all.

Every day over the next fortnight was worked, mostly until midnight, including making carefully measured drawings of the plates and train of the machine. On 28 September planning was complete and the plates were sent off to Buck the next day. 5 October (notebook marked 'Flint's funeral') sees Gould return to the job, making notes on the wheel train and motion work counts, and starting to dismantle the balances.

Mercers

It was at this point that Gould decided, for the first time, to employ the services of his friend Frank Mercer. On 6 October 1932 Gould visited St Albans: 'Took the 2 chain barrels, wheel, its broken part of arbor & the pinion driving the motion-work, to Mercer, & arranged with him that he should tackle the winding gear (& also the balance springs) as soon as the plates come back from Buck'.

He was back the same day continuing the dismantling of the balances. The 10th saw Gould over at Hopgood's in Blackheath with a long list of jobs needing doing, accompanied with the usual neatly typed and illustrated instructions, covering 4½ pages. The next two weeks saw work continue steadily, often well into the night, and with jobs now returning from Hopgood and Buck, Gould notes on 20 October: 'We Progress! The butterfly is beginning to emerge from its ten-year chrysalis'.

Gould spent the next day 'drawing hard all day, with short intervals . . . Knocked off Midnight', so that the following day 'after 2 days of extremely hard work, finished a set of 6 detailed drawings . . . for Mercer's use in making the new winding gear. Wrote & bound up some 2,500 words of notes in connection with these'. He was then able, on 23 October, to take the main frame, relevant parts and his drawings and notes, all typed up and home-bound in his familiar brown paper covers, to Mercers at St Albans, to get the making of the winding gear started.

On 25th the remaining parts, now polished, were collected from Buck, and Gould arranged with his friend and fellow naval officer,

Captain Jauncey of the antiquarian horologists Clowes and Jauncey, to have a new seconds hand (and later an hour hand) cut for H1. Two days later he was visiting Greenwich to leave more jobs for Hopgood and collect one from him, and things were progressing well.

Gould now struggled to understand the action of the compensation of H1, which he needed to draw and explain to Mercer's if they were to make the missing parts. On 2 November he noted in the book 'I think I now fully understand all details of the compensation . . .' and evidently feeling pleased with himself remarked: 'Going to pictures, so knocked off 7.30pm'. Five days later he wrote to Mercer enclosing the parts relating to the balance springs and a model (out of wood and iron after scrapping one in copper wire) he had made of what he needed to be constructed by them. He also sent a separate set of elaborate bound drawings and notes on the springs, all carefully copied for the notebook as usual (See plate 15).

Mercer replied the next day saying that most of the first part of the work was done, though 'the hairspring question is going to be a little difficult'. Gould's work continued in the evenings, with the dismantling of the gridirons and the drawing and description of the compensation. On 13 November the children were at Downside so 'not able to do much', but two days later he travelled to St Albans to collect the winding work. He described it as 'a most admirable job' and, leaving the parts of the compensation, promised to write shortly with instructions. The 18th was 'A busy day . . .', Gould calling first at Hopgoods in Blackheath to take the escape wheel for trueing, and to collect finished work, then on to the Observatory to see Bowyer, after which he travelled over to the Science Museum in South Kensington to re-start H3 which had stopped a few days previously 'probably through case being jarred'. After this he continued on west to Bucks to drop off all the balance parts for polishing. The notebook records that the day was rounded off with a quick visit home and off out again to hear Gilbert & Sullivan: 'Mikado in evening'.

As the end of November approached Gould began making a model of the escapement in cardboard to ascertain exactly what form the pallets should be. The 26th saw the pallets coming to a final size, but a social call to neighbours (for billiards) interrupts: 'It is my evening for the Lowthers, & I shall turn in when I get back'.

The 28th was another busy day out. A journey to Blackheath to see Hopgood's foreman about repairs to H1's balance bearers, was followed by a visit to Bowyer and Dyson at the Observatory to show them parts and drawings, after which he 'Lunched with Palliser @ Naval College'.

Gould wrote to Mercer again on 1 December with a third set of notes, this time on the parts needed for the compensation, and explaining how he believed it works, asking 'If there is anything else you or Godman want to know . . . please let me know . . . don't forget my telephone number, Ashtead 192'. Bill Godman was Mercer's man who mostly carried out the jobs the firm did for Gould. Copies of all three sets of notes for Mercers, on the winding gear, the balance springs and on the compensation, were then bound up together by Gould and form notebook No.10 in his series.

H1's compensation

For the record, it should be pointed out that Gould's reasoning behind the action of H1's compensation, which is admittedly very complex, is at least partly mistaken in its assumptions. He states that the greater the tension of the balance springs the *slower* the balances vibrate, noting that this conclusion came as a surprise to him. This it certainly should, as the opposite is in fact the case. Presumably he drew his conclusions by noting that the balances vibrated slower when he applied additional force to each balance spring at the moment it is most relaxed (i.e. when it is collapsed and its opposite number at the other end of the balance is at its most extended) which is the moment when the compensation acts. This of course *will* cause the balance to vibrate slower as the strengthened spring resists the return of the balance (i.e. it reduces the restoring force acting on the balance). So it is true to say that increasing the tension of the balance spring, at the moment when it is most collapsed, will cause the balances to vibrate slower, but only at that point. Having said this, the reconstructed parts he instructed Mercers to make were all correct, and all function just as they should.

Both pallets for H1 were finished on 4 December, but a visit by the children caused him to stop work for most of the day. Then, just when things seemed to be well on the way, another accident set Gould back. Noticing some of the (*lignum vitae*) rollers in the (oak) third wheel were tight, and knocking the pins out as carefully as he could 'the whole wheel suddenly split clean in two, along a diameter with the grain'. He supposed it was probably weak anyway and may even have broken of its own accord in the clock had this not happened. Besides, he reasoned, it allowed him to see how Harrison attached the oak wheel to its oak arbor, via angled pins knocked through. He repaired the wheel by attaching wooden ties (pieces of a wooden ruler!) across the break, a most unsatisfactory mend which would be repaired properly later.

Writing on 16 December, Gould asked Mercer for the work to be completed, if possible, before Christmas 1932 and on 22 December sent his final instructions about the winding work. Mercer replied on Christmas Eve that work was progressing well, but they had had considerable difficulty with getting wire for the balance springs. Nevertheless, they were hoping to have everything done before the end of the

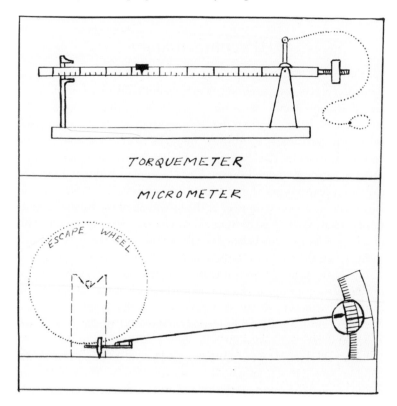

48. Gould's sketch of two of the special tools he made to facilitate the restoration of the timekeepers. The *torquemeter* was a highly sensitive instrument which could be attached to a tooth on any of the wheels to check the evenness of the driving force, or *torque* at that wheel. (© Sarah Stacey and Simon Stacey, 2005)

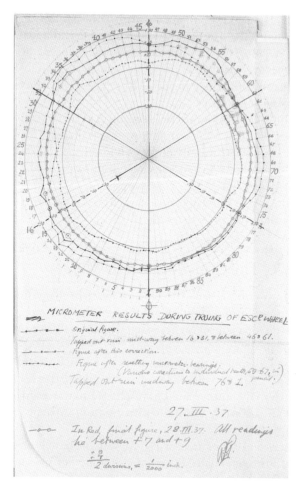

49. The *micrometer* was used for checking the concentricity of the teeth-tips on the escape wheels in the timekeepers. This diagram plots the gradual improvements to the wheel (H2's in this case) as Gould carefully adjusts each tooth individually and rechecks the wheel on the micrometer. (© Sarah Stacey and Simon Stacey, 2005)

year, Mercer noting 'rest assured we are pushing on with it'. By this time Gould had also completed trueing up the escape wheel; a task he managed by using the special micrometer he had made for the RAS regulator and H3, to measure very accurately the position of each tooth.

Work on H1 had to be put on hold over Christmas 1932, and the New Year holiday. Cecil and Jocelyne were staying at Downside for this period and Rupert had an opportunity to spend some time with his children. On the 3 January they all engaged in a practical project of a rather different nature: 'the children & I have been very busy erecting one of the *Underground* toy theatres...'.

Over the holiday period, on 30 December, Mercer wrote to Gould to inform him that, true to his word, the work was all due to be completed that week, and on 4 January, once the children had returned to their mother, Rupert drove to St Albans to collect the parts. The notebook records of Mercer's job: 'All his work is complete and well done'. Returning to Ashtead the same day, Gould worked on until 11.10 p.m., making a baseboard and dust cover to test the timekeeper on.

The Contribution by Mercers

Frank Mercer now needed to decide how much to charge for his work and, apparently unknown to Gould, asked the Astronomer Royal, Frank Dyson for a note of the costs and schedule of H1's project so far which, surprisingly, Dyson consented to. The handwritten note, dated 7 January, and marked 'Confidential' and 'To F.Mercer', lists the costs so far[294] Mercer replied a few days later that 'after reading your notes on this, my brother and I have decided to waive any claim for payment for our work, naturally it has been of extreme interest to ourselves...'

In this second restoration of H1, Mercers were one of the more significant firms who helped Gould by carrying out specific tasks. In summary these were chiefly:

(1) Making and fitting the winding wheel and pinion winding key;
(2) Making and fitting the balance springs and their connections for the compensation;
(3) Supplying the two mainsprings.

It was extremely generous of Frank Mercer to decide, at the end, not to charge the Observatory for this work and this was fully recognized by Gould. At the beginning of notebook 10, he has annotated the front (in August 1941), concerning these parts: 'while I could plan this portion of the work, I greatly doubt whether I could possibly have executed it. Mercers did this for me—magnificently—and refused to take a penny!'.

Unfortunately this generous remark has been misinterpreted in later years and it has been suggested by Frank Mercer's son, Tony Mercer, in all three of the books he has written, that it was in reality Mercers who principally carried out the full restoration of H1. This is wholly untrue and both Gould and Frank Mercer would certainly have been horrified at this suggestion.[295]

Although the winding gear, compensation and balance springs were now complete, there was still a great deal to be done to finish the restoration of H1. The pace of Gould's work dramatically increased through January 1933, working every day, usually well into the night to get things done. As more parts went off to Hopgood for repairs and adjustments and to Buck for polishing, Gould records on 8 January 'The pace of the work is getting too hot for keeping detailed notes', though most of it does in fact appear to have been recorded.

After many hours spent trueing and balancing the balance bearers and the balances themselves, Gould notes: 'The big push begins' on 18 January. A number of small jobs with the balances and escapements now caused problems and many hours were spent trying to solve them. But on 24 January, with the escape wheel being driven by a small weight on a cord, the notebook records 'After being almost in despair, the "big push" has ended satisfactorily. I have had the pleasure of seeing the machine run the weight down its full height, banking slightly. Rang Greenwich up & arranged provisionally to have machine collected on Tuesday!' Gould also alerted *The Times* newspaper so they could cover the return of H1.

It was a fatal mistake of course as, predictably, another problem arose. With the machine together with its new mainsprings for the first time on 29 January 1933, he found there was insufficient power to drive the clock at a safe amplitude. That day the distinguished clock and watch collector and historian, Courteney Ilbert (1888–1956) came to tea, along with Gould's neighbour Lowther. Ilbert and Gould got the springs out to inspect them, putting them back again and setting them up as much as they dared, but still without success.

There was nothing else for it but to let the Observatory and *The Times* know, and postpone H1's triumphant arrival at Greenwich. A visit to the spring makers Cottons of Clerkenwell the next day saw new springs ordered, with delivery promised for the very next day. Gould did at least have the services of Clerkenwell's horological industry to support him. Today the supply of a special order, first-rate chronometer spring in 24 hrs would be quite unthinkable.

On 31 January, the springs ('a splendid job') were duly fitted and after a little experimentation, at 5.20 p.m. Gould noted 'No.1 is going very sweetly, swinging about 7° or a little over'. The last job, and the one Gould considered the most tricky, was the adjustment of the check-pieces in the balance springs. He would later (in his 1935 lecture) describe it as 'like trying to thread a needle stuck into the tailboard of a motor-lorry which you are chasing on a bicycle'. With the timekeeper now running, Gould made the packing case for its transport and on the morning of 3 February 1933 the timekeeper was taken down to the awaiting transport.

Gould had had a restless night worrying, because he realized overnight that the case wouldn't go down the narrow winding staircase out of his workshop. The solution, which Dodo was never allowed to know, was to cut the banister off where the turn in the stairs was required, and refit it, hopefully invisibly. Rupert's son Cecil recalled that, as far as he knew, Dodo never did find out!

H1's arrival in Greenwich

They arrived at Greenwich at 12.45 p.m. to find the *Times's* photographer had arrived early and, according to Gould's notebook, had 'left in a huff at 12.30!' Another was quickly called, one 'Mr Barker, efficient & courteous'. After setting the timekeeper up, Gould walked down the hill 'to the Gloucester [a pub at the lower, west gates of Greenwich Park] for bread, cheese and beer'. Returning after lunch to see Dyson, H1 was then got into the showcase in the Octagon Room in Flamsteed House, 'going, & looking really magnificent' the notebook proudly records. Later that day Gould added 'So ends this log, and a long chapter in my life . . .' a few days later (8 February) telling the whole story of H1's restoration in a lecture to the B.H.I.

But it is not the last entry in the notebook as H1 needed further adjustments, and Gould visited several times over the next week or two, to adjust the balance bearers, which tended to get displaced. The 12 February 1933 saw Gould visiting the Observatory in company with his children and Courteney Ilbert, to attend again to the balance bearers. The notebook also records that day that Gould 'Brought away original glass "penthouse" cover of No.1'. This extremely interesting comment seems to refer to the remains of Harrison's original mounting case for H1, but alas, nothing is known of its whereabouts today.

Meanwhile, Gould began a series of articles and talks on this latest restoration. An article for *The Times* was written on 5 February, then published in the Horological Journal in June (Gould had given a lecture to the BHI on H1's restoration on the 8 February), and an article on the same subject appeared in *The Observatory* the same month.

Following the delivery of H1 to the Observatory, Gould asked Frank Dyson if there would be any objections to a short description of his work on the timekeepers, along with a small photograph of him, being put up in the Octagon Room, next to H1. Dyson, who was soon to hand over to the new Astronomer Royal, Harold Spencer Jones, had a word with him and wrote to Gould that they would be very happy to do so.

This, for Gould, was an important psychological achievement. Not only had he now completed the restoration of all the Harrison timekeepers, but this was now officially recognized and approved of in the most conspicuous and appropriate way: a notice at the Royal Observatory itself.

Dyson also sent a summary of the costs incurred (totalling £40/6/6), and a short report of all the work done on the Harrison timekeepers by Gould, to the Hydrographer (by now Admiral J.A.Edgell). Along with his report, Dyson added the suggestion that the Admiralty Board might consider making Gould some ex-gratia payment for all his trouble. Edgell and the Admiralty Board agreed and Gould was voted £100, Edgell suggesting that H1 should at some stage go on show at the Admiralty for the Board to see 'in person'.

In fact, during the following year, 1934, H1 began to prove somewhat unreliable again and, after several visits to try and ensure its reliability, even including fitting larger mainsprings, Gould was obliged to ask for the machine to be returned to Ashtead for another service, H1 arriving there on 1 August. After many repairs and adjustments to the fusee assembly (Hopgood made a new stop for the maintaining power), to the balance springs, to the escapement (new pallets), and to the anti-friction rolls, work which took the whole of August, the timekeeper was taken back to Greenwich again and set up, running, on 3 September 1934. Finally H1 began to settle down to reliable running, and this phase of Gould's Harrison timekeeper restorations was truly complete.

R.J. Cyriax and Sir John Franklin

It was about this time that Gould was first contacted by a researcher who would remain a correspondent and friend for many years. In the early 1930s Richard J. Cyriax had begun to research the story of the Franklin expedition, with a view to publishing a book on the subject and had contacted Gould after reading the chapter on Franklin in *Oddities*. Gould soon realized Cyriax was a meticulous researcher and they became firm pen friends, corresponding for over 10 years.

In May 1933 Cyriax had suggested they meet in London and they agreed to rendezvous at Waterloo Railway station, Gould offering a brief 'pen picture' describing himself at the time: 'As we are unknown by sight to each other, let me say that I stand about six feet four or a little over, and am clean-shaved with a fresh complexion. I always wear either a brown or blue suit, much the worse for wear, and in hot weather a flannel shirt. I am just going grey at the temple, otherwise I have a good deal of dull-coloured hair, intermediate between fair and dark. I am 42'. Always one for covering every eventuality, in his next letter Gould was even more specific: they would meet on 3 June (1933) at 6.27 p.m., 'I shall be wearing a dark blue suit, flannel shirt, blue tie & grey Homburg hat & shall be carrying a Burberry and a small brown attaché case'[296]. Gould would soon recognize that Cyriax's total dedication to his subject and his attention to detail marked him out as the leading expert in the matter of Franklin's expedition and, as will be seen later, the correspondence between the two scholars continued off and on until the end of Gould's life.

16

The Loch Ness Monster 1933–1934

Although at the beginning of the 1930s Gould had had another minor nervous breakdown, life at Downside in the early and mid 30s was generally as stable and happy as Gould had known since childhood.

Rupert was now the only family Dodo had; her other son had died during the war, her husband had died in 1923 (both deaths usually commemorated 'In Memoriam' in *The Times* in early April) and her brother Hilton Skinner had passed away in 1928. For Dodo, looking after Downside, socializing with friends in the locality, and involving herself in charity work kept her busy, and having Rupert around and seeing her grandchildren on a reasonably regular basis was a comfort to her. As for Rupert, his socializing still included the monthly dinners with The Sette of Odd Volumes.

Alexander Keiller

At the O.V. dinner he attended on 25 April 1933, when S.R.K. Glanville ('Brother Ushabti') spoke on 'Human Personality in Ancient Egypt', Gould had with him, as one of two guests, his close friend Alexander Keiller (1889–1955).

Keiller was heir to the family's famous marmalade business in Scotland, and was himself a noted archaeologist, having been excavating the celebrated dig at Avebury for some years. The talented and ambitious Keiller had many interests in common with Gould, from early motorcycles and technology to witchcraft and the occult. How they met is uncertain but it is most likely to have been the publication of *Oddities* which brought them together.

They had more personal things in common too: Keiller's marriage had been on the rocks since the late 20s, and they doubtless understood each other's personal situations. A big difference between them however was in their personal wealth and Keiller would soon become something of a benefactor for Gould. Gould, in turn, was trying to get Keiller admitted to The Sette of Odd Volumes in 1933. The Candidates

50. Alexander Keiller (1889–1955). One of Rupert's closest friends during the 1930s, archaeologist Keiller sponsored his expedition to Loch Ness in 1933. Rupert failed to get him admitted to The Sette of Odd Volumes. (Alexander Keiller Museum/National Trust)

book reveals Keiller was lined up as 'Brother Antiquarian', but he was eventually obliged to withdraw under By-Law 4 (failing, after four attempts, to get Council approval), owing to objections from the secretary, Lord Moynihan that Gould, as an Emeritus Odd Volume, was no longer entitled to propose new members.[297] There was however nothing in the rule book which made this clear and Gould was furious with Moynihan, demanding to have his Emeritus status removed (he could just about afford the annual fee by this time), but it was something he still felt badly about 2 years later when, to everyone's sorrow, Moynihan died suddenly; the first time a President had died while in Office.

That evening, alongside Keiller's signature in the attendance book, Gould signed himself simply 'X', a cryptic mark, the significance of which was probably known only to Gould and Keiller at the time but, within a year, would be the subject of Gould's next book.

The Loch Ness Monster

The fact was that in the Spring of 1933 rumours had begun of sightings of a large creature, apparently of unknown species, living in Loch Ness. Having published *The Case for the Sea Serpent* just over 2 years before, Gould soon got to hear of this and needless to say was intrigued by the descriptions, sounding, as they did, remarkably 'Serpentine'.

In late October, the highly sceptical opinion of one of Gould's fellow Sette members, E.G. Boulenger ('Brother Shark') the Director of the Zoological Society's Aquarium, appeared in *The Observer* newspaper,[298] in which he dismissed the reports as simply a case of mass-hallucination, and concluded that 'it is safe to dismiss the creature as mere "moonshine"'.

By November, the reports had appeared in most of the national newspapers and questions had been asked in Parliament. But it was becoming clear that no professional zoologist was prepared to investigate, the sightings being generally dismissed with a variety of 'explanations', and Gould felt something had to be done.

Investigation

It was at this point that, as the restoration of all the Harrison timekeepers was now effectively complete, and no major project was outstanding in Gould's itinerary, Alexander Keiller proposed to him that he personally should go to Scotland and investigate Loch Ness. Anticipating Gould's objection that he could not afford such a project, Keiller offered to fund the whole exercise. There had of course been several journalistic enquiries made already at Loch Ness, but as yet there had been no systematic attempts at impartial research, and Gould gladly accepted the challenge.

Leaving Downside in early November 1933, Gould's first stop was Edinburgh, to research the already published material, chiefly from the files of *The Scotsman*, and to make a list of eye-witnesses for interview. Then on to Inverness to start work. After establishing the virtual

51. Gould astride *Cynthia*, the Matchless motorcycle he had bought in Scotland to get around. At the time, he hadn't ridden a motorcycle for nearly 20 years. (© Sarah Stacey and Simon Stacey, 2005)

impossibility that any large creature had entered the Loch via the canals feeding into it, Gould concluded that there was a remote possibility that something might have made its way, against the outflow, into the Loch from the River Ness. And so the interviews began, on the assumption that this would reveal the identity of some well known sea-creature.

Gould now considered the best way to get around the Loch (somewhere between 40 and 50 miles in all) during his investigations, deciding in the end to buy a second hand motor-bicycle 'But I had been out of the saddle for nearly twenty years, and I looked around for a mount with some trepidation'. Nevertheless, enquiries in Inverness soon turned up a 2-year old 246cc Matchless, 'fast, comfortable and very easy to handle . . . the best bargain I ever made in my life'. He christened her 'Cynthia' after the heroine in the film *Christopher Strong*, released that year.[299] So pleased was Gould with his new mount that, after returning to London to write up his project he submitted an article about his adventures to *The Motor Cycle*.[300]

The interviews

From 14 to the 25 November, Cynthia carried Gould twice round the Loch, enabling him to interview some 50 eye-witnesses. His interview technique and record taking was scrupulous and intelligent: 'If witnesses could draw, I got them to make a sketch of what they had seen. If they could not, I made an outline sketch under their inspection and direction. I put no leading questions, and offered no opinions or theories of my own; nor, unless this was unavoidable, did I mention to any witness what I had heard from another. If two witnesses, in company, had both seen the creature, I did my best to obtain independent statements from each of them . . . I eschewed the evidence of children—they may be accurate observers, but the tendency to "go one better" is always present. Lastly, if a witness had used a telescope or binoculars . . . I compared the instrument, for power and definition, with my own binoculars'.

Always making careful notes on ordnance survey maps and recording the altitude and bearing of the sun at the reported time, it is difficult to see what more Gould could have done to get the facts straight. And the witnesses themselves impressed him 'they were quiet and cautious, tending to minimize the importance of their respective experiences and to understate, rather than to exaggerate, what they had seen'. The press, on the other hand, did not impress him much at all, mostly misquoting witnesses and glamourising the sometimes rather pedestrian descriptions.

The sightings

This is not the place for detailed descriptions of the large number of pieces of evidence he gathered, but a summary of typical sightings might give a flavour of what was being described. Often the witnesses first noticed a strange disturbance in the water, followed by the appearance of part of a large grey-brown body, resembling 'an upturned boat'. In some reports this suddenly disappeared under the surface, in others it was seen to move off at some speed, just on or below the surface when, in some cases, the body appeared as two humps in the water, creating a considerable wake. One or two of the sightings were more specific and extraordinary. For example in the case of Arthur Grant, a vet and thus a supposedly well-qualified observer, a complete animal was seen, 15–20 ft long and resembling a cross between a giant sea-lion and eel, loping across the road and into the Loch. Another very remarkable sighting was that reported to Gould in the following year by Professor A.W. Stewart, a witness of the most impeccable scientific credentials, whose impartial and sensible description goes a long way (at least in the mind of the author of this biography) towards evidence of something real and unusual in the Loch at that time (see p. 272 in Chapter 18).

There is not space here to cover Gould's clear and logical thought processes, as described in his published account. Suffice to say that his startling conclusion, reached only after carefully processing the evidence he had gathered, was that the witnesses were not imagining things. Neither were they having a joke at his expense nor that of the press. Although he had not had a sight of the creature himself, Gould was now certain there was something unusual in the Loch and that in all probability it was a fresh-water specimen resembling the creatures he had described in his earlier work *The Case for the Sea Serpent*.

Anticipating the most likely reaction of readers to this, he comments: 'Of course, if anyone chooses to assert that I went to Loch Ness with the intention (conscious or subconscious) of identifying the "monster" as a "sea-serpent" and points for confirmation to the fact that I have already committed a book about such creatures, and am an avowed believer in their existence, I have no means of disproving his assertion. But if I am any judge of what I think, and of how I form my convictions, I can—and do—contradict it most emphatically'.

Gould was very much aware that the press was following his progress. They were actually quite helpful to him at various times during his investigations, and before leaving Inverness on 25 November 1933, he decided to issue a communiqué for general distribution to the Press Association, giving his preliminary findings. The press release concluded that what had been seen was in all probability 'a large living creature of anomalous type, agreeing closely and in detail with the majority of the reports collected in his book *The Case for the Sea Serpent* (1930). In his opinion, no other theory can be advanced which covers the whole of the facts. He hopes to publish his results in book form at an early date'.

The Reactions

In fact the communiqué appeared in only a couple of Scottish papers and when, three days later, Gould, calling from Glasgow, enquired of the

Press Association in London as to why nothing had appeared in any National, he tells us: 'the voice at the other end—distant in more senses than one- remarked 'Oh, we don't want that-we don't believe in it'.

Needless to say, Gould was furious that his three weeks hard work was being dismissed without their even bothering to discover how the evidence had been put together. Nevertheless, on his return to London he found to his delight that *The Times* itself, no less, was interested in doing something on his research. On 9 December 1933, a three-column article appeared under his own name, with an Editorial Leader to back it. Predictably, this produced a flurry of correspondence of all kinds, mostly (as Gould himself described it) of the 'busybody' kind, and ranging from reports of mermaids to outright condemnation of his findings. Various notes and letters continued off and on until the new year.

An old adversary reappeared in the form of the geologist F.A. Bather, pouring scorn on Gould's plesiosaurus theory as he had done when *The Case for the Sea Serpent* had been published 3 years before. But this time, Bather was not alone in his scepticism. Even some of Gould's friends, having read his statement to the Press and assuming he was not wholly serious, asked him what he *really* thought. Such questions only prompted feelings of having been insulted, Gould noting angrily: 'I have never written a page of fiction: and I have never, in my life, printed a statement which I had not previously done my best to verify, or an opinion which I did not sincerely hold'.[301]

The other predictable result of the press release was that all manner of challenges and offers were made for securing a live specimen of the monster, and a similar number of pranksters produced fake evidence and sightings of monsters, all of which naturally came to nothing.

The book

Work began on writing the book, as promised in the press release, as soon as Gould returned to Ashtead, and on 27 February 1934, he gave a talk to the Sette of Odd Volumes[302] on the Loch Ness Monster. Of course, his fellow brothers in the Sette could not resist the temptation of ribbing him mercilessly over the whole affair.

The cover of the menu that night contained a large blank space where a picture should be, titled: 'Drawing of the Monster to be inserted when details of his appearance have been duly testified'. Inside, a footnote reads: 'Note: Brother Hydrographer wishes it to be known that he has not had a drink since October. It is not however, yet known whether his observations of Gaelic phenomena were anterior or subsequent to this surprising act of temperance. The guests are therefore respectfully requested not to refer to the subject, as it is one upon which Brother Hydrographer is inordinately touchy'.

The evening thus appeared as jolly as usual, but there were at least two sour faces in the company. Lord Moynihan ('Brother Incisor') and, of course, E.G. Boulenger ('Brother Shark') both lived up to their Sette titles, and were highly critical of Gould's views and made no pretence about it, much to Gould's annoyance. But Gould would have his say when the book was published.

Another lecture followed soon after the O.V. talk—on 17 April 1934 Gould spoke to the group known as 'The Anchorites'[303] on 'Sea Monsters', the lecture being exceedingly well received. Indeed, so much was said of its success that it led to the request for a rather more significant lecture, on the subject of John Harrison, to be given to the Society for Nautical Research the following year.

The Loch Ness Monster and Others

The book that Gould now completed to summarize his research, titled *The Loch Ness Monster and Others*, was released in June 1934, published by Geoffrey Bles. Bles had recently taken over Philip Allen's, but had kept on its Director, Gould's friend, J.G. Lockhart ('Brother Investigator').

Gould knew the book would be pulled apart and heavily criticized by the likes of Boulenger and Bather, wherever arguments were weak, and he covers the evidence very carefully and comprehensively. The Introduction describes the circumstances surrounding the research trip to the Loch and its immediate aftermath. The evidence of 47 sightings of the monster by 69 people is then given, simply stating the facts as told to the author. To avoid the constant references to 'the monster', 'the creature' etc, etc., Gould decided it should simply be referred to as 'X', a title he had coined some months before, in discussions with Keiller.

The next chapter considers the various explanations for what was seen. Here Gould considers Boulenger's claim, which is supported by the opinion of the distinguished palaeoanthropologist, Sir Arthur Keith (1888–1955), that the whole affair is the result of mass-hallucination.[304] In an article in the *Daily Mail*, Keith ends 'I have come to the conclusion that the existence or non-existence of the Loch Ness monster is not a problem for zoologists, but for psychologists'. To this remark, Gould comments: 'Possibly. And if the psychologists have any time to spare after solving the problem, they might usefully employ it in studying the mental processes by which this remarkable conclusion was reached', going on to expound on the substantial evidence and the very sane and respectable people responsible for providing it.

Another chapter analyses the evidence clinically in an attempt to discover common threads and anomalies in what was seen and described. This part of the book ends with the summary, itself beginning with a summary: 'On the evidence there can, I suggest, be little question that Loch Ness contains a specimen of the rarest and least-known of all living creatures'. The exact nature of this creature remained uncertain, but Gould, in spite of his press release of some months before, on reflection found the evidence for a sea serpent of the plesiosaurus type (as described in the earlier book) less convincing and concluded that one possibility could be 'a vastly enlarged, long-necked, marine form of the newt . . .'.

A second part to the book looks at 'Some Supplementary Cases' and really forms an addendum to his earlier book, considering as it does further sightings of Sea Serpents, reported since the publication of *The Case for the Sea Serpent*. The chapters end with a look at three carcasses of sea creatures, not in an attempt to suggest them as the remains of Sea Serpents, but to discuss how the evidence can be, deliberately or mistakenly, misinterpreted. Gould illustrated the book with no less than 46 figures and drew for the dust jacket a simple map of the Loch, Ordnance Survey style.

It is interesting that *The Loch Ness Monster and Others* was not dedicated by Gould to Alexander Keiller, who he nevertheless describes in his Preface as the book's 'onlie begetter', but Keiller's wife, Veronica. Alexander and Veronica were estranged and by this time living apart, Keiller having apparently treated his wife rather scandalously,[305] and it is interesting to speculate, though it would be only speculation, whether Gould's dedication was based only on friendship, and pity for her situation, or on a deeper affection for Veronica. Cecil stated that during the latter part of his life Rupert had 'a number of girlfriends', two others being dedicatees of later books, though Gould stated (in 1935) that there had only ever been one woman in his life.[306]

Reviews

The book was announced in *The Times* on 26 June 1934, price 10/6, a non-committal editorial review appearing on 29th. A full and complimentary one, covering one and a half columns, appeared in the *Times Literary Supplement* on 12 July, as well as reviews in most of the other national dailies. The publishers dedicated an advert to the new book in *The Times* on 6 July with the, perhaps germane, heading 'Do we believe?' and quoting Harold Nicholson in the *Telegraph's* review 'In spite of overwhelming human and photographic evidence, there are still many people in this country who attribute the Loch Ness Monster to mass hallucination. I beg such people to read Commander Gould's book'.

'Brother Shark'

However, one who numbered among 'such people' was E.G. Boulenger, who had also been asked to review the book, and this time the gloves were off.[307] Dismissing the complete book, and Gould's integrity with it, Boulenger stated that he 'repudiated the whole business as a stunt, foisted on a credulous public and excused by a certain element of low comedy'. Not surprisingly, Gould was again very offended at this slur on his character and was thereafter never on good terms with Brother Shark.

Boulenger would give two talks to the Sette of Odd Volumes in the coming year, the first on 'The Zoo Aquarium' on 30 April 1935. The menu for that night's dinner remarked jovially: 'Brother Hydrographer will no doubt have something to say about the L*** N*** M******'. But he would not. The ill feeling was too great and this was one of the only dinners Gould didn't attend at this period. On 25 February 1936 Brother Shark was again speaking, on 'Family life and courtship in the Zoo', and again the regular attendee Gould was not present.

It must be said, that while today's specialists in this field much admire Gould's seminal work on the subject, they do recognize the limitations of his knowledge and point to one or two significant failings in the book. Dr Mike Dash, erstwhile publisher of *Fortean Times*, singles out one of Gould's major weaknesses: that he had no zoological experience or qualifications. This was a weakness Gould himself was all too aware of, and he was very ready to admit where he felt he had made mistakes. In a typically frank annotation in his own copy of the book, Gould notes on page 211, footnote 2, on his remark about the swim bladder of the basking shark, 'This is a "howler" for which my scanty knowledge of marine (or any other kind of) zoology is responsible. Dr Calnan, in conversation with me some time after this book appeared, informed me that the basking-shark hasn't got a swimming bladder! RTG 2-II-42'.

Of another significant case, the Spicer's sighting of something crossing the road by the Loch on 22 July 1933, (p. 43), Gould annotates: 'Were I rewriting the book, I should have omitted this case. I think the Spicers saw a huddle of deer crossing the road. RTG'. This author confesses to being rather perplexed at this remark. In the book Gould takes some trouble (three and a half pages) to discuss what, on the face of it, seems a highly implausible account, saying that he then 'became, and remain, convinced that it was entirely bone fide', only to condemn it after all.

Gould dismisses the question of a lack of food supply for any large animal almost as an irrelevance, but this question really needed answering and, in commenting on the book, Dr Dash also points out that while three photographs of X are produced as evidence, there is no discussion of them, save only remarks such as (of Hugh Gray's photograph) that it is 'both interesting and undoubtedly genuine', with no further comment.

And Gould's tentative conclusion to the whole phenomenon, that X (which he evidently believed was of the same species as that he described as the Sea Serpent) may be some form of long necked newt, would also seem highly improbable in the light of modern zoological knowledge. Newts are incapable of the vertical undulations characteristic of many of the sightings of X, and amphibian species such as the newt cannot survive in salt water, meaning that if X is of the newt type, it cannot be the same as the sea serpent.

Nevertheless, some believed the 'newt theory' at the time, and it seems Gould's explanation was the inspiration for the first movie film made about the monster. In his comprehensive study *In Search of Lake Monsters*, Peter Costello reminds us[308] that in 1934, Wyndham Productions made *The Secret of the Loch*, written by Billie Breston and Charles Bennet. It starred the popular Seymour Hicks, with the monster in the form of a giant newt. Curiously enough the film happened to be the first 'talking picture' made in Scotland, though Costello considers it 'a pretty bad film', and Halliwell's Film Guide dismisses it as 'Mildly amusing exploitation item following the 1934 rebirth of interest in the old legend'.[309] Gould was a keen aficionado of film, going at least once a week to the local cinema and sometimes one in town too. He does not appear to refer to *The Secret of the Loch* in surviving correspondence, but one can safely assume he would have seen it and would perhaps have been amused and flattered.

As noted in Chapter 12, the year following publication, *The Loch Ness Monster and Others* was incorporated into a new edition in German, including much of the material from *The Case for the Sea Serpent* and some additional data supplied by Gould's joint author, Georg-Gunther Freiherrn von Forstner, himself a sea serpent witness.

A few years later, Gould had two chances, in the most public manner possible, to provide the evidence again and answer the doubters, when he gave national broadcasts on the subject, firstly on BBC radio, and then on television, as described in Chapter 20.

Today, Nessie, as the monster is familiarly named, is a well-known phenomenon and *The Loch Ness Monster and Others*, as the first book on the subject, is probably the most important source for the dozens produced since, the more notable amongst these being listed in the sources given in Appendix 2. In Constance Whyte's *More than a Legend* (1957) the author is so respectful of Gould's work she includes a brief biographical piece on him in an Appendix, notes from which were included in another primary work on the subject, Bernard Heuvelman's *In the Wake of Sea Serpents* (1968).

Another important work on the subject, and one which discusses Gould's contribution in some depth, is the 1983 book *The Loch Ness Mystery Solved*, by Ronald Binns, with R.J. Bell.

Binns includes a few, rather curious personal notes on Gould, describing him as 'a six foot tall seventeen stone Falstaffian eccentric . . .' and claiming that he 'died in obscurity in 1948, as forgotten as his cherished monster'. However, Binns goes on to say that in the following decade, thanks to Gould's books, there was a great resurgence of interest in the Loch Ness Monster, and concludes that: 'without Rupert Gould the monster would, one suspects, have died a quiet death in the summer of 1934 and never been heard of again'. But he is critical of Gould's naivety and failure to recognize his own shortcomings, and his failure to face up to the evidence which ultimately pointed to 'normal' explanations for the sightings of the alleged 'monster' phenomenon.

Whether in fact that phenomenon actually has physical, living form, or is, as E.G. Boulenger might have said, the result of a considerable amount of wishful thinking on the part of devoted followers of things Fortean (and of commercial interests), is still just as much a matter of opinion as it ever was. One thing is certain, Binns is correct in suggesting that had it not been for Gould's interest, his hard, disciplined, intelligent research in December 1933, and the resulting publication of this book, the whole subject may well have returned to obscurity, with any further sightings going unreported for fear of ridicule. As it is, interest in the Loch Ness Monster and Sea Serpents has never diminished since, and whether one is a believer or not, the phenomenon has provided us with much food for thought and scope for debate.

The Harrison Timekeepers and the NMM 1934–1935

Harrison's timekeepers H2 and H3 were on show together at the Science Museum from June 1932 and, naturally enough, the curator there Frank A.B. Ward, had been keeping in touch with Gould over H1's restoration, with an eye to a loan once it was finished. Gould had visited the Museum in December 1933 to restart the two timekeepers after they had been allowed to run down, and Ward would certainly have heard all about H1's completion earlier that year.[310] So, prompted by Ward, the Science Museum's Director, E.E.B. Mackintosh, duly applied, on 18 January 1934, to the new Astronomer Royal, Harold Spencer Jones, at the Observatory for permission to have H1 join its brothers on display there, pointing out that Dyson had tentatively promised this back in 1930.[311]

To his disappointment though, Spencer Jones, answered[312] that he wished for H1 and H4 to remain at the Observatory. Spencer Jones explained: 'Harrison's timekeepers Nos 1 & 4 are the two which were actually tried at sea and which have therefore particular interest from the point of view of navigation. It seems therefore appropriate that they should remain here, in view of the close connection of the Observatory with marine chronometers'. This was reluctantly accepted by Mackintosh, who then requested photographs instead.

There was of course a hidden agenda in Spencer Jones's refusal; one which would soon become very clear to the staff at the Science Museum, but which they probably already suspected: there were plans being discussed by the Society for Nautical Research to establish a new National Maritime Museum in Greenwich.

The Society for Nautical Research

The SNR (Society for Nautical Research) had been founded in 1910 with a view to supporting maritime historians and helping to improve the various naval museums in Britain. Its journal, whose first editor was Gould's friend Leonard Carr-Laughton, was dubbed *The Mariner's Mirror*, for which, it will be recalled, Gould redrew the cover in 1924 (Gould was a life-member of the Society). The title of the journal was an anglicised version of the first 'modern' sea atlas, Waghenhaer's *Speculum Nauticum*, (1584).

Although it had not been an original aim of the SNR's to found a maritime museum for the nation, the idea was proposed in principle, by the Navy League as early as 1913. In 1920, Geoffrey Callender (1875–1946), a naval historian and one of the teachers at the Royal Naval College at Osborne, was appointed Honorary Secretary and Treasurer of the SNR, and from the outset it had been Callender's ambition to see a national naval museum founded in Greenwich.

The National Maritime Museum

During the late 1920s and early 1930s some of the elements which were to make the proposed museum a reality slowly began to fall into place. In the early 1920s Arthur Smallwood, the new Director of The Greenwich Hospital (the Naval equivalent of the Army's 'Chelsea Hospital', and founded as a charitable institution to care for retired seamen and their families) had decided that the Hospital's school for boys in Greenwich should relocate to new buildings at Holbrook in Suffolk, where an estate had recently been bequeathed to them.

The collections of marine artefacts and paintings at the Greenwich Hospital also happened to need more space, as the Hospital buildings were now occupied by the Royal Naval College. Amongst other champions, a passionate and wealthy benefactor was now supporting the idea of a national naval museum. The Scottish shipping millionaire, James Caird (1864–1954) was already a generous supporter of naval charities and causes.

After considerable political and administrative debate, unnecessary to relate here, the SNR was able to claim, in 1927, that the concept of a National Maritime Museum—the name chosen by Rudyard Kipling—was now definitely on the cards. Trustees had been nominated by the Admiralty, the Honorary Secretary being the Admiralty Librarian, W.G. Perrin. James Caird was nominated a Trustee the following year and between then and 1934, Caird, guided by Callender, spent over £300,000—a truly gigantic amount at the time—on acquisitions for the new museum.

In 1933 it was agreed that Professor Callender would be the first Director, and on 26 March 1934 the Prime Minister announced the introduction of a Bill for the Establishment of the National Maritime Museum at Greenwich, the 'National Maritime Museum Act' receiving Royal Assent on 25 July.

Meanwhile, during the early 1930s, Callender, temporarily based at the Royal Naval College, had been busy seeking other appropriate acquisitions and loans for the new displays. Relations between the fledgling National and its older brothers was predictably patchy on the sensitive matter of acquisitions, and there was considerable tension at various times, particularly between the NMM and the Imperial War Museum and the Science Museum over collecting policies.

So, when the Director of the Science Museum, E.B. Mackintosh, had his request for the loan of the newly restored H1 refused by Spencer Jones at Greenwich in January 1934, he must have feared the worst. In fact it seems the Astronomer Royal and the Admiralty Board had secretly agreed with Callender, at least a year or two before, that eventually the Harrison timekeepers would go to the NMM. In 1932 Gould had been asked to write a short history of the timekeepers for Callender's information when the decision was being made as to their final destination.[313] But there was never any real doubt, and by June 1934 Callender was planning where in the old school buildings would be the new gallery to display them.

Writing to Gould on 4 June 1934, he told him 'Speaking in confidence, I have got as far as selecting Harrison's Room in the Museum'. The main purpose of his letter to Gould however was to ask him to give a lecture to the SNR in early 1935 on the subject of The Marine Chronometer, or Harrison's Timekeepers. He pointed out that, with the opening of the NMM imminent 'there could be no riper moment for arousing interest in the subject'.[314]

As noted in the previous chapter, 12 months before, Gould had given a lecture to the London dining club 'The Anchorites' on 'Marine Monsters'. One of the members, Admiral L.R. Oliphant, who was also a Council member of the SNR, had enjoyed it so much he now strongly recommended Gould as a speaker.[315]

The SNR's lecture was to be given in the Drapers' Hall, 'with all the attendant hospitality . . .'. Drapers' Hall was the ceremonial hall of the Worshipful Company of Drapers, one of London's great livery companies, and an exceptionally imposing venue. Of course, it was absolutely the perfect opportunity for Gould to announce the completion of the monumental task he had engaged in for fifteen years, and he gladly accepted. There was, it seems, another slightly devious ulterior motive in giving the lecture on the Harrison timekeepers at this time, though whether this was planned from the outset, or just a happy result of the way things turned out, is only conjecture.

While a loan to the Science Museum of H1 and H4 was now definitely not on the cards, H2 and H3 were already there and there was no suggestion, at this time, that they should not remain there on 'permanent loan'. Callender knew however that sooner or later they would have to bite the diplomatic bullet and gently break the news to Mackintosh that these too would be taken from them. This would not be easy, but might be made at least slightly easier if H2 and H3 were, at the time of breaking the news, not actually at the Science Museum, on the basis that 'possession is nine parts of the law'.

So the decision was taken to have all the restored timekeepers displayed at Gould's lecture at Drapers Hall, now programmed for 21 February 1935. The Secretary of the Admiralty duly wrote to Mackintosh at the Science Museum on 15 January requesting that H2 and H3 be released for loan to the SNR at Drapers' Hall, adding 'that on conclusion of the lecture they should be removed to the Admiralty for examination and any adjustment necessary'.

Of course, Mackintosh smelled a rat, and replied on 18 January that he 'shall be happy to accede to your request, and will afford every assistance possible for the removal and return of these historical objects'. Then, getting straight to the point, he continued 'In order to minimize the risk of damage to these valuable time-pieces, may I suggest that at the conclusion of the lecture they should be returned here, where the necessary examination and adjustment could be undertaken in their *fixed positions*, without risk of subsequent damage in transport'. In his reply, on the 26th, the Hydrographer simply ignored this direct point, stating that the SNR would be making arrangements with the Science Museum directly for the removal of H2 and H3. Callender then wrote asking if Gould could collect them on Monday 18 February, to allow time to set up for the Thursday lecture, and offering tickets to any Science Museum staff who may want them.

On 14 February, just seven days before the big SNR lecture, Gould gave a talk, 'Ways of Measuring Time—Modern Clocks', the third of a series presented to Bedford College for Women. The text was later published in the *Horological Journal*. Announced in *The Times* on 31 January, this was a prestigious lecture and no small duty just a few days before his important presentation to the SNR.

The SNR Lecture

'John Harrison and his Timekeepers' was the title of the lecture Gould gave that afternoon at 2.30 p.m. on 21 February 1935. It was a lecture which was, of the many he gave in his career, unquestionably the most important and significant. The Admiralty and the SNR probably had something of a hidden agenda that day relating to the timekeepers and the new National Maritime Museum. But Gould also had a very definite agenda of his own; this was the best possible opportunity to summarize what these timekeepers had cost him and stake his claim to fame.

The Times had published notice of the lecture back at the beginning of December, a leaflet had been printed by the SNR for circulation advertising tickets,[316] and the usual notice appeared in *The Times* on the day in 'Court and Social', as well as there being a notice of this Nautical Research Lecture under 'The Services'.

Presided over by Admiral Sir George P.W. Hope, Chairman of the new National Maritime Museum Trustees, the lecture was attended by nearly 200 guests, many of them important figures in the naval and horological worlds. The lecture included 43 lantern slides and was also illustrated by having the timekeepers themselves present, including K1, and H5, borrowed from the Clockmakers Company. All were running, set up on the table in front of Gould in the hall.[317] It must have been an incredibly impressive sight and a considerable task in itself to arrange, but one which must have thrilled Callender. Seeing the complete series in a row like this could have left no doubt in anyone's mind that the collection as a whole (excluding H5 naturally) should be permanently displayed in one place.

For Gould, the lecture was probably his best, perhaps his only, opportunity of fully describing to such a large and influential audience, the passion he felt for his subject. The subject of Longitude, the work of early pioneers in attempts to solve the Longitude problem and then, of course, the extraordinary story of John Harrison, and his trials and tribulations, were all described eloquently and accessibly. Gould felt very keenly that John Harrison's work was wholly unappreciated and he spoke feelingly of that great man's achievement. The lecture ends with one of his typical 'rousing appeals', calling for recognition of Harrison's greatness. It is ironic though, that in his lecture Gould continued to damn Harrison with faint praise, and gave the laurels to Pierre Le Roy as the 'father of the chronometer'.

But equally important for Gould was to reveal the agonies and ecstasies he himself had experienced in the restoration work. Two-thirds of the lecture focused on the Longitude, Harrison and the later chronometer. The remaining third looked at the work Gould himself had done to restore the timekeepers. In short, it was a public justification of the sacrifices he had made during the previous 15 years, and an opportunity for him to gain the proper recognition he craved. In discussing the fate of the Harrisons he remarked: 'I must apologise if, in so doing, I seem rather egotistical. It so happens that through force of circumstances I have had more to do with those machines than anyone else except their maker'.

There were probably a number of other aims one could identify too. Although Gould is careful not to apportion blame to staff at the Observatory, he gives painfully detailed descriptions of the state the timekeepers had got into. Remarks like 'No.1 . . . looked as though it had gone down with the Royal George and had been on the bottom ever since', and (of the same timekeeper) 'as Sir Boyle Roche might have put it, there were enough missing parts to fill a bucket', left no doubt as to how neglected they had been.

Needless to say, this was not a point picked up by the Astronomer Royal, Sir Harold Spencer Jones. In publicly congratulating Gould on his lecture at the end of the afternoon, Spencer Jones chose instead to jovially bring up the question of Maskelyne's alleged 'conflicts of interest' in his dealings with Harrison. The account of the meeting in the *Mariners' Mirror* reported that Spencer Jones 'professed himself anxious to do penance for the misdemeanour of his predecessor'.

Gould was also anxious not to claim all the credit for these restorations for himself, but makes it abundantly clear, in the course of describing the work, that he had the help of several others, naming Buck, Hopgood, Ilbert and Mercer (all present in the audience) particularly singling out mention of Mercer as he had generously refused to accept payment for the work done at St Albans.

Such was the impact the lecture had, that members of the audience were still referring to it years later. It even inspired a letter to *The Times* a few days later from one of the audience, who noted 'All who had the privilege of listening to Lieutenant-Commander Rupert Gould's lecture . . . must have come away with a feeling of gratitude for justice being done to a great man'. The SNR annual report was equally ecstatic: 'The lecture, which gripped his hearers from the first moment, was listened to with rapt attention by a large audience . . .'

Keeping the momentum, and interest in Harrison going, Gould wrote a piece for the *Illustrated London News* in March, and gave a similar lecture to the BHI in October that year, published in the *Horological Journal* in December.

John Harrison and his Timekeepers

The SNR lecture was to have a very long lasting effect. The full text was published as an article in the *Mariners' Mirror*, with a list of lantern slides and a record of some of the more notable attendees headed 'Among those present were . . .'. This article, simply titled 'John Harrison and his Timekeepers' was immediately produced as an off-print. That same year the noted horological bookseller Malcolm Gardner announced in his catalogue: 'Just out: *J. Harrison & his Timekeepers*—25 copies only, signed and numbered by the author, 7/6'.

The SNR then (probably just after the war) reprinted it in pink covers, with some very slight alterations to the text (presumably made by Gould) and without the list of attendees or lantern slides. It was then reprinted by the NMM in 1958 and was virtually never out of print until the mid 1990s. Seventy years later, it is still highly regarded by antiquarian horologists and will undoubtedly remain one of the major landmarks in the horological literature.

Callender was happy too. The Harrison timekeepers were now all out of the Science Museum (they were now moved to a temporary exhibition in the Admiralty Library) and they could discuss their future with advantage. At the Science Museum, FAB Ward (who had attended the lecture, but without Mackintosh) noted on the Harrison loan file six days later that he had written privately to Gould to ask him if he knew: 'a) Where the Harrisons now are, and b) When they are likely to be returned here'. A file note the following day records that Gould telephoned him 'to say that the Harrison chronometers are now at the Admiralty and that no immediate decision as to their future location is to be expected'.

On 12 March 1935, the Hydrographer, by this time J.A. Edgell, decided the time had come to put the Science Museum out of their misery. He wrote to Callender 'I think the time has come for you to unmask your batteries and put in a request to the Admiralty for the loan of the Harrison Chronometers . . .'. This would be the formal rubber stamping exercise, after which there could be no prevarication or pretence; the machines were bound to go to Greenwich.

Hearing nothing, and worried at the silence, Mackintosh wrote to the Admiralty on 23 March asking when they would be returned to the Science Museum 'They occupy a very important place in the Time Measurement Collection of the Museum, and since they were removed from the special cases which were constructed for their exhibition here I have received numerous enquiries as to the date of their return'. A note of 3 April in the museum's files, by a clearly irate Mackintosh, then records that he had been telephoned by Sir Vincent Baddeley (of the Admiralty Board) informing him 'that the Admiralty had been asked by the Trustees of the Maritime Museum at Greenwich to lend them all the Harrison chronometers for exhibition & that the Admiralty (i.e. V.B) were inclined to agree. I gave him my views on this question. I suggested the division as hitherto. He then said Greenwich wouldn't be able to exhibit them for sometime, & offered us the whole lot temporarily. I accepted & said we'd exhibit the whole lot (it may possibly help us); so please be prepared'.

With the wisdom of Solomon, the Admiralty Board managed to offer the Science Museum some consolation, on the strict understanding that they would all go to the NMM the next year. This they reluctantly accepted, and Gould set up all the large timekeepers, H1–H3, in South Kensington on 9 May 1935. H4 and K1 did not go to the Science Museum straight away, as Gould decided to have the two watches sent to Ashtead first to overhaul them before setting up at the Museum.

As an aside, it's worth noting here the terms on which the timekeepers were loaned to the Science Museum. It was stated on the form, completed by the lender (Edgell as Hydrographer) in May 1935, that there were no objections to the Science Museum photographing the Harrison timekeepers, no objections to those photographs being sold, and no objections to those photographs being published.

However, to the question: 'Have you any objection to the permission being given to private persons to copy, photograph or measure the objects', the Hydrographer's definite reply was: 'Yes'.[318] This statement marks the beginning of a policy which has pertained ever since, that the Hydrographer, as owner of the Harrison timekeepers, does not wish to allow measurements or copies of the timekeepers to be made. The several copies of H1 made over the years, and those of the other timekeepers which are also currently under construction (2005), have thus all been made without the approval of the Hydrographer and without comprehensive measurements being made.

H4's second overhaul and the first for K1

The work on H4 took from 30 April until 12 June 1935 and Gould made the usual meticulous notes and drawings, filling quarter of a large notebook. At the end of the notes is an 'Analysis of the Work' which states, rather frankly: 'I have made a much better job of it than I did in 1921. *How* I got it to go then I don't know—it must have largely been fool-luck. I have gone about it much more thoroughly now, and I think I can say that no manipulative problem has offered itself for which I could not provide a solution . . .'. A later note states 'if the movement is to be on view dial down at the NMM—as it is at present—I shall have the plates gilt', a very controversial decision to make without consultation. Fortunately no such action was taken.

The work on K1 followed straight after, taking from the 16 June to the 18 July 1935, though the middle part of the overhaul was interrupted by tennis, when Gould was umpiring at Roehampton and then Wimbledon. The same meticulous notes record the work on K1 in a continuation of H4's notebook, the work progressing smoothly until 7 July. On that day Gould records: 'Starting with the hope that today might see K1 assembled, I had to discontinue work soon after noon with a broken pivot (5th wheel). I cannot imagine how it can have happened'. The wheel was taken to J.F. Cole in Clerkenwell, who promised to have it re-pivoted in a couple of days. Once the wheel was back Gould continued, but: 'during operations my glass fell from my eye, knocked the 5th wheel out of the plate & bent the new pivot. Luckily, I managed to straighten it'. Several days were then spent adjusting the action of the remontoir before finally getting it right, the watch then settling down to reliable running by the 18th, in time for its delivery to the Science Museum on 22 July 1935.

The usual notices in the press announced the arrival of all the timekeepers at their temporary home in South Kensington, and a new set of photographs was taken of the whole collection, copies being presented to the Admiralty and to Gould. For the time being the Science Museum staff were content.

18

Professor Stewart, the BBC
and Tennis 1936

It often happens with the publication of research, that as soon as a book is available, letters from interested readers provide more information. Gould found this to be the case with just about all his books, but a contact which came from the publication of *The Loch Ness Monster and Others*, in 1934, was of particular interest to him.

A year or so after publication, Gould received a letter from A.W. Stewart, author and Professor of Chemistry at Queen's University, Belfast. On 4 September 1935, while holidaying in Scotland with his wife and daughter, and in common with a considerable number of strangers (some nearby, others out of sight), Professor Stewart witnessed the appearance of something unusual in the Loch. On his return home, he sat straight down and wrote to the author of *The Loch Ness Monster and Others*, to describe what he had seen. Stewart's last book, *Alias J.J. Connington*, a partly autobiographical work published after his death, contains the description of the sighting.

Several hundred yards out in the Loch they saw, quite distinctly, a great splashing commotion on the surface. Stewart wrote 'My wife exclaimed, rather contemptuously: "It's only a school of porpoises"; which described the appearance exactly. Speaking for myself, I saw four (or five) "coils" or "humps" in motion among the foam. These remained in sight for perhaps a minute, and then vanished so far as I was concerned. The foam patch remained however. It was not a "white cap", for there was very little wind at the moment, and the foam remained in one spot and did not drift as a "white cap" does'. Stewart wished to be very clear on a number of points: The air was clear and the light was good; they had not been thinking of sightings of the monster and certainly did not have 'expectant attention' as Gould had termed it.

What they saw was something large and active, emphatically not a tree trunk or suchlike. All three of them saw it and were in complete

52. Professor A. W. Stewart (1880–1947). Stewart and Gould were regular correspondents during the mid 1930s. Gould used the letters to put on record many of his feelings and thoughts at the time (Courtesy of Miss I. Stewart).

agreement afterwards, and they did not, at any time, discuss it with anyone else who had witnessed the occurrence. As it turned out, the sighting had been made, entirely independently, by others from another part of the Loch, and Stewart ruled out having been subject to any kind of 'mass hallucination'. As for what it was he had seen, Stewart had no answer, preferring to echo the words of his 'sagacious friend' Arthur Machen who, when asked for his opinion on a difficult question, replied 'Indeed sir, I don't venture to have any opinion'.

In his field of Chemistry, Alfred Walter Stewart (1880–1947) was in fact a very distinguished academic, having lectured at Belfast and Glasgow (before returning to Belfast to take the chair in Chemistry). He wrote many advanced textbooks and was particularly noted for his clear and accessible lecturing style. Gould could not have wished for a more sober and scientifically scrupulous witness for a sighting at Loch Ness.

But in addition to his scientific credentials, Stewart had interesting literary talents too which immediately appealed to Gould. In his spare time Professor Stewart wrote 'whodunnit'-style detective stories, often using his professional knowledge of chemistry and physics to inform the plot of his mysteries. He was, in his day, very well known by his pseudonym: J.J.Connington.

So Gould and Stewart soon discovered they had much in common. As well as being exceedingly well read, Stewart was a great lover of music and, like Gould, had the best obtainable gramophone and a large collection of records. Not surprisingly, they became good 'pen friends' and for a year or so, from late 1935, they began a regular and wide ranging correspondence. Although they never met (or perhaps because of it) Gould felt able to discuss almost any subject that arose, including the highly sensitive judicial separation, his close friendships, likes and dislikes, and all manner of personal foibles that one could not hope to have found elsewhere, so the letters have proved a valuable source of intimate biographical detail.

It seems Gould himself recognized this and, reminiscent of the personal asides in his Harrison restoration notebooks, it's almost as if he intended the correspondence to be used in this way. At one point he asks Stewart (in jest) whether they might ensure the volumes are left to the British Library. Soon after they began, he started binding up copies of his letters with Stewarts replies, presumably with the intention that they should be preserved and that others may refer to them one day. Early on, as the correspondence was proving so interesting, Gould suggested to Stewart that they might share letters with his friend Alexander Keiller, saying 'He is a man of my own age, and my own type of mind . . .' By December 1936 this correspondence ran to seven slim volumes, but unfortunately volumes one and three are lost[319] and the first few letters, which would be most interesting, have not been seen.

Abductions

In volume one, in discussing the kind of fiction they enjoy, Gould apparently mentioned deriving sexual gratification from accounts of abductions, and Stewart then refers to a work entitled *The Man-Stealers*. In reply to Stewart, Gould notes:

Re abductions. I may have misled you. The only abductions which interest me are those of young women. I take it that mentally, I have a streak of the sadist in me. Not otherwise—I have never been cruel, physically, to any woman in my life (by the way, there has only been one) and I believe myself incapable of so acting. But there is no escaping the fact that, so long as I can remember, I have always been interested, and have taken pleasure in, accounts of girls being kidnapped, gagged, blindfolded, bound hand and foot, handcuffed, fettered and otherwise reduced to a condition of complete helplessness. Also in illustrations and / or photographs of the same

Following this clarification, Stewart is able to recommend another title, Fowler Wright's *Island of Captain Sparrow* which he said he hoped 'fits your needs'. Embarrassment never seems to have been an emotion experienced by Gould; he was happy to discuss his own predilections as a matter of fact, something which was neither right nor wrong but simply there, to be analysed along with everything else—a tendency which goes some way to explaining another extraordinary series of letters he kept among his possessions.

At about this time, or soon after, Gould engaged in correspondence, copies of which were found in an unmarked brown envelope by his son Cecil among Gould's papers, after his death. The contents were copies of type-written letters to Alexander Keiller, along with Keiller's replies, concerning the merits of group sex activities the men engaged in, including one other (identity unknown), with a young woman. The letters, perhaps themselves written partly for gratification, described aspects of the proceedings, which were apparently conducted with ritualistic precision.

The men would meet with the woman at the Café Royal in the afternoon, and would then proceed to an address in South London where the young woman, who had been sent in advance to have a bath (an important part of the ritual apparently), would be waiting for them. They would then take it in turns to have intercourse while the others watched. Cecil was astonished that his father would have kept such letters, describing them as 'absolute dynamite' at the time. The survival of this strange little file of letters is interesting as much for what it says about Gould's view of himself, as it does about his activities at the time. Not knowing what on earth to do with the letters, Cecil and Jocelyne had a quick giggle as they rapidly looked through them and he then

53. *Fetters and Chains.* Pen and ink drawing by Gould, *c.*1910. Although never knowingly cruel to women, Gould had always taken pleasure in accounts and images of girls being 'reduced to a condition of complete helplessness'. (© Sarah Stacey and Simon Stacey, 2005)

promptly tied them to a weight, took them to Westminster Bridge and threw them into the Thames.[320]

During interviews for this biography, three people, two of whom had met Gould, claimed that he was in fact homosexual. On pursuing this question, it seems that the first had simply heard this from a secondary source, and the other two (brothers) stated that their father, who knew Gould professionally, had told them so, but were unable to say on what evidence he had deduced this. Whether, in the light of the 'group activities' mentioned, one can infer bi-sexuality in all of the men who took part in those rituals (there is some evidence to support that supposition in Keiller's case) is highly debatable. Gould certainly had an interest and sympathy with the subject as in later life he was planning to co-author a book on Homosexuality (see the last Chapter). But it must be said that there is no other evidence that this author can find, which supports the suggestion that Rupert Gould was entirely homosexual, and there is considerable proof of his sexual and emotional interest in the opposite sex. It is possible that rumours of his wife's lesbian lover, mixed with his now having returned to live with his mother, might have led to confusion on this score, but after so many years it is impossible to come to a definite conclusion on the matter.

BBC Broadcasting

With five full scale books to his name, and further writing, in the form of continuing articles and letters to the press, Lt Cdr R.T. Gould was slowly but surely creating a public name for himself. But it was in radio broadcasting that Rupert Gould would first become a truly household name—a 'media star' in modern parlance—during the 1930s.

This aspect of his career in fact began back in June 1927, right in the middle of the action for judicial separation. On 24 June 1927 the BBC Producer Lancelot de G. Sieveking (1896–1972) wrote to the broadcaster A.J. Alan (Pseudonym for L.H. Lambert, 1883–1941) that he was 'glad to have secured your friend Rupert Gould to speak on Big Ben'. The short talk was to be called 'Why Big Ben' and focused on the origin of the great bell's title (it was named after Westminster's Chief Commissioner at the time, Sir Benjamin Hall).

The live talk (all broadcasts were live until the introduction of magnetic recording in the 1930s) went out from Savoy Hill in July, Gould reading from a script he had prepared and which had been edited. Gould evidently enjoyed the experience and offered Sieveking the synopsis for his new book, *Oddities*, to be published the following year. He wondered if Sieveking might find any of the subjects useful for further radio talks by the author. Sieveking could find 'no opening at present . . .', but kept the synopsis on file for a further 3½ years.

It was in January 1934, just as work had begun on writing *The Loch Ness Monster and Others*, that Gould was approached and invited to give

occasional talks on the BBC's *Children's Hour*. The programme was nominally an hour but more often than not actually lasted 45 min, usually
broadcast between 5 p.m. and 5.45 p.m. By 1934, *Children's Hour* was five
years old (the BBC itself had only been in existence for about thirteen
years) and was undergoing a reorganization.

Although it is commonly believed that many Britons, having
survived 'the war to end all wars', discovered new found freedoms and a
determination to 'live life to the full' in the 1920s, in truth this mostly
applied to a small, relatively well-off sector of society (including, briefly,
the Goulds for example). For many, it was a period of terrible strife,
followed by the effects of severe economic depression. By the early
1930s, the Government was striving to consolidate the family unit and
one strategy was to use the BBC, by bringing the household closer
together round the wireless set. Part of this determination to unify
younger families was to make improvements to the style and content of
Children's Hour.

Uncle Mac

A new *Children's Hour* organizer (London area), Derek McCulloch,
famously known as 'Uncle Mac' to the young 'listeners in', had recently
been appointed as part of this new strategy. McCulloch had ambitious
plans for making the programme more professional and informative,
whilst keeping the all-important personal touch. Gone were the daily
birthday readings and in was a new signature tune in the form of *The
Blind Seller's Song*, played on an antique musical box.

Also in were a number of well-known light entertainers and the talks
area was expanded considerably. It was in this section that Gould was
employed, along with a small number of other authoritative speakers,
including 'The Zoo Man' (David Seth-Smith), 'The Farmer' (John
Morgan), and Stephen King-Hall on world affairs.[321]

The Stargazer

Gould's title was to be 'The Stargazer', on the assumption that the majority
of his talks would be on astronomical subjects. But on his pointing out
that more than a few talks on this subject would soon become boring, he
was told that, within reason, he could range over any subjects he chose,
'an outlet of which I took full advantage', Gould recalled.

Remembering that his earlier broadcast had been scripted, he asked when they needed to see these for his talks, only to be told that on *Children's Hour* they wanted an informal talk and that he should only bring along a few notes and speak from them. Gould recalled: 'And . . . that's exactly what I did. I used to walk into Broadcasting House with a postcard on which I'd scribbled a few notes. From these I'd extemporize, at the microphone, a fifteen-minute talk. I've never 'dried up' while talking; and I've always felt very grateful to the B.B.C. for reposing enough confidence in me to let me hold forth to, perhaps, a million people without any previous indication of what I was going to say'.[322]

In fact not all the talks were extemporized in this way. All those given during the war were carefully scripted and vetted, and some in the late 1930s were read from script too. The very first talk was given on 14 March 1934 and was on 'The Indian Rope Trick'. Gould had been collecting notes on magic tricks, and children's toys of all kinds, since childhood and had a large collection of 'penny toys'.

In company with many across the nation, Muriel and Miss Gurney, Dodo, and Muriel's mother ('Grandma') had all bought their first radio set in the early 1930s and were able to listen in to the new media star, R.T. Gould. Remarkably, over the next 10 years Gould would give over 100 of these talks, and such was his fund of knowledge and interesting topics that it was never necessary for him to repeat a subject, the talks ranging over an incredibly wide area of subject matter. He was always careful to ensure that he didn't overrun his time, and developed the uncanny knack of winding up his talk, in the most natural way, just as the fifteen minutes was up. On closing his talks he was asked to say 'good-bye', then pause for a couple of seconds and then say 'good-bye' again. When he asked why this was necessary, he was told 'to allow the children to say good-bye, of course'.

A whole generation of young listeners were inspired by the encyclopaedic knowledge and storytelling abilities of 'The Stargazer'. During preparation for this biography, a few people especially made contact to say they still remember talks by 'The Stargazer' with affection. Mr.G.P. Cole recalled the talks containing 'sound interesting facts', and were 'thrilling to listen to'. Mr J. Black 'used to dash home from school to listen to him, . . . five miles cycle ride, Mum furious'. Mr Alan Partridge wrote to say that the 'talks by "Star Gazer" were among my favourite items on Children's Hour. It was probably these talks which aroused my interest in astronomy, leading to my becoming for a short time a Scientific Officer at Edinburgh Royal Observatory . . .'

Gould's speaking voice was quite distinctive too. As well as speaking rather quickly (in the first broadcasts he was constantly having to remember to pace his delivery) he spoke with what might be termed a kind of lisp, or more correctly perhaps, a 'retrusive R', where the 'R' sounds similar to a 'W'. Mr Wallace Grevatt, author of an excellent book on the *Children's Hour*, remembered 'a compelling voice who introduced us listeners to an entirely new world . . . most certainly the Patrick Moore of his day'.

In fact, Patrick Moore himself met Gould. Moore recalled: 'I came across Commander Gould aeons ago, when I was young and he was the BBC *Children's Hour's* 'Star Gazer'. I had joined the British Astronomical Association at the age of eleven (a record then: fifty years later I was President!) and my first meeting with the Commander was at a BAA meeting. He towered over me (even though I was tallish even then) and went out of his way to make himself very pleasant. I was immensely impressed with him. I met him on several occasions after that, and I remember talking about his Macaulay-type memory! I doubt if he ever forgot anything . . . we also talked about the Harrison chronometers, and his collection of old typewriters (. . . all my books are done on the 1908 Woodstock I have had since the age of eight). I do remember him saying that when he started repairing the Harrisons he was acutely conscious of his responsibility'.[323]

McCulloch observed, during an interview in 1934: 'The B.B.C. knows that the children of today are the adult listeners of the very near future . . . In this department we work on the principle that a child wants to know things and we endeavour to tell them in the most interesting way possible. We talk to our children as equals. We believe that children have intelligence and we do not insult it'. A laudable principle and, in the 1930s, one which worked admirably.[324]

'The Stargazer' in particular was a great success from the moment he stepped up to the microphone, being voted a place in 'request week' in December 1935, 'despite the grousing of one or two non-appreciators'.[325] Ironically though, in spite of his popularity among the younger 'listeners-in', he admitted to Professor Stewart that he was never at ease with children. Cecil's account of his father concurs: he was only ever affectionate in a very formal way, more like a kindly teacher than a father to his children.

Needless to say, the money he was paid for the BBC talks was of real importance to the perpetually impecunious Rupert. So short of cash was he that it became a standard request of his that he be handed the BBC's cheque, in payment of that day's talk, before he left Broadcasting House, and he would, on occasions be very irritable if for some reason this was forgotten.

Having quickly proved himself to be an accomplished broadcaster, Gould was invited to give other talks on the BBC at various times, occasionally suggesting topics himself. One such was a talk on John Harrison, proposed by Gould in March 1935, on the occasion of the 200th anniversary of H1's completion and immediately following Gould's notable lecture to the SNR. But on that occasion the offer fell on stony ground, the Producer ('GNP') not being able to see how such an 'out of the way subject' could be fitted in (in fact it was accepted soon afterwards).[326]

In September 1936, Malcolm Brereton, a Producer in the BBC's General Talks, got in touch via Geoffrey Bles (apparently not realising Gould already broadcast on *Children's Hour*) to ask Gould if he would do a 5-min talk on Sea Serpents, as part of a series called *They May be Right*. Over lunch at the BBC, Gould persuaded Brereton that he should not only do Sea Serpents, but could take on a whole range of occasional talks (outside of *Children's Hour*), and lent him copies of *Oddities*, *Enigmas* and the Harrison lecture from the *Mariners' Mirror*. Brereton was duly impressed and talks began on a range of subjects at various times, often as part of the series titled 'The World Goes by'.

Talks were usually arranged by post, occasionally by telegram if a commission was last-minute. The request gave the subject and length of talk, followed by Gould submitting a draft script, after which a contract would be sent for him to sign. The fee paid was usually one guinea per minute, a little less if the talk was longer than 10 or 15 min. A date would then be fixed for a reading of the edited script to be rehearsed in a studio booked for the purpose, sometimes then rehearsed a second time on another occasion in a studio, after which the broadcast would go out at the appointed hour.

One of the first of these was the twenty minute talk 'The Man who Discovered The Longitude 1736', now commemorating the 200th anniversary of H1's successful trial to Lisbon. The broadcast went out at 6.20 p.m. on Christmas Eve 1936. Although 'The Stargazer' had already spoken on Harrison to younger listeners a few weeks before, here was a chance to introduce the adult world to that remarkable story, and tell them of his work too.

By the mid 1930s Gould was becoming something of a public figure and with the regular (if not very remunerative) employment at the BBC, he was now just about managing to keep his head above water financially. And the BBC was the place to be if one was interested in networking, meeting the influential people of the time and making one's way. For example, Gould's introduction to Vita Sackville West, to whom (as Mrs Harold Nicholson) he presented a signed copy of *Enigmas*[327] was probably through the BBC; she too did many broadcasts during the 1930s.

Writing continues

Between broadcasts, Gould continued writing. A letter of his to the *Times Literary Supplement* in 1934 helped settle the question of the pronunciation of the surname of the celebrated diarist Samuel Pepys. It so happened Pepys's cousin had lived in Gould's very own village of Ashtead and he knew of a contemporary map of the area with lands shown belonging to the cousin, whose name, he informed readers, is uniformly shown, phonetically, as 'Peeps'. Towards the end of 1934 Gould was busy preparing his lectures for Bedford College and the big Harrison lecture for the SNR (due in February). But he was also busy writing the text for his next book.

Captain Cook

Although he had been obliged to pull out of his contract with Argonaut to produce the comprehensive work on Cook and his journals, the preparation he had started for the biography was now put to good use. Signing a new contract, this time with Duckworths, Gould produced a smaller life of Cook as part of their 'Great Lives' series, *Captain Cook* being published in May 1935.

In six chapters (144 pages), with six illustrations by Gould, the book succinctly relates the life of this great navigator, including the question of how he met his death. In an article in the *Mariner's Mirror* in 1928, Gould had published the latest thinking on what happened that fateful day, 14 February 1779, on Hawaii. The book was reviewed in the *T.L.S.* on 30 May 1935, the reviewer considering that 'Lt. Cdr. Gould, whose little book displays the admiration that a seaman feels for a master of his profession, has written a worthy sketch of a great man'.

Always difficult to find secondhand, the book was produced in a new edition by Duckworth's in 1978, including an Introduction by Gavin Kennedy, an admirer of Gould's who included some biographical details about the author. Kennedy began: 'Of the hundreds of biographies of Captain James Cook, three stand out above the rest: those by Andrew Kippis, J.C. Beaglehole and Rupert Gould'. Comparing Gould's little work with the contemporary account by Kippis (1788), and the magnificent biography by Beaglehole (1974), the result of 40 years' work and still considered by many to be the best, Kennedy notes 'The third outstanding biography of Cook is this modest volume, written by Rupert Gould and first published in 1935. It is shorter than Kippis and Beaglehole by several hundred pages; yet on any comparison it matches them in excellence. Apart from his conciseness and meticulous approach, Gould was uniquely fitted to write the book by the circumstances of his life . . .'. Kennedy added a couple of pages of biographical notes on Gould.

At the head of Chapter Two in his personal copy of the original book,[328] Gould has annotated: 'To my mind, this is the best chapter in the book: and the best concise account of a complicated subject that I have ever written. RTG 16-I-40'. He also sent a copy to the historian Hugh Mill, noting 'I am vain enough to think that although it is so short, it is not the least valuable life of Cook that has yet been written'.[329] He goes on to say that, in spite of his debacle with Argonaut, he is soon to be signing a contract for a more complete and definitive (125,000 words) life of Cook with another publishing house, though this never came to pass either.

By contrast, he also wrote about a somewhat lesser explorer at this time. Always keen to research and publish on polar matters, he contributed an article to the *Bulletin of the Geographical Society of Philadelphia* in 1935, describing the extraordinary, and half-baked proposal by commander J.P. Cheyne to undertake an Arctic exploration by hot air balloon in 1880, an attempt which, needless to say never actually got off the ground.

Tennis

Rupert Gould's lifelong interest in the rules and game of tennis was put to good use during the 1930s when he was invited, from about 1930, to umpire at several of the great Championships. Although generally good at sports when a schoolboy, Gould had never been much of a team player, and his preference was for games of the solo variety. Billiards was a favourite, but above all, tennis was his great passion—'I think there's no finer game,' he told the nation's children in one of his talks—and tennis was seldom far from his mind. Even restoration on the Harrison timekeepers would be called to a halt if there was an opportunity to get out on the court for a game.

As with all things that interested him, Gould had studied the history of this subject since childhood and was exceedingly well informed on tennis, including the modern game's predecessor, real or royal tennis. When his friend, Evan Baillie Noel, Secretary of the Queen's Club and himself a great tennis expert, lectured to The Sette of Odd Volumes (he was 'Brother Paulmier') on 'An Odd Survey of the Royal Game of Tennis', Gould illustrated the menu with a detailed plan of the highly complex court and with a caricature of the speaker, unusually, bearing the cost of three guests on that occasion.

Captain Wakelam

In February 1933 the radio sports commentator, Captain H.B.T. Wakelam joined The Sette of Odd Volumes, almost certainly at the prompting of Gould. Wakelam was already something of a celebrity himself, being the first British broadcaster ever to give a running commentary (on the England versus Wales rugby match at Twickenham in 1927). During the 1930s he was the BBC's principal commentator at Wimbledon, the Davis Cup (Wimbledon and Paris), at all the major rugby matches and at all the cricket Test Series, including the first television commentary on the test match between England and Australia at Lords in 1938. Predictably, he took the Sette title of 'Brother Commentator'.

It was Wakelam's turn to step into the Odd Volumes spotlight on 23 May 1933, when he spoke to The Sette on being on 'The Other Side of the Microphone', Gould doing the caricature of Wakelam for the cover of that night's menu. Naturally Gould and Wakelam were good friends. When Professor Stewart mentioned in a letter to Gould that a tennis commentator he had heard on the radio had mentioned Gould's name, Gould noted 'I expect the man . . . talking about me when broadcasting from Wimbledon was my friend Brother Commentator of the Odd

Volumes, Capt H.B.Wakelam. He generally works in something about me sooner or later'.

'The Stargazer' told his young listeners,[330] in a talk simply called 'Umpiring at Wimbledon', that he had been umpiring there for 7 years (and that's what he was busy with at the time), adding 'I'm not quite a centenarian yet, and I hope to have the pleasure of umpiring for some of my younger listeners before I finish' (alas, this was not to be; although The Stargazer was just 47, he would be dead within ten years). Good eyesight, good hearing and an ability to keep your head, were The Stargazer's recipe for a sound umpire. Usually, umpires are asked to attend a second match every day as a line judge, and in this role watching the ball, never the lines, is the essential thing to remember, especially given that 'in all matters of fact affecting that line your decision is absolutely final—even the umpire can't overrule them', not a regulation which applies today.

As for players challenging decisions 'As a matter of fact, your first class player hardly ever queries a decision—he knows jolly well that there must be a small percentage of wrong decisions, and if one goes against him the next one, most likely, will be in his favour. If you have trouble—I've had very little myself—it's generally from a second-class player, or even a "rabbit". Well, It's best to be tactful with such people, but be firm if you have to'. In fact, Gould was capable of getting pretty angry where bad behaviour was concerned and his son Cecil, who deplored the modern tendency for tennis players—even the great ones—to question decisions, and to generally 'behave badly', remarked that if his father had been umpire at a match when, for example, John McEnroe was giving voice to one of his famously rhetorical outbursts, Cecil reckoned 'my father would have made absolute mincemeat of him'.

On once being asked who he regarded as the best player he had seen (and he had umpired many at Wimbledon, The Wightman Cup, The Davis Cup, the Beckenham Tournament and others, including matches with Fred Perry and other greats) Gould was in no doubt. He stated unequivocally it was 'Reggie Doherty, the older and taller of the two famous brothers who, between them, held the Singles Championship for 9 years, and the Doubles Championship for ten. Reggie . . . was a master player . . . at his best he could beat any opponent on any kind of court, and if only he'd had decent health—he was a semi invalid most of his life—I don't think he'd ever have been beaten . . . He died in 1910 when he was only thirty six'.

The umpiring duties were by no means as easy and stress-free as one might imagine. In a letter to R.J. Cyriax[331] he mentions that he is currently umpiring at Wimbledon 'and my days are pretty full and long', having to leave first thing in the morning and not arriving back home until late. And accepting the role meant accepting a full three weeks work, umpiring in the qualifying rounds for the first week, followed by the Wimbledon fortnight itself, working full days, everyday. Acting as a foot-fault judge he said was particularly tiring, especially when having to stare into the sun.

On 14 July 1936, at the conclusion of an exhausting Wimbledon fortnight, he noted to Professor Stewart 'Wimbledon has taken all my time and left me flat as a pancake. It is really rather a strain. I left here every morning at 12.30 [presumably actually pm] for about a fortnight, worked or watched all day, and got back, usually about 9.30 or so, tired to death'. Nevertheless, he was all set to carry on umpiring at the Davis Cup the following two weekends. Besides, umpiring had its compensations, and some amusements; Gould continued to Stewart: 'I had the honour . . . of taking the chair for the opening match of the meeting on the Centre Court—Perry v. Stratford. It was also my duty to inform Stratford, about half way through the second game, that the majority of his fly-buttons were undone. Having thanked me effusively, he did them up, *coram populo*, with no apparent embarrassment . . . the match was just a preliminary canter for Perry'.

Of the many matches he had officiated at, he especially recalled having the privilege of umpiring the final match of the Davis Cup tie between U.S.A. and Germany in 1937, when Budge defeated von Cramm after being led 4–1 in the fifth set. 'It was a very wonderful exhibition of modern, all round tennis'.

There was no payment for umpiring, the privilege being considered rewarding enough, but officials were sometimes presented with tokens of appreciation from the authorities, and Gould was given a set of three silver spoons engraved: '1934 Davis Cup Great Britain—Holders—beat USA Challengers—For Umpiring R.T. Gould' the second marked the same but '1935' the third (now parted from the set) probably marked '1936'. In all, he umpired officially during 8 years of tournaments and Championships (the last season he umpired was that in July 1939), totalling dozens of matches.

It was at a Wimbledon match in July 1936 that, by chance, Gould saw his wife again. He told Professor Stewart: 'I was just going on to the court to Umpire a match between Fru.Sperling and a Miss O'Connell, when I caught sight of my wife. She didn't see me at the time, but afterwards she came up and spoke to me just outside the Centre Court and I provided her with tea in the Umpires' refreshment-room. I was very glad to see her again, not having done so for years. I have altered in form far more than she has; she has hardly changed at all, and remains, as she always was, the most beautiful woman I have ever seen in my life. It is difficult to describe my frame of mind at the time, and subsequently; perhaps Milton's 'calm of mind, all passion spent' best expresses it. Time was, soon after we separated, when I both loved and hated her simultaneously. Now, I think, I do neither—at least, not actively'.[332]

❧ **19** ❧

Many Projects 1936–1937

Rupert's son Cecil had by this time left school and, wishing to study the History of Art at the Courtauld Institute, had been recommended to visit Germany for 6 months, learn the language and take in the more important collections.

In March 1936 Hitler had broken the terms of the Treaty of Versailles and marched his troops illegally into the Rhineland, and the outlook was increasingly bleak. But the general consensus among the Gould family was that war was not imminent and it was agreed he should go. In pessimistic, misanthropic mood Gould wrote to Professor Stewart:[333] 'Anyhow, if another general European War comes, I hope it will make an end of civilized Man altogether. He is obviously incompetent to form a stable society, and should abandon the job to the bees or the ants'. Around this time Gould had probably been reliving the First World War in his mind, and in excruciating detail, as he was contracted to do nineteen maps for A.C. Delacour de Brisay's new book, titled: *And Then Came War, An outline of the European Tragedy*, the book seeing publication in 1937.

Although Rupert and Muriel rarely met, discussions between the parents about their children's future had apparently been carried on by proxy, both grandmothers having a say too, and there was no issue with communication on this score. Cecil set off for Germany in April 1936, Rupert observing to Professor Stewart that even if war does come 'he will be no worse off interned there than fighting here'. No doubt then, when Rupert and Muriel met at Wimbledon that year, they had much to talk about.

Another busy year

The year 1936 was another very busy period with Gould's literary and lecturing commitments. The year had got off to a sad start. Gould was billed to speak at the Odd Volumes dinner on 21 January 1936, on the subject of 'Abraham Thornton and his Wager of Battle', a subject he would publish in his second edition of *Enigmas* in the 1940s, but the evening was cancelled at short notice, owing to the death of George V at 11.55 p.m. the previous night. The talk was eventually given at the OV (Odd Volume) dinner on 26 May 1936.

Cecil recalled that after hearing the radio announcement 'The King's life is moving peacefully towards a close', his father, in typically pedantic mode, simply remarked: 'the word "drawing" would have been better than "moving" '. After the funeral, Gould wrote to Stewart with a sketch he had done, from the top of the Admiralty Arch (access provided by courtesy of his ex-employers in the Hydrographic Office), of the funeral procession on its way to Westminster Hall on 23 January.

Amalgamated Press

At the end of the previous year, Amalgamated Press had asked Gould to write a series of articles on exploration for an unnamed publication (this would eventually form part of the two-volume set *Shipping Wonders of the World*), resulting in an order for five articles, the first of which was to cover Shackleton's expedition of 1907–1909, in 4,000 words. He told Professor Stewart 'I sent it in—and got it back like a boomerang! However, the Ass.Editor—quite a young fellow—took the trouble to point out more or less why it didn't suit—it wasn't dramatic enough for them, and was too quiet in tone . . . I rewrote it accordingly . . . packing it full of superlatives, laying on the local colour with a trowel. When I read it through, it seemed the most awful tripe—still, it was that or nothing, so in it went. And they were entirely satisfied with it!'.

An article on Scott was next, but this time Gould was to hold back on publishing the true drama of Scott's ill-fated expedition, as he saw it. He felt very strongly that, out of hubris and with no feeling for his men's welfare, Scott was determined to get to the pole primarily with man-hauled sledges. As a consequence, Gould felt, he killed his Polar Party, a view which he told Stewart 'I should have no hesitation in publishing in a book of my own, but which there is no need to make public via the Amalgamated Press'.

While *Shipping Wonders* was still totally occupying his time, Gould was nevertheless committed to another book, *Three of a Kind*, the deadline being 1 August. This was to be a discussion of three Polar explorers who had all falsely claimed to have reached the north pole, Louis de

Rougemont in the late nineteenth century, Dr F.A. Cook (21 April 1908), and Admiral R.E. Peary (6 September 1909). However, in April 1936,[334] Gould read in the press that Dr Cook was now trying to sue the Danish author Dr Peter Freuchen for libel, Freuchen having published that Cook's claim was unsubstantiated; and Gould was now thinking he may have to cite the claims of George Psalmanazar, instead of Cook, for fear of another libel action.

A few weeks later and he was complaining to Professor Stewart that the pressure was mounting: 'I have three urgent articles to write for *Shipping Wonders of the World*, and three of a kind is due by August 1 and not a line of it is written. Gosh, what a life! I wish there were 48 hours in the day—time seems to go much faster with me now than it did even five years ago'.

As it turned out, owing to overwork, Gould would again have to withdraw from the contract and, although the publishers reassigned the deadline for the coming year, *Three of a Kind* never saw the light of day. But in spite of this workload, Gould still hoped to get the larger biography of Cook done soon and through his literary agent,[335] he had already signed another contract with Methuens. This was to write a 90,000 word book to be titled *Nine Days Wonders*, on the subject of famous hoaxes and deceptions (this was another of those projects, it may be recalled, he had begun several years before). The manuscript was due for delivery in August 1938, but like many of these projects, it was never completed.[336]

A Book of Marvels

Meanwhile, as a means of getting something from him, in December[337] he signed an agreement with Methuen for yet another publication, to be called *A Book of Marvels*, and part of their Fountain Library series. This was in fact simply a re-publication of seven of the chapters from *Oddities* and *Enigmas*, mostly just reprinted, but occasionally with footnotes added. The book, which was released in 1937, was dedicated to 'Miss J. Bower', about whom nothing is known. Gould's son Cecil could only surmise that she was 'probably one of my father's girlfriends'. As it turned out, *A Book of Marvels* did not sell well and was remaindered in 1940, there still being a large number of copies unsold. These may well have eventually been destroyed, as this is one of the Gould titles which is more difficult to find these days.

Children's Hour Annuals and *Radio Times*

Owing to his great popularity on *Children's Hour*, Gould was invited to contribute articles to the *Radio Times*, several appearing in 1936, and one in each of the *Children's Hour* annuals in 1935, 1936, and 1937. All these articles, (each illustrated by Gould of course) begin with virtually the same little potted biography of *The Stargazer*; the children had been writing in to ask for more about their favourite broadcaster.

In the 1935 article, *In Quest of a Monster*, Gould tells the story of Sea Serpents and the Loch Ness Monster, assuring the reader that there really is something in the Loch. In 1936 he reveals *Some of my Hobbies*, a

54. Jocelyne and Cecil in the garden at Downside, *c.*1936, with the *Graf Zeppelin*, Rupert's huge box-kite. It required several people to launch and was mostly flown when on holiday in Norfolk. (© Sarah Stacey and Simon Stacey, 2005)

piece he told Professor Stewart he was rather pleased with. In it he describes his love of tennis and billiards and his fascination with mechanical things like typewriters, clocks (naturally) and mechanical toys. The clockwork model submarine 'K2', which is 3 ft long and goes 100 yards at one winding, first on the surface, then submerging and resurfacing again automatically, is a particular favourite. He also describes his own, home-made model helicopters, constructed with cardboard and elastic bands with which he had been experimenting.

Kite-flying was also a great passion, and being the man he was, he had to construct the biggest and best: the largest was a nine foot box kite which he dubbed the *Graf Zeppelin* after the famous German airship! For the 1937 annual, the article was titled *Make your own Puzzles*, with instructions, for example, on how to make several kinds of wooden blocks formed from several ingeniously dovetailed and tapered parts. Among a number of other amusing tricks, Gould illustrates Sam Loyd's wonderful 'get off the earth' puzzle, where a number of figures (in Gould's example drawn as deep-sea divers) are drawn over the edge of the world, the disc of the world being rotatable, half of each diver being drawn on the disc, half on the background. In one position the number of divers is seen to be twelve, but on rotating the disc slightly the figures miraculously become thirteen in number, a very effective and amusing puzzle.[338]

Churchill

On 10 April 1936, Gould attended the annual Navigating Officers' Dinner as usual, telling Professor Stewart rather pathetically that he 'pretended to myself, for a few hours, that I was still a Naval Officer. Our guests of honour were Edgar Britten, of the Queen Mary, and Winston Churchill who was expected to make the speech of the evening. But Lord what a lamentable performance it was! He followed the old tradition that "to er-er is human, to hm-hm divine," his delivery was bad, his matter generally dull and trivial, and he seemed to have no concentration at all. In fact, after he had been speaking for five interminable minutes without saying anything, my next-door neighbour . . . remarked to me in a loud whisper "Is he blotto?." I replied that I didn't think so, and felt rather sorry he wasn't'.

The Blattnerphone

Gould's BBC broadcasts continued almost without break during the second half of the 1930s, and in June 1936 he had his first opportunity to hear his own voice. He told Professor Stewart: 'I had a curious experience the other day. My talk on "*Comets*" was canned on the Blattnerphone for reproduction in an Empire broadcast; and one day last week I went to hear it. First of all I heard the reproduction of McCulloch's voice, which I recognized instantly. Then came Stuart Robertson, and again I should have known the voice immediately. Next came a strange voice—one which I could have sworn I had never heard before. It was only by the words, and the blurred "r"s, that I made sure it was mine. I conclude—since I am convinced, by my recognition of the two other voices, that the machine reproduces a voice with utmost fidelity—that no one, in general, has any idea of what his voice sounds like to other people; probably because part of it comes to him, but to no one else, via the bony framework of his head'

One or two historical landmarks are described by Gould in his letters to Professor Stewart at this time. On 5 December 1936 he noted: 'We live in stirring times. Last Monday evening I had walked down to the level crossing, intending to pay my usual weekly visit to an Epsom cinema. I noticed a bright red glare in the sky, obviously from a big fire, and diagnozed, from its bearing that it was either the Epsom gas-works—in which case one would hear about it in a minute or two, or else at Sutton. I never dreamed, until I got home, that it might have been the Crystal Palace'.[339]

Then changing the subject to the abdication of Edward VIII, Gould noted: 'And now there is this hideous tangle about the King's marriage. I'd heard rumours (as most people, I suppose, had) for some weeks; but the details were a complete surprise to me. Personally, I think the harm is done; and if the alternatives are marriage or abdication, I am for marriage every time . . . From the little I know of Constitutional Law, I incline to think that the Government have no real power to prevent the King from marrying whom he pleases; but as to the utter inexpediency of his choice I think that there can hardly be two opinions'.

The National Maritime Museum

In April 1936, the fitting out of the new National Maritime Museum at Greenwich was well underway and plans were complete for its opening to the public. On 18th of that month, the new Director, Professor Callender, wrote to the Hydrographer, Admiral Edgell, formally asking him to arrange for the transfer of the Harrison timekeepers to Greenwich. So, as the year's loan to the Science Museum was now nearly up, Edgell duly wrote to E.E.B. Mackintosh, the museum's director, asking for their return. Gould noted to Professor Stewart at the time that the Harrisons 'are going to leave the Science Museum next month—but not if the authorities there can do anything to stop it—and will find a permanent home in the new National Maritime Museum at Greenwich. I shall be glad when they are finally shifted. I have always managed, in the past, to safeguard them from damage while being carted about in lorries—but I have a haunting fear that one day my run of luck will end. It is very difficult to get all the heavy, fragile parts chocked up so that they are safeguarded from shocks'

Just as Gould suspected, with such sensational artefacts as the Harrisons, Mackintosh felt he had to have one last try at keeping them on loan, and he replied in passionate vein to Edgell on 5 May: 'I had hoped that, when their Lordships appreciated the important gap filled by these chronometers in the history of Time Measurement in the National Collections here, they would waive this intention and allow them to remain on loan here and thus avoid a gap in this history which of course is otherwise unfillable . . .'. He went on to *'earnestly beg'* their lordships to reconsider the question of transfer, in view of the importance of these chronometers in the Time Collections at this museum, and allow them to remain on loan and available for study to the interested portion of the metropolitan public', a hint, perhaps, that having them at Greenwich would be making them somewhat inaccessible.[340]

Of course it was to no avail, the Hydrographer replying that the decision had been made carefully and that 'they therefore regret that they are not able to agree to their continued exhibition at the Science Museum',[341] Mackintosh noting on the museum's file 'An unfortunate decision in my opinion, but it must of course be accepted'. Gould was then asked to oversee the collection on 9 June, travelling with them and setting them up in a temporary home at Greenwich, pending their display in the new Navigation Room in the south west wing of the Museum.

Later, Gould wrote to Frank Ward, the curator at the Science Museum, to thank him for having sent a copy of the new Time Measurement catalogue, which he was very complimentary about. He took the opportunity to remark: 'I hope you don't bear me any ill-will in connection with the removal of the Harrison machines from the Science Museum to Greenwich. As I told you, since they were once got together in going order my chief concern has always been that, wherever they were finally housed, they should be together. On the other hand, I will freely admit that if my opinion had been sought (it wasn't) as to whether they should stay at South Kensington or go to he National Maritime Museum, I should have recommended the latter course. They stand right apart from the history of horology in this country, and the development of its ordinary clocks and watches; they are of essential nautical interest—and the world's largest and completest nautical museum is the ideal place for them, in my judgement'.[342]

NMM Curator of Navigation

Gould had been aware of the proposals for a new maritime museum since the 1920s, being a life member of the SNR, and as plans progressed for the museum's staffing in the early 1930s he naturally made discreet enquiries as to the possibility of becoming the museum's first Curator of Navigation. It was a post for which he was supremely qualified and one which he desperately wanted to fill. But again, the opprobrium associated with the judicial separation came back to haunt him, and he was politely informed that such an appointment would not be appropriate at the time.

Instead, the job was given to one Captain Maxwell, 'a little man with a moustache and whistling dentures', Cecil remembered.[343] Maxwell was evidently far less well qualified and had little idea on horological matters, but Gould and he got on reasonably well and he would take Gould's advice when needed. On one occasion, when Maxwell sought Gould's advice on a number of very simple matters concerning the history of chronometers, Gould replied: 'The questions . . . which you raise are all answered in a work which expounds the subject with minute accuracy, marvellous fullness of detail, and superb literary style—in fact, words fail me to describe its many and varied excellences. It is entitled *The Marine Chronometer, its History and Development*, and I believe that the Museum possesses a copy'.[344]

The Harrison cases

On 13 June 1936 Gould wrote to Callender concerning the new display of the Harrisons, evidently not yet finalized, suggesting opening doors for new showcases. The existing ones had glazed tops which lifted over, onto the bases.[345] Gould noted again that the Observatory still has Harrison's original glazed cases for H2 and H3 and suggests they consider having these cleaned and displayed alongside the clocks, or at least stored underneath so as to be available if a visitor wished to see them. By agreement with the Astronomer Royal and the Hydrographer, these cases were sent down to the museum, but it is sad that the idea of having them cleaned and displayed wasn't acted on at the time, as H2's case is now lost and H3's case, unrecognized at the time, only just avoided destruction in the early 1970s.

H2 Polished and Lacquered

Gould also pointed out that H2 needed its movement cleaning as it was never lacquered and, after over 10 years, it had become covered in fingerprints. He offered to do this after the autumn holiday, in October 1936, pointing out that H4 and K1 will need cleaning again before the museum is to open the following year. Both the Astronomer Royal, Harold Spencer Jones, and the Hydrographer, Admiral Edgell agreed that H2 should be polished and lacquered, or, as Edgell put it, 'that Cinderella should be provided with a new coat'. Writing on 8 September to Maxwell from the Grand Hotel in Sheringham, Gould described and drew a specification for new showcases, proposing a wheeled trolley be made of exactly the right height for the cases, so that the timekeepers could be slid in and out of the cases without difficulty (this was not done at the time, but on the Harrison's redisplay at the Observatory in 1985 the idea was finally introduced). H2 arrived back at Downside for dismantling and polishing on 3 October 1936.

Half of a large notebook (the second half of Book 6, used for H4 & K1's overhaul the previous year,) and the whole of a somewhat slimmer book, recorded the dismantling, cleaning and readjusting work, which actually started on the 18 October 1936. The work continued until 12 April the following year, involving Gould in far more trouble than might be expected for a second overhaul. Complete dismantling of H2 took just four hours, Gould noting that in 1923 it took him a week! Work was suspended for the first two weeks of November, and much of December. Gould explained in the notebook that he had 'a lot of drawing and writing to do'. The BBC had, in the end, asked Gould to give a broadcast on the Harrison timekeepers, and the script had to be carefully drafted for the twenty minute talk. 'The Man who Discovered the Longitude 1776', went out at 6.20 p.m. on Christmas Eve 1936, having been announced, *inter alia*, in *The Times* and the *Horological Journal*.

Work on H2 was again briefly interrupted as Gould visited the National Maritime Museum early in the New Year of 1937. The NMM's new showcases for the timekeepers were supplied, and he saw to the fitting of the bases personally, to ensure the timekeepers were safe and correctly mounted.

H2's polishing and reassembling was well underway by February, after another breakage (of the pivot of the 3rd wheel—this time caused by Bucks during polishing) had been put right by Hopgood. Gould himself had a terrible problem to overcome at this point, spending many hours struggling with cleaning out surplus lacquer which now filled all the holes in plates and other parts.

But the way he cleaned the lacquer out of the screw holes was frankly most unsatisfactory. He decided to cut grooves ('flutes') down the threads of many of the screws, so they *cut* their way through the lacquer as they were screwed in! He noted: 'As Panhard said, "It is brutal, but it works" '. When even this method failed on some of the smaller screws on the anti-friction wheels, he resorted to opening out the holes and using the screws as *rivets* to secure the wheels to their centres.

Then there was an even more prominent misdemeanour, committed on 10 March 1937 when, after long hours finally getting the timekeeper together again, Gould realized the stop work was incorrectly set up. 'Things looked gloomy, because to reset the count wheel would mean getting off the front plate—a huge job! Turned the matter over in my mind during breakfast & decided to attack the front plate round the pivot of the count wheel arbor, with a view to giving the arbor enough side-play to let the wheel clear its driving pinion on the fusee. In this way the setting of the count wheel could be adjusted at any time without parting the plates'. What he actually did was chain-drill round the pivot hole and remove a rough plug of the plate so the wheel could be moved out and into its correct position. He then covered the hole with a small square plate, screwed to the front plate, with a small brass block riveted to the back, which had a pivot hole for the wheel: a truly dreadful botch for which there can be no real excuse. The square plate was later reshaped into a regular octagon, so at least it matched the fusee pipe, in an attempt to reduce the effect of its appearance.

Poor Workmanship

One can only reiterate the comments in Chapter 5, that there are quite a few aspects of Gould's restoration work which would be utterly condemned today, and which one could really have hoped for better, even at the time. Much of this poor workmanship had to be put right by the Ministry Of Defence chronometer workshops when the time-keepers were overhauled during the 1950s, 1960s and 1970s, and it must be said the contribution of those staff to the preservation of the Harrison timekeepers, mentioned in the last chapter, is often overshadowed by Gould's pioneering work.

H2 was finally completed and delivered to Greenwich on 12 April 1937, Gould setting it up in its new case after having lunch at his old haunt, the Royal Naval College. In fact, several more visits to Greenwich were required up to the end of April before the timekeepers were finally all running reliably, but on 17 April Gould wrote to Spencer Jones that all the Harrisons were installed in the new Navigation Room at the Museum.

Of H2, necessarily turning a blind eye to some of the repairs, he noted: 'It looks brand new, and I think that I have rendered its mechanism as nearly perfect as I am ever likely to get it'. He then broaches 'a matter on which I should greatly value your opinion. Unless some one drops a bomb on the museum, there is little doubt that the Harrison machines will, or should be, going long after I am dead—anyhow, I want to do what I can to ensure at least the possibility of this. And the only way in which I can do this is to make sure that my knowledge of the machines does not die with me. The ideal plan would be, I suppose, to find some young mechanic, with a permanent job either at the museum or near it, to whom I could give personal instruction—as Harrison did to Kendall. If I may say so without vanity, I think that the chance of someone turning up in after years who would do the work for love as I did, is not very great. But failing any prospect of my instructing some one else, the next best plan—I think—would be to put such written material as I have into some sort of order, and augment it wherever necessary'. He goes on to offer all his notebooks to the Observatory, in the hope that they would be preserved in the manuscript collections, and adds 'I hope one day to put together a full monograph on Harrison, with an appendix giving details of all his timekeepers'.

The Astronomer Royal, the Hydrographer and Callender all took Gould's point about a Harrison trainee, and it was agreed in due course that a new member of the Observatory chronometer staff (Mr Holborn Warden) would be given instruction under Gould. However, as often happened, events overtook this intention, as they did, sadly, with Gould's intention to do a full monograph on Harrison.

At the time, Spencer Jones replied accepting Gould's generous offer and, just in case he was now looking for a new project, wondered if Gould might be interested in looking at the Shepherd electric clock system for them which was 'one of the earliest applications of electricity to horology and the distribution of time' but which '. . . is of no use in its present condition'. The alternative was to offer it for loan to the Science Museum, who might restore it. In fact Gould had to decline, claiming modestly that 'I know very little about the historical side of electric clocks', though he then cites some of the Shepherd clock's history. As it turned out, the Science Museum did not get involved in a restoration either. In fact, apart from a rather botched attempt to restore the circuitry in the 1970s, nothing has been done to the system since, and a full restoration is now underway (2005).

The Arctic, Sea Monsters, and the Antarctic

There had been other interruptions to work on H2 in early 1937. On 26 January, another OV dinner and talk had engaged Gould's attention. Taking his old school friend H.C. Arnold-Forster with him as a guest, he spoke that evening on Louis de Rougement, one of a number of explorer–imposters who claimed to have reached the North Pole, while apparently having done no such thing (De Rougement, it will be recalled, was to have been one of the *Three of a Kind*, the book which at the time Gould was contracted to write by Methuens, but which never appeared).

Then, while he had been busy working on H2 at Downside, Gould had heard a talk on the radio, on 28 February 1937, by his old adversary E.D. Boulenger, titled: 'Sea Monsters: Do they Exist?' in which he evidently gave a decidedly one-sided view. Gould wrote immediately to Malcolm Brereton at the BBC asking whether he could reply to Boulenger and air the subject of sea monsters by giving a talk of his own. He needed to redress the balance, as the evidence for the existence of such creatures was, in his view, 'overwhelming'. 'Sea Monsters—A Vindication', an eighteen-minute talk, was thus broadcast at 9.10 p.m. on 31 March (the BBC's Scottish Region broadcast a recorded version of it at 10.25 the same night), the script having been edited for critical references to the Press, and the removal of expressions such as 'the stunt press' and 'enterprising journalists', which the BBC felt may reflect on them.

It was by all accounts a fine talk and in the BBC's regular 'broadcast extracts', which were production team discussions and summaries kept on file for future reference, the notes on this talk (and others later in the 30s) comment that Gould is 'an excellent example of a 1st class broadcaster'.[346]

The NMM Opens

The 27 April 1937 saw the opening of the National Maritime Museum. Such a high profile national event would inevitably have a political element at such a tense time, both nationally and internationally. The Spanish Civil War was constantly in the news and the vexed question of the appeasement of Germany's sabre rattling (Hitler had by now occupied the Rhineland and had signed the pact with Italy) was doubtless on everyone's mind. At home, with the controversial subject of the abdication, and the new King's impending coronation, it was seen as vital that an event such as the opening of the new National Maritime Museum was a success and tended to unify the nation.

The soon-to-be-crowned monarch, George VI and his wife Elizabeth would perform the opening ceremony, in company with Queen Mary and the young Princess Elizabeth. Gould's now-celebrated O.V. friend, the M.P., A.P. Herbert, proposed a river pageant, an idea eagerly accepted and, on the day, the Thames was lined with public keen to take part in the auspicious occasion. Such were the crowds of well wishers it took the Royal party *eighty minutes* to travel the six miles back to town.[347]

Ownership of the Harrisons

On looking round the museum after the opening that day, the Astronomer Royal, Harold Spencer Jones noticed that some of the labels credited the Harrisons as loaned by the Observatory and some loaned by the Admiralty. He spoke to Gould about it while still at the museum, and duly dropped Callender a line to say (of the Admiralty's ownership): 'The latter statement is incorrect. All of the Harrison time-pieces belong to the Observatory and the costs of their renovation and repair have been paid out of Observatory votes. In order to comply with the requirements of the Exchequer and Audit Department, I have to obtain each year a receipt from you for their loan'. In fact it had been Gould who placed the old Science Museum labels in the cases (much of the displays and labelling was incomplete at the Royal opening) to provide some information about them.

Realising this may be a sensitive issue however, Callender contacted the Hydrographer, and Gould was asked to prepare notes on the historical facts as to ownership. In May, Gould submitted his findings 'in connection with the recent transference of the Harrison timekeepers from the Science Museum to the National Maritime Museum' to the Hydrographer, copying Spencer Jones and Callender.[348] The six-page document[349] titled: 'Notes on the Ownership of John Harrison's Timekeepers' gave a statement of the facts, from Harrison's parting with them in May 1766, up to Gould's 'discovery' of them in 1920. The conclusion was that they must belong to either the Admiralty or the Royal Observatory, the former having the stronger claim. His view was that they should, in a sense, be regarded as working instruments and simply be 'issued' to the NMM from the Observatory by direction from the Admiralty, the audit paperwork being the same as with ordinary marine chronometers serving on ships. Gould's report was accepted and this was the chosen practice thereafter.

As for the labelling, there was still some sensitivity with the Astronomer Royal, Spencer Jones feeling that he had responsibility and rightful custodianship for them. So, adopting his usual wisdom of Solomon, the Hydrographer, Admiral Edgell, proposed that the labelling thenceforth read: 'Lent by the Admiralty from the Royal Observatory, Greenwich', which satisfied everyone.

Typewriters

On 9 June 1938 Gould gave one of his 'Stargazer' talks on the subject of 'The Story of The Typewriter'. Like so many interests of Gould's, the esoteric, some may say eccentric, fascination he had for these most useful office instruments goes back to his childhood; he first used one when 8 years old. His study of the history of the typewriter was sufficiently well developed by 1928 (as will be recalled from Chapter 11) for him to

55. The 'Sholes-Densmore' typewriter as redrawn in 1943. Gould had one of the largest collections of typewriters in existence and wrote the first independent history of the instrument. (© Sarah Stacey and Simon Stacey, 2005)

give the paper to the Royal Society of Arts on the subject. He had been actively collecting them since the previous year, and by the time of his talk Gould had one of the largest collection of typewriters in existence.

The 'Stargazer' talk was typically wide-ranging: from the instrument's earliest origins in Henry Mill's patent of 1714, to the 'early "blind" platen type, where the operator could not see the result until the exercise was complete. Gould even discussed the possibility of a dictating machine, able to interpret the human voice and produce text from it automatically. This, Gould says 'might possibly be made—but it would have to be immensely big and complicated, and could never be more than a curiosity'. He would no doubt have been fascinated with today's electronic word processing capabilities, and with the voice recognition software now available. The talk was followed by an article in the *Radio Times*, illustrated with a drawing of the 'state of the art' Sholes–Densmore typewriter of 1874.

Cecil recalled that at Downside, on the workroom door, his father, in unusually facetious mood, had placed a sign reading: 'HOME OF REST FOR AGED AND DECAYED TYPEWRITERS SUPPORTED ENTIRELY BY VOLUNTARY CONTRIBUTIONS NO DESERVING CASE REFUSED ADMISSION'. By 1938 Gould had managed to amass a collection of over fifty typewriters, a few on loan to the Science Museum but most on shelves in the workroom, and Dodo was becoming concerned that the collection might continue to grow. If he was obliged to refer to them while in Dodo's presence he would use the code word 'lobsters' when a new consignment was due, fearing a total ban on further acquisition!

The collecting did indeed continue and by 1943 he had accumulated no less than seventy-one of them. After the Second World War he wrote a series of articles for the journal *Office Control and Management*[350] which built up into *The Story of the Typewriter, From the Eighteenth to the Twentieth Centuries*, a slim book produced posthumously by the same publishers in 1949. Mr Bernard Williams, for many years an avid researcher and collector of typewriters, has very kindly made a few comments on this pioneering work of Gould's, as well as casting an eye over the text of his earlier lecture to the Society of Arts on the subject. One of the most surprising things that he has noted is that, unlike Gould's published work in antiquarian horology and in scientific mysteries, his contribution to the history of the typewriter is very little known in the small but enthusiastic world of typewriter collecting. Apparently none of his work has ever appeared in any of the typewriter collecting journals and most collectors are unaware of it.

Nevertheless, well over half a century after they were published, the book and article have very little in them which Mr Williams could correct, and he is highly complimentary about the style and content of Gould's essays: 'I am very impressed by the quality and accuracy in all his work researching & collecting Typewriters. He has obviously read & researched much of the available literature . . .' While there were of course earlier publications on typewriters, these were almost all commercial pamphlets which were naturally biased towards the story of that particular company, and Gould's was the first comprehensive and impartial history.

The editor of *The Story of the Typewriter* was Dudley W. Hooper, a friend of Gould's who wrote a short Introduction to the book as a homage to the late expert. Hooper had actually met Gould while on holiday at Sheringham in 1934 and had been one of the group of hotel residents which followed Gould's rallying call for help when the great box kite was 'going up'. As it turned out, Hooper lived near Gould in a town close to Ashtead and he was a regular visitor to Downside, playing billiards and studying the vast collection of typewriters in the workroom.

The NMM

By the early Summer of 1937 the Harrison timekeepers, all running and looking spectacular, were the star turns in the new National Maritime Museum's Navigation Gallery. At the end of May, Gould had been called in to start H4, which had unaccountably stopped, but there was little else technically required of him at the museum. Gould was of course as busy as ever with writing and broadcasting but, lacking the gratification of getting his hands dirty, he had kept himself busy making up a dedicated travelling case for the Harrison special tools. Maxwell was asked to post all the tools to him at Ashtead, but on their arrival Gould found that 'some of my gadgets seem to have been rolled on by elephants . . . but . . . the new case will ensure that it doesn't happen again'.

The First Orrery Restored

Another fine restoration opportunity arose at this time, when Gould had the chance to overhaul an important example of the scientific instrument known as an orrery (it may be recalled he had already overhauled one, back in 1924, see p. 137). An orrery is in fact what was otherwise known as a planetarium; not the kind at Madame Tussauds, but a smaller, table top sized instrument which is a three dimensional, working representation of the solar system. Full orreries include all of the known planets, each orbiting the Sun at their correct relative rates and positions.

The first such instrument, showing only the Earth, Moon and Sun, had been made by the great watch and clockmaker George Graham, in 1705, for Prince Eugene of Savoy. However, about, it was studied by the celebrated instrument maker John Rowley, who then made his own version of it c.1712. This instrument then belonged to the fourth Earl of Orrery, and the Earl's friend, the author Sir Richard Steele stated that its makes, Rowley, had dubbed the design "an Orrery". And it was this very instrument, the world's first orrery, that Gould was now to overhaul for its owner, the current Earl.

Every detail of the work was recorded in a notebook, just as he had done with the Harrison timekeepers. The notebook, and indeed the Orrery itself, are now in the care of the Science Museum at South Kensington. The book is titled 'Lord Orrery's Orrery', then subtitled facetiously 'The Brompton Orrery', a punning reference to the famous church, the Brompton Oratory, just round the corner from the Museum in South Kensington.[351] The instrument was taken from the Earl's London home at 46 Ennismore Gardens, to Ashtead on 13 June 1937 and work on dismantling began the same evening. Over the next month, many evenings were spent dismantling and note taking, brass parts were pickled in a strong ammonia solution and many were then sent to Bucks for polishing. Careful records were kept of the whole process, along with numbers of teeth on wheels and pinions, and notes on the re-correction of an error introduced into the orrery in 1876 by a well meaning but ill-informed repairer. The whole job was completed on 31 July and the instrument delivered to Admiralty House, in Portsmouth, the Earl's place of work as Commander in Chief, Portsmouth. Later in the year, Gould wrote the restoration up for the *Illustrated London News*,[352] just over a full page being devoted to the work and a description of this and other Orreries.

The notebook was left with many spare pages, and typically, Gould used it again, starting at the back, for notes on some experiments he was doing later in 1937. These tests were on toy helicopters he had designed and made from cardboard and rubber bands, as described and illustrated in one of the *Radio Times* articles. At the time, some of these tests were carried out with Cecil at Kew Green, outside Muriel and Vivian's house, and Cecil remembered one of helicopters, rising high up into the air and landing on the roof of the church, where they had to leave it! Another section of the back pages was used for notes on skyscrapers and a brief note on the violinist Nicolo Paganini, probably both examples of the 'notes on a postcard' which Gould took with him to the microphone when giving his 'Stargazer' talks at the BBC.

H3 and the RAS Clock

In September 1937, when Gould was on holiday with Dodo at Sheringham, he received a note from Maxwell to say that H3 had stopped. Apparently a group of boy-scouts 'had been dancing a jig around the case' and the timekeeper had objected. In fact, Gould had never been confident about the cross wires on the balances, and it was now going to be necessary to fit new ones, as well as new axial wires for the balances, which would unfortunately mean transporting H3 back to Ashtead for a while.

After a visit to pack the timekeeper on 14 October, Gould wrote to Maxwell, one of the more pressing issues being some form of payment, in advance, for the work to be done. He said 'I hate asking for money in the ordinary way before I've earned it; but at the moment my need is *urgent*. Take it that, whatever happens on this occasion, I shall not make such a request again'. H3 arrived at Ashtead again on the 19 October 1937 and began the next stage in the extraordinary saga of its repairs and overhauls. As it turned out, the timekeeper would not be put back on display at the National Maritime Museum for almost ten years.

Another 'Harrisonian' commitment at this time was a visit, on 1 November 1937, to the Royal Astronomical Society to strip and service the Harrison regulator, which had been proving somewhat unreliable, the whole job taking just 4 hours flat. Later in the month he would return to fit a new stop detent so the clock could not be 'overwound'.

Torrens and 'Nail and Cork'

On 10 September Gould lectured to the British Horological Institute on 'The Chronometers of the Clockmakers Company', a lecture which was reported in the *Watchmaker and Jeweller*, and the *Horological Journal* (HJ) in December 1937. The latter report, which extended to five pages, prompted a letter in the February issue of the HJ from the noted amateur horological historian David Torrens who was Professor of Medicine at Trinity College, Dublin. Torrens, whose horological knowledge was by then encyclopaedic, pointed out very politely, that Gould's claim, supported that evening by Frank Mercer, that the early chronometer makers 'must have worked by trial and error, and by rule of thumb . . .' was rather misleading.

In a two-page commentary titled 'Nail and Cork' Torrens simply pointed out that these early chronometer makers were primarily watchmakers and that the hand crafts used for making chronometers were basically very similar to the well established systems used in watch making where a multitude of special tools were used. Torrens also stated what was already evident to those who wished to face facts: the chronometer industry was rapidly shrinking; this we know had been the case since the First World War.

The chronometer was, he noted, wholly unsuited to mass production, as the market is too small, the materials used do not suit this kind of production and much of the materials and tools have to be bought from abroad, making the financial viability unlikely on a large scale. He ended with an affectionate look back to the times when craftsmen could take their time to do a job to the utmost of their abilities and produce first rate work, something he implies, not seen today in chronometry. Torrens, who like Gould, knew the world of the chronometer maker intimately, was fully aware that though some makers had gone out of business, there was at least one company, Thomas Mercer Limited of St. Albans, which was 'bucking the trend'. Not only was Mercer's staying in business, but they were just then beginning to adopt something like mass-production techniques to chronometer production.

Frank Mercer saw Torrens' note as a direct criticism of his business, and immediately submitted a scathing and verbose report, published the next month in the HJ. So sarcastic and vehement was Mercer's reply to what he called 'this onslaught', which he described as 'grossly misleading', that one is tempted to suspect that he 'protesteth too much', and that Torrens had touched a raw nerve. Mercer ended with a snipe at Torrens' amateur status as an historian: 'if I were to infer in a like manner [sic] upon my medical friend's ground, I would suggest that in this—our age—we are smothered with a flood of patent medicines and cures . . .'. Probably because of the tone of Mercer's letter, Torrens decided not to respond and the question was not further discussed. One would normally have expected Gould to have joined in such discussion, but there was a very good, but sad, reason he did not.

Leaving Downside, Leaving London 1937–1939

Dodo

Work had been underway on H3 at Downside for just over a month when, on 22 November Gould notes as an aside: 'Mother is very ill, but is, I hope, on the mend', work continuing 'without sawing or hammering' to ensure peace for Dodo. But two days later his worst fears were confirmed.

Writing to the Arctic specialist, R.J. Cyriax, on the 29 November, Gould explained 'I am sorry not to have written sooner, but calamity has suddenly come upon me. My dear mother died here on Wednesday of pneumonia'. She was buried in Ashtead churchyard on Saturday, 27 November, an announcement going out two days before in *The Times*. H3's notebook entry for 14 December continued from three weeks previously: 'Mother is dead and buried, and at long last I resume my work, thankful to have something apart from business matters to occupy my mind'. The business matters were of course seeing to the administration of Dodo's will and liasing with her executors, his cousin Edwin Kennedy Hilton in Manchester and Dodo's friend Margaret Hunton.

Rupert must have always known that his mother's death, apart from leaving him grief stricken, would also place him in a difficult situation. Dodo's will left the main body of her estate in trust, just as her mother and her grandfather had before her. The will provided for Rupert only in that he was to have a small number of Dodo's personal possessions and that the income from the trust fund would go to him for his life. This was nowhere near as large as it had been in previous generations, but did represent a slight improvement in Rupert's income from that point on. Most of Dodo's valuable possessions were left to her grandchildren and other relatives and friends and the Executors were instructed to sell all the real estate 'when they think fit'.[353]

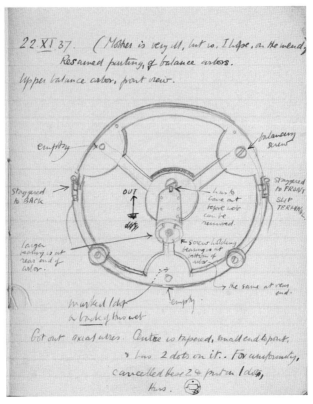

56. The entry for 22 November 1937, in H3's notebook, No.8. Gould notes that his mother, 77 years old, was very ill. She died two days later and there began a new, unsettled phase in Gould's life. (© Sarah Stacey and Simon Stacey, 2005)

Although she left the decision to them, Dodo clearly did not intend that Rupert should continue to live at Downside. The Trust fund itself was then to be held in trust in equal shares for Cecil and Jocelyne. So, for Rupert there was nothing for it but to move out and find accommodation elsewhere.

The death of Dodo thus marks the end of Gould's 'halcyon days' at Downside, and the beginning of the last ten years of his life, a period of very mixed fortunes. Dodo's death also meant that H3 would have to continue on its travels; there was no way Gould could complete work on it before he was obliged to leave. On 3 March he records pathetically in the notebook: 'Finished the last job I expect to do in my beloved workshop . . . I can clearly see that it is hopeless to dream of finishing No.3 here. Tomorrow I go to the N.M.M. & ask Callender if he can give me a temporary workshop there', and the next day he records: 'Callender has given me a room in the Queen's House . . .'

All of H3's pieces were then transported by Gould in Dodo's car from Ashtead on 25 April 1938 and work began again with the timekeeper in the Queen's House,[354] at the National Maritime Museum in Greenwich, on the 13 May, continuing off and on for the next year and a half.

The Red House Hotel

On 30 April 1938 Gould left Downside for the last time, and moved in to residential accommodation in The Red House Hotel, in Leatherhead. Here he had two rooms, one each for his bedroom and library/living room. His large collection of typewriters (about 60 from Downside) were now accepted on loan at the Chiswick Polytechnic in West London, which relieved him of that considerable burden of storage. Unfortunately though, he had been obliged to sell most of his books, along with almost all the other furniture from Downside. Nevertheless, he was reasonably comfortable during his time living at the hotel in Leatherhead, which lasted for the next 17 months. One of the few things of Dodo's that Rupert was able to keep, was her motor car (a Standard 9 saloon of about 1930) so, as long as he could afford the petrol, he was at least able to get about.

Muriel

Although he did have a slightly larger income now that Dodo's will provided some interest from the Trust fund, one third of this was immediately claimed by Muriel as alimony. But this increased income for her produced an unexpected bonus for Gould.

Up to that time, Muriel had always been dependent on Vivian Gurney for support (putting it more bluntly, Cecil described his mother as 'under her thumb'). When Miss Gurney found she could no longer 'call the shots' in the relationship, there were, according to Cecil, tremendous arguments. He recalled: 'Early in 1938 Miss Gurney stormed out of the Kew house, and took a flat in London. I could not believe our good fortune. Her pervasive hatred of Grandma and of my father, and their hatred of her, which had contributed to poisoning the background to my childhood, now receded into the past'.[355]

At this point Rupert made a determined attempt to turn his fortunes round and make another go of his life. First, he wrote to the Admiralty to ask if there was any chance he might now be employed again. The reply thanked him and informed him that, while there was nothing for him at present, his offer had been 'noted'. This was perhaps not quite the brush-off it sounds, as it did eventually lead to something, but only a few years later.

Second, knowing that Vivian had departed, he now approached his wife and tried for a reconciliation. But she too refused, having, as Cecil put it, 'no wish to exchange one form of subservience for another . . . I imagine she would have realised that my father's wish was largely dictated by his convenience. Now that his mother was no longer there to keep house for him, why not his wife?' This is probably only part of the story, as Rupert felt, with some justification, that he had been very badly treated by his wife over the separation, and hoped for some form of retraction by Muriel of the very exaggerated claims about his behaviour before the court case[356] and it was no doubt partly for that reason that the reunion failed.

So, with his attempts to redirect his life having come to nothing, Rupert was now very much on his own. Indeed, over the coming years his life would become increasingly solitary, and though he would find two more close friendships with women in the next few years, one has the strongest impression that, from the late 1930s, he had become philosophically more of a loner.

Although very depressed at the loss of his mother, and busy dealing with the administration, Gould kept going with the BBC broadcasts. One he was involved with on 19 February 1938 appealed particularly to one of his pet hates. The programme was to be a discussion on Astrology, to be called 'Can the stars fortell?' and Gould was to be one of five taking part in the debate, the programme to be rehearsed and then roughly scripted before going out live a few days later.

R.H.Naylor of the *Sunday Express*, was a believer, and cited John Flamsteed's horoscope, cast at the foundation of the Royal Observatory Greenwich in 1675, as an example of serious scientific use of Astrology, unaware that Flamsteed's horoscope was intended as a joke. In fact Flamsteed had written next to it the sarcastic annotation in Latin: '*Risum teneatis, amici?*' ('Can you keep from laughing my friends?'). Naylor did not know of this important annotation but, Gould says, 'I did—and flattened him with it'.[357]

Glasgow Empire Exhibition

From May to November 1938, the Empire Exhibition was being held in Glasgow, and the Hydrographer decided it would be a good idea to have some of the Harrison timekeepers on display in the Navigation section within the Government pavilion. It was agreed that the large timekeepers were too difficult to transport, but that No.4 and K1 could go. Gould was asked to overhaul them, but naturally had to decline owing to his own upheavals, though he did offer to deliver them to Glasgow, an offer which was not taken up, as a member of the Observatory's staff escorted them instead.

The subject of sea monsters appeared again during the year. Although fully committed with Dodo's administration business, Gould had agreed to give the SNR a lecture on Sea Monsters, at The Drapers' Hall again, on 23 February 1938. According to the Annual Report of the SNR, the talk was 'listened to with the closest attention from the first word to the last!' A broadcast on BBC Regional on 8 June brought the listeners up to date on the Loch Ness Monster, still very much alive and swimming in the Loch, Gould told them.

A supporting article then appeared in *The Listener* on 16 June. In the autumn Gould lectured on Sea Serpents to The Ghost Club, with the Chairman, the famous 'ghost-hunter' and investigator of the paranormal, Harry Price presiding. The meeting was held at the Royal Societies Club, in St James's on 20 September and the lecture was duly reported on the following day in *The Times*. Gould had followed the work of Harry Price (1881–1948) and his National Psychical Laboratory with interest over the years and occasionally corresponded with him. That same year Price had approached Gould asking his advice about a possible investigation of the 'paranormal' sightings at Loch Ness, though nothing came of Price's plans.[358]

Work on H3 had ceased during the summer, but on 15 September Gould was back at the National Maritime Museum, and in pessimistic mood. The notebook reads: 'Amidst rumours of war, I had planned to come here today to re-stow & secure every part of No.3 in the 4-handled wooden case, so that if the Museum has to clear for action the case could at once be put into any place of comparative safety assigned to it. But for the moment the outlook is less black, owing to Chamberlain's going to Germany. I shall leave here a parcel of cloth & newspapers ready for packing & if any mobilisation takes place I will come here I hope, at once and pack the machine'.

'Peace in our time', as Chamberlain promised, was not sufficiently convincing for Gould though, who became increasingly pessimistic in outlook, and a few days later he decided the time had come to write his own will. He got his cousin Edwin K. Hilton to draw up something for him on 29 September 1938. Appointing Muriel as his sole Executrix, there was precious little else for him to say. His typewriters were all to go to the Chiswick Polytechnic, and his Harrison notebooks to the Royal Observatory. Anything and everything else was to go to Muriel.

Navigation articles

Meanwhile, a little writing kept his mind occupied. *The Daily Telegraph* produced a special *Clocks and Watches Supplement* on 7 November 1938, with a long article by Gould titled 'Finding the Way at Sea', and headed 'With a chronometer, Columbus would have realised America was not Asia'. Naturally, Harrison and his amazing timekeepers are the focus of the story.

With war becoming a distinct possibility, the production of books celebrating the strength of Britain's armed forces were bound to be popular and reassuring. Gould contributed an article on *The Navigator at Sea* to a 6d (sixpenny) book titled *The King's Navy*, which declared it was 'Published Especially for the People of the British Empire'. Gould's friend, the maritime author Frank Bowen, was one of the Associate Editors.[359]

Work on H3 at Greenwich continued into the New year, but Gould made very slow progress. On 21 February 1939, he was back again, but got little done as on this occasion he was showing the noted horologist Courteney Ilbert H3 while it was in pieces, and took him for a tour of the new museum galleries.

The entries in H3's notebook now start to sound increasingly worried about things generally. Typical entries read: 'serious financial crisis' (the NMM were no longer paying his expenses); 'Harrassed by many worries'; 'Maxwell predicts crisis in Mid February'; 'Attended but did little. I am worried. Chamberlain meets Parliament today' and so on. It seems that, for Rupert, the notebooks doubled as means of pouring out his heart to an unseen audience, as well as for recording his practical work. Perhaps, as with his macabre drawings, he found that getting his fears down on paper in some way eased his mind.

OV Dinners

OV dinners continued however, and on 24 January 1939 Gould took Arthur Hinks of the RGS, and George Naish of the NMM to hear him speak on 'The Siberian Meteorite', the story of the meteorite which devastated huge tracts of forest in Siberia when it came to Earth on 30 June 1908. On 2 June Gould travelled over to Bath with his daughter Jocelyne to an unusual OV dinner held in the Assembly Rooms in the City. The Mayor of Bath was Adrian Hopkins, 'Philatelist' to the Sette. There was no menu, but a souvenir booklet for the occasion was issued to the Sette in 1949. Sadly, the Assembly Rooms were destroyed by enemy action in April 1942 and this was deemed an appropriate memento of the Sette's evening held there.

Broadcasts continue

On 1 January 1939 Gould asked the Producer of Talks if he could do a broadcast on the Antarctic discoveries of James Weddell. There was a specific reason: an article had recently been published by one Professor William H. Hobbs of the University of Michigan, in the *Transactions of the American Philosophical Society*, entitled 'The Discoveries of Antarctica within the American Sector, as Revealed by Maps and Documents'. Hobbs's article sought to discredit the British claims for priority of discovery in the region and, with a great deal of very misleading information, including some charts which actually had coordinates *deliberately falsified*, attempted to prove a prior claim for the United States.

Gould, who had of course specialized in just this subject—no one alive knew more about it than he—was apoplectic that such falsehoods were being expounded in a serious professional journal. This was the kind of academic source which Governments might rely on in international negotiations over sovereignty, and Gould desperately wanted to refute, or at least comment on, these claims, in a BBC talk. Naturally at the time, the BBC were cautious of anything so politically sensitive and only allowed a general talk about Weddell's expedition in the 1820s, carefully editing the script. But Gould would have his say in the *Mariners' Mirror* in due course.

R.T. Gould was a particularly popular speaker at this time and was very much in demand. In February and March 1939 he had advised on, and taken part in, a 'docu-drama' radio play on three famous mysteries, The Comte de Saint Germain, The Disappearance of Mr Bathurst, and The Borden Murders, Gould taking the role of Narrator and Commentator during the plays, the programme being repeated later in the year.

Television

Then, on 28 March and again on 2 April 1939, Gould was invited to take part in a television broadcast from Alexandra Palace on the subject of Sea Monsters, entitled *Leviathan*. Notes of Gould's on the back pages of his copy of *The Loch Ness Monster and Others* includes: 'Television Programme March 28 [9.40–10.20], April 2 [3.20–4.0]; Morning Dress'.

Gould had already done at least one other television presentation— Cecil remembered being with him at Alexandra Palace on 4 August 1938—and Gould was evidently as good on TV as he was on radio. On 30 April 1939 he was at Alexandra Palace again, taking the role of Master

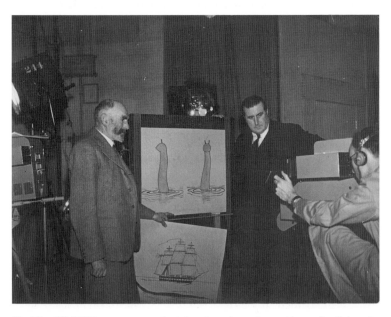

57. The BBC TV programme *Leviathan*, broadcast from Alexandra Palace in March and April 1939, featured Gould talking about Sea Monsters. (© Sarah Stacey and Simon Stacey, 2005)

of Ceremonies at the *Puzzle Party*, a programme chiefly for children. The fee, 8 guineas, was, as per his usual request, to be paid at the stage door before leaving.

Big Ben

The great clock at Westminster, the very first subject which had brought Gould to a BBC microphone, was now the subject of yet another literary contract, another of those contracts which unfortunately he would fail to honour. The agreement, with the horological book dealer and antiquarian Malcolm Gardner, was to write a small book on the history of this important clock. The date for delivery of the manuscript was 1 May 1939, and two advances, totalling £10 10s (ten pounds, ten shillings) had been made to Gould, who was, as usual, desperately short of money. In fact it seems Gould got no further than research for the text and it would be several months of chasing before Gardner was able to get recompense for his advance. Gould did eventually associate himself with a publication on Big Ben, by writing the Foreword to the book on the clock by Alfred Gillgrass, some years later (See Appendix 1).

R.J. Cyriax

The name of the Franklin expert, Richard Cyriax, has been referred to already in this story. Since their first correspondence in the early 1930s Gould and Cyriax had been writing, off and on, pretty well continuously about the book on Franklin's ill-fated expedition, which Cyriax was writing, and in 1933 Gould had agreed to draw relevant maps for the book. Cyriax lived in Leamington Spa, and in the later 30s Gould was doing small pieces of research in London for Cyriax too, receiving a nominal fee, usually half a guinea. The drawing of the maps would be another of those long, drawn-out projects, and it was only in mid 1939 that things were nearing completion. In a letter of 8 May 1939 Gould apologises to Cyriax, saying 'I have been having rather a bad time of physical and mental worries lately. Post-influenza depression was complicated by a return of another old trouble—eczema—which affects my sleep (I average about 3 hours a night at present) and I've had a flood of financial troubles as well, chiefly brought on me by my wife's solicitors . . .'

With eyesight suffering under the strain, Gould finally got the maps completed in August 1939, writing to Cyriax on the 22nd that he hoped

to leave them with the publishers, Methuen's, the next day. The letter
ended rather melodramatically: 'The news is not too good to-day. If, by
force of circumstances we shouldn't meet again, accept my very best
wishes, and my assurance that I shall always be pleased to have met you
and worked with and for you'. Writing again on the 25th, he said he had
delivered the maps, and hoped to see proofs, but that now 'my plans are
entirely uncertain. Good Bye and best wishes from, yours ever . . .'.[360]
They didn't correspond again until 1944.

Preparations for War

Earlier that same day, 25 August 1939, Gould had been very busy: 'Owing
to the exceedingly grave international situation, I spent yesterday in
safeguarding, so far as I could, my work on the Harrison machines. At
the request of the R.A.S., I went there, [and] packed the movement of
their Harrison clock . . . The movement, pendulum, keys and relevant
papers will go to Oxford, the remainder will be stowed in the basement.
I then went to Greenwich and saw the Astronomer Royal. I suggested
that if the Harrison timekeepers were removed from the . . . National
Maritime Museum it would be a good thing not to put all the eggs in one
basket . . . He suggested . . . transferring the three big timekeepers to
Cambridge. I then went to the National Maritime Museum'.

Here, in the Queen's House, (the notebook, No.9, records), he packed
'No.3's' pieces in the travelling case, and then secured the balances and
trains of 'No.1' and 'No.2', and wedged the balance of 'H4' and 'K1', in the
Navigation Gallery. This would appear to be the very first use of the term
'H4', as opposed to 'No.4', by Gould. The note ends: 'Told Maxwell that
the machines could now be removed from the cases without much
likelihood of damage. A hurried job, done by a worried man—I hope it
has been well enough done. RT Gould'. That same day, 25 August, the
Astronomer Royal, Harold Spencer Jones wrote to Professor Sir Arthur
Eddington at Cambridge Observatory and Professor H.H. Plaskett at
Oxford University asking for safe haven for some of the Harrison
timekeepers. Still worrying about the matter a few days later, he was
down at the National Maritime Museum. The Director, Sir Geoffrey
Callender's diary for Monday, 28 August, reports: '3.45 p.m. Astronomer
Royal created disturbance by fussing about Harrison chronometers'.[361]

As it turned out, Eddington replied first, saying he was 'willing to take
the Harrison Time—Machines', and so H1, H2, H4 & K1 were sent to him

at the Observatory in Cambridge. H3, packed up but still in pieces in its box, remained at the N.M.M. for the time being. Plaskett replied soon after, saying Oxford was happy to help, but by that time it was a *fait accompli* and only the RAS regulator was in Oxford for the duration.

Leaving London

What exactly happened next, as the certainty of war approached, is somewhat unclear. We know that on 26 August, Gould was driving west and passing his friend Alexander Keiller's estate at Avebury, as Keiller notes in his diary that Gould visited 'on his way to take up his duties under the Admiralty at Bath'.[362] This was almost certainly Gould telling Keiller what he hoped might come about. There is no evidence at all that there was any official work on offer to him at the time.

But four days later he was indeed in Bath, staying at the Pulteney Hotel, as he wrote to the Astronomer Royal from there: 'I came here on Saturday last, having reason to think that, in the event of hostilities, I should be appointed to somewhere in the neighbourhood', though what, and where this can have been is difficult to say. He may have had thoughts of helping at the Royal Observatory's chronometer workshops during wartime, but these were initially sited in Bristol, some miles away (after enemy bombing, they were soon moved to Bradford on Avon). Perhaps it was suggested to him that, should he be required, this would be the part of the country he would need to be in, but it is now difficult to be clear.

Gould continued, in a slightly panicky tone, to Spencer Jones 'But I am uncertain whether I may not have to leave at very short notice with a minimum of baggage'. He had borrowed three volumes relating to Big Ben from the Observatory (it seems he had indeed been researching for Malcolm Gardner's book), but now returned them by post.

He also took the trouble to state, for the record so to speak, that he had deposited the Harrison notebooks 'in the keeping of Alexander Keiller Esq . . . and are left by my will, to the Observatory'. Two things seem to be uppermost in Gould's mind at this point: First that he felt there was a chance he would be killed during the coming conflict, and second, he was concerned about the survival of the Harrison notebooks. This, naturally enough, would have been partly to ensure those valuable records of the timekeepers survive as technical records, but perhaps also because he felt that in them rested his strongest claim to posterity.

Gould had in fact planned to take a week's holiday with his children in Sheringham, in Norfolk, from 1 September, and now, at the last moment, he decided he would stick to the arrangement. Cecil, who had been on a hectic pre-war tour of Europe's museums and galleries (fearing he may not get another chance) had just returned to England and was preparing to join the fire brigade.

Having heard nothing from their father, the children had assumed the holiday was off. Cecil's diary for the 31 August takes up the story: 'Mad, crazy mad! Today has broken all the laws of reason and sense as far as we're concerned, but its very incongruity and unexpectedness has been a welcome change . . . On coming home I was astounded by Mama's news that Papa had just rung up from Bath and that Jocelyne and I were to go to Sheringham immediately. My instructions from the fire brigade had been to report immediately on general mobilisation, but as that had not yet occurred I thought I could probably get back in time, and even a day at Sheringham was worth it'.

When they arrived at 7.15 p.m., their father was nowhere to be seen—neither had he even booked the hotel—but he arrived later that evening, having driven all the way across country from Bath to Norfolk.

After just two nights in Sheringham, Cecil felt he had to return to London 'to hold himself in readiness' and was at his mother's house in Kew on the morning of 3 September when Neville Chamberlain made his fateful speech on the wireless, announcing that England was at war with Germany. Rupert too returned south with his children and, by now slipping again into a state of nervous depression, made his way back to the Red House Hotel.

What happened to Rupert next is described by Cecil: 'On the outbreak of war he had got into his car and driven aimlessly from Leatherhead in a westerly direction. He got as far as Shaftesbury by nightfall, settled into an hotel there and drifted into a state of near imbecility within a few weeks'.[363]

❧ 21 ❧

Upper Hurdcott and The Brains Trust 1940–1945

When asked whether his father really did just 'go west' on hearing Chamberlain's announcement, Cecil confirmed: 'Yes, my father did drive aimlessly in September 1939. He was on the verge of another nervous breakdown. I suppose subconsciously he chose west as being farthest from German bombs (we all thought there would be an air raid on the night war was declared)'.[364]

Having hastily put all his books and possessions in store at the Red House Hotel in Leatherhead that same day, the 3 September, Gould took the car and headed out. By nightfall he was approaching Shaftesbury in Wiltshire and, almost out of petrol, decided to check in to an hotel.

The Coombe House Hotel was his choice. 'Private and Residential . . . Standing in own sheltered grounds of 50 acres . . . Recommended by Doctors as Good Health resort, Quiet and Bracing' is how the contemporary advertisement described it, 'Terms from 4 to 7gns, according to season'. One of the best in the district, it boasted 'Magnificent Ballroom', tennis, 9 hole golf course, billiard room etc, etc. It is certainly a beautiful place, today housing St Mary's School, Shaftesbury, but having lost none of its charm and peaceful grandeur. Surrounded as it was by some of England's finest and most idyllic countryside, The Coombe House Hotel must have seemed a haven in a storm.

But there was no escaping his predicament, and the fact that war had arrived. Indeed, this area too was soon a hive of activity with army and airforce personnel everywhere. In fact, as the war progressed, Coombe House Hotel itself became a recuperation centre for wounded and exhausted American bomber crews. So, 'quiet and bracing' as the place was said to be, Gould's mental state deteriorated, ultimately reaching 'near imbecility' (as Cecil described it), including loss of speech and memory. After a few weeks at the hotel, Gould was admitted by the hotel staff into Grove House, a local nursing home, primarily for expectant mothers, where he was treated for a nervous breakdown and began a recuperation. It seems he was in care here over the autumn and early winter, by which time he was slowly making a recovery and during which he was able to travel up to London at least once, visiting the Royal Geographical Society in Kensington.

Cecil went down to visit him in late December 1939, when he was back at the Coombe House Hotel.[365] This was probably only for the Christmas period however, as in January 1940 he was back in Grove House, from where he was writing to his friend Arthur Hinks at the RGS that he would only be there a few days longer. Evidently now pretty well recovered, he remarked that he would soon start work on a reply to the dreadful article by William Hobbs (mentioned in the previous chapter), but only, he said 'If I get no Naval employment'.[366] Gould had offered his services to the Admiralty back in 1938, and now, under the exigencies of wartime, he would soon be asked to take on some research for them.

Grace Ingram

Gould's very satisfactory recovery was in part due to the emotional support he had from one of the staff at Grove House. Veronica Grace Ingram (known as Grace), a young divorcee who was employed at the Home in a secretarial capacity, had taken a personal interest in Rupert. This interest was encouraged by the Matron who, by Cecil's account, considered it to be beneficial as a form of 'sex-therapy'. Cecil noted, rather unkindly, that 'I remember her [Grace] as a small, thin and pale woman, suffering from epilepsy and very over-sexed. She would have been in her mid thirties . . .'.[367] This was rather unfair of Cecil ('typically waspish' remarks his niece Sarah). Grace had been divorced from her husband in rather tragic circumstances. Their child had died during birth and the husband had left Grace, unable to put up with her regular epileptic fits and partly in the belief that her illness had caused the death of the child.

Rupert and Grace now continued to see each other and had soon, to use current terminology, become 'an item', though not yet living together. Meanwhile, given his financial circumstances, Rupert could not easily return to life at the Coombe House Hotel, and now settled into less grand accommodation at the Grosvenor Hotel in Shaftesbury itself.

Many of his friends and professional colleagues had of course been trying to get in touch with Gould during the past few months, and having now informed the Red House Hotel in Leatherhead of his forwarding address, he slowly began to get his correspondence in order again.

Big Ben

One of the more unpleasant tasks would have been dealing with Malcolm Gardner, who had been sending letters to him, via the Red House Hotel, to announce that the contract for the Big Ben book was now null and void, and that he wanted his advance payments back. By 31 December 1939 this had grown to £11.10s, the extra pound added was 'due on R.D. cheque destroyed'. Determined to get his money back, Gardner wrote again on 4 January, observing 'It is rumoured in horological circles that you have obtained an excellent war time appointment under the Admiralty, whereas you will doubtless appreciate that my present position as an antique dealer is far from happy, my sales being reduced to about a quarter of what they are in normal times. My bank balance is so low at the moment that I shall have difficulty in meeting my quarterly expenses, so that I should greatly appreciate a cheque in payment of the above account, or at least something on account.' On the 10th he wrote to Gould's solicitors (his cousin's firm) in Manchester, saying that Gould had told him he was on a list of creditors submitted to them for payment by the Trustees of his Estate. It seems that in desperation Gould was hoping to be bailed out from the residual Trust fund, though whether his relatives agreed to do this, given that the funds were being held in trust for his children, is not known. One way or another, Gardner appears to have got his money as the flow of letters ceases after this.[368]

BBC

Correspondence with his old friends at the BBC began again in early 1940. They had apparently tried to get in touch in December and had sent him a cheque, apparently an *ex-gratia* payment of some kind, as Gould wrote to say how much he appreciated the generous action of the Corporation. Evidently fully recovered at this stage, Gould now began accepting regular contracts for broadcasts again, both as *The Stargazer* and as a speaker under the heading of general talks, sometimes doing appeals on behalf of the forces. One regular broadcast was as part of the slot specifically aimed at the Anti-Aircraft personnel, on the radio programme known as 'Ack-Ack, Beer Beer' (the code names for the letters 'A' and 'B').

With finances worse than ever, Gould even had no bank account at this stage, and had to be paid by cash. The request for him to speak often came at fairly short notice, and many of his commissions were sent by telegram to him in Wiltshire. Broadcasts were mostly sent out from London—just about the only reason he travelled there at this period—but occasionally there would be broadcasts from Bournemouth or Bristol. These did not always go according to plan, one long round trip to Bournemouth being a complete waste of time owing to cancellation caused by air raids, but where possible, 'the show went on'.

Invariably the talks were scripted, or at least were discussed carefully beforehand to avoid any sensitivity, though Gould did apparently overstep the mark on one occasion. During a broadcast on 7 August 1940, when speaking about the hoax perpetrated by the American explorer Dr. Cook, who claimed to have reached the North Pole, Gould apparently made a controversial remark of some kind. It is not stated how he upset the censors, but he was evidently ticked off pretty hard. Gould apologized again in a letter to the Producer, Pringle, on the 9th, to which Pringle replied that they should 'say no more about the Cook business', and that he was 'glad you have no lingering ill-feeling' about it.

Upper Hurdcott

On 3 September 1940 Gould wrote to Pringle at the BBC from a new address: c/o Mrs Harding, Upper Hurdcott, Barford St. Martin, Nr Shaftesbury, saying that he would be there for the next four weeks, after which his address would be uncertain. What seems to have happened is that the money was finally running out, and Gould could no longer afford to live even at the Grosvenor.

Mrs Harding was in fact Grace Ingram's mother, and had just purchased a short lease on Upper Hurdcott, a small stone-built farmhouse just off the Barford to Shaftesbury Road. Taking pity on Gould, Grace persuaded her mother that he might be taken in as a paying guest, the four weeks perhaps being a trial period to see how everyone got on. Evidently things worked out well, as Gould would spend the rest of the war years at Upper Hurdcott, and would find a kind of peace here similar to that he had known at Downside.

The house itself, which had been known for a long time as 'Green's Farm', was entirely as built, well over 100 years before. There was no electricity or gas and heating was by fire and range in the kitchen, the lighting entirely by oil lamp and candle, and no telephone, the mail and telegram from the Post Office in Barford the only means of contact. There was at least enough room for the couple and Mrs Harding, and in due course Gould was able to set up his books and files in a study of his own. As it turned out, Mrs Harding died within a year of having settled into the house, and Grace and Rupert then had the place to themselves.

Upper Hurdcott had a large garden with many fruit trees—a small orchard in fact—and one of the several duties Rupert undertook as a working member of the household was harvesting the crop of apples each year, for storage in the loft. For fruit that even he couldn't reach, he invented and made a special 'apple picker', with a remotely controlled claw on a pole, for getting to the topmost boughs. An old stone outhouse in the form of a 'horse box' was turned into a useful, if rather damp, workshop and Gould was naturally also the household's 'handyman'.

The Green Dragon

Always fond of a drink, especially after a good bout of work, Gould was a regular at Barford's local pub, The Green Dragon (today called the Barford Inn). Some of the older locals in the area today can still just remember him walking down alone to the village from Upper Hurdcott at midday for a 'liquid lunch' and to replenish his supply of pipe tobacco—he was seemingly never parted from his 'Sherlock Holmes' style curly pipe, which was constantly alight.

The Marine Chronometer

Apart from preparing for broadcasts, his spare time during these few quiet years was spent reviewing and annotating the books he had previously published, most notably his first major work, *The Marine Chronometer, its History and Development*, which he very much hoped could now be published in a second edition. The war years were when the two annotated copies ('Copy A', the advance copy, & 'Copy B', the special interleaved copy, see p. 122) received most of their additions.

By 1925 he had evidently decided that Copy B, with its interleaved blank pages, should be the primary copy to be used for producing the new edition and all notes in A were copied to B in February that year. When he left Leatherhead in September 1939 it seems he left both books behind, and strangely, when he did start annotations again (still initialling and dating every note he made) it was in copy A. It is possible that Copy B had been left with Malcolm Gardner just before the War, 'on account', as Gould was always short of money. The 9 January 1940 sees the first new annotation in A, and this was probably when he returned to Leatherhead. An annotation in his own copy of the biography of James Cook is marked 16 January 1940, 'The Bull, Leatherhead', so he was back by then. The autumn and winter of 1940 reveal intense activity in annotations in Copy A, with other brief bursts of activity in 1942 and early 1943, but on 22 July 1943, annotations cease in copy A and begin again in Copy B. Evidently he had collected the book from Gardner, or from wherever it had been, at that time.

Hobbs

In November 1940 Gould was ready to tackle the matter of the Hobbs article (see previous chapter), along with a new travesty published by another American scholar, Lawrence Martin. Gould wrote to Hinks at the RGS 'My hand hasn't quite lost its cunning, and I should like to have a crack at Hobbs and his cronies myself—but I'm handicapped here by being short of data—especially charts', following it up, on the 10 January he notes to Hinks: 'As you say, Martin is nearly as slippery as Hobbs. Their methods of discussing geographical problems in print approximate to those of solicitors in a lawsuit—if there is a document in existence which tells against them, they ignore it so far as is humanly possible . . .'. He then added topically (about the Royal Observatory at Greenwich): 'By the way, I saw in the press recently that the Huns had managed to hit the Christie Altazimuth. Judging by what I've heard of its peculiarities and defects, I should imagine no one at Greenwich will lose much sleep over it'!

The result of Gould's thorough investigation into Hobbs and Martin's published pieces, which took up most of his spare time during the first half of 1941, was an important article entitled: 'The Charting of the South Shetlands 1819–1828', published in the *Mariners' Mirror* in July 1941. Arthur Hinks also published a critical review of Hobbs' article, so there was a twofold rebuttal.

Antarctic Sealers

Ostensibly Gould's article is an introduction to the pioneer surveys of the Antarctic by the early 19th century 'sealers'. Their charts produced the first delineations of that part of the Antarctic regions, and were immensely important both hydrographically and politically as they naturally had a direct bearing on national claims to sovereignty. Gould goes on to say that a résumé was needed, not just for interest, but as a corrective: 'In 1939 there appeared at Philadelphia an astonishing monograph by an American geologist, Professor W.H. Hobbs . . .'

Then, after finding (with some difficulty) a few things to compliment in the article, Gould continues 'Unfortunately, it is accompanied by a most regrettable commentary, acridly bitter in tone, painfully wrong in its facts, and scurrilous in language' and 'I do not propose—space, time, and patience would alike fail me—to follow Professor Hobbs closely through the whole bulk of his monograph; although I am quite prepared to point out a distinct mistake, misstatement, or suppression of fact upon every one of its seventy-one pages, including the title page . . .'. By simply relating all the historical facts, chapter and verse, Gould then proves that the Englishman Edward Bransfield, Master, R.N., was the first man to discover and chart a portion of the Antarctic Continent.

Gould continues: 'That is his offence—in Professor Hobbs's view, quite unpardonable. An American, Capt. N.B. Palmer, saw the same land nine months later; therefore the honour of first discovery must go to him. At whatever sacrifice of candour, of common fairness—even of common sense—Bransfield cannot, must not, *shall* not have gone southward . . . or even entered the strait which now bears his name'. Not leaving Hobbs a leg to stand on, Gould shows that, incredibly, Hobbs was so determined to prove the unprovable, he had actually falsified charts by cutting off the titles and reassigning them to other navigators, in one case even superimposing a border so he could alter the longitude of the chart by five degrees, to suit his case. Gould notes 'It is an unfortunate feature of Professor Hobbs's slips, and even his misprints, that—like the mistakes in a restaurant bill—they are almost invariably in his own favour', and as a parting shot: 'In a subsequent article Professor Hobbs modestly plumes himself upon "whatever of reputation I may have *for thoroughness of research*". The italics are mine—and the operative word, I fancy is "may" '.

The article was illustrated by Gould. 'I had to take to glasses to draw the charts for it—but I've turned fifty before needing them' he noted in a letter to Geoffrey Callender the following year. As soon as the article was published, Gould sent a copy to Hobbs, and eventually received the following reply:

(From the University of Michigan)
August 14 1941

Dear Sir,

Lest you should feel impelled to send me another of your precious reprints of the 36-page article published in the Mariners Mirror of July 1941, I hasten to acknowledge the receipt yesterday, together with your letter enclosed within its pages. Of my monograph to which your aspersions are directed, it seems to have achieved the distinction of bringing out two reviews of perhaps quite unparalleled proportions—21 and 36 octavo pages, respectively. It is unnecessary here to deal with their quality, but the large number of reprints mailed of the earlier article was probably also unique. One does not as a rule make reply to reviews of one's published articles, even when they are brief, but when they take on such prodigious proportions, such a proceeding lies outside the realm of the possible. In the case of the article by Hinks I have replied only to his charges of falsification, and this in a paper of four pages, which I herewith enclose. Perhaps you will not understand this, but for a long time I have refrained from publishing even this brief reply, and I have sent out only perhaps 25 copies of it, and these to persons specially concerned. My reason for this is that in the great conflict in which we are now engaged, my heart and my efforts are so given over to supporting Britain in the struggle, Anything tending to produce bitterness between the two peoples I think should be avoided. I have already glanced over your paper, and am now wondering whether duty should impel me to read it throughout. Its quality does not lure me. I note in the first pages that you go somewhat out of your way to support the great faker, Cook. One wonders if you know that the Explorers Club, an international group including Cook's closest friends, after a full investigation voted unanimously to expel him from the Club for the faking of the ascent of Mount McKinley. For the greatest oil frauds in history, he was later sentenced to the federal penitentiary for a term of fourteen years and six months, and after appeals to higher courts, ending in the United States Supreme Court, had utterly failed. The opinion in the United States Supreme Court was rendered by William Howard Taft, afterwards President of the United States. There were other frauds and fakes to his credit, but you would not be interested to learn of them.

Please believe me, Very Truly Yours, Wm.H.Hobbs

Sending a copy of this (and his reply) to the polar historian Dr Hugh R. Mill on 6 October, 1941, Gould remarks that Hobbs's extraordinarily hypocritical claim to sensitivity on the question of international relations during the war, (given that he published these false claims in the first place) 'has given me one of the heartiest laughs I've had for a long time . . .'. It goes without saying that Gould had no anti-American tendencies personally. One of his closest friends was the American Paul Chamberlain and he would have been just as critical of anyone, of whatever nation, who produced poor (let alone fraudulent) scholarship such as this.

Gould's reply reads:

(From Upper Hurdcott)
6.X.41

Sir,

I have your letter of August 14, and its enclosed pamphlet. These would have reached me sooner if sent to the address heading my letter of July 28 (and this) instead of to the Admiralty—which I left in 1927. I must thank you for sending me your 'Early Maps of Antarctic Lands' which I have read carefully. I do not find anything material in it which is not dealt with in my 'Charting the South Shetlands'. And you must forgive me if I remain a little sceptical of your professed motive for restricting its circulation. The avoidance of 'anything tending to produce bitterness between the two peoples' is a most laudable aim at all times, in war and peace alike; but how far this was likely to be furthered by your monograph is a question upon which I need not give an opinion—though it is not difficult to form one. On the other hand, and speaking for myself, if I could not have made some more effective and convincing reply than yours to pertinent and damaging criticism, I would not even have printed 25 copies of it. As Bentley said long ago, 'No man is ever written down, save by himself'.

There is much to be said for your view that your monograph has received a quite disproportionate amount of critical attention. But you should remember that an author's incompetence to handle his subject, however manifest, is no measure of his capacity to do harm—to impose his views, by sheer weight of assertion, upon an uncritical public. That is why his writings need critical attention, although undeserving of it. You are incorrect in assuming that my article is simply a review of your monograph. Primarily, as its title states, it is a summary—concise but, I think, accurate, of the true facts relating to the charting of the S.Shetlands in 1819–1828. Interwoven with this summary there runs, I readily admit, a secondary commentary upon a few of your more glaring errors; but these were not of my making or seeking, and I conceive myself fully entitled to find some useful purpose for them—just as the Spartans, long ago, were accustomed to inculcate temperance by letting their children view the antics of a drunken man. As for my tendency—in your conception—'to support the great faker, Cook', if you will read with rather more attention the few lines which I incidentally devote to his work you will find that I have expressed one opinion, and no more, concerning it. And that is, that I do not believe he reached the North Pole—one of the few matters upon which I should have imagined, *a priori*, that you and I were likely to find ourselves in agreement. Admittedly, I also remarked that his claim to have done so was more plausible than Peary's, but that it might very easily be, and yet wholly or in part untrue. For your information though it may be as well for me to remark that I have done my best during a good many years past, to make myself fairly fully acquainted with all the extant information relating to the lives and exploits of both Cook and Peary. I hope one day to produce a book of a kind which, in my judgement, has not yet appeared—a temperate and dispassionate examination, in detail, of the whole Cook—Peary controversy as it presents itself to a neutral investigator. Naturally, such a work is not to be undertaken lightly or effected hurriedly. It shall never be said of me, if I can help it, that I blundered into controversy upon a subject of which I knew little, and would learn nothing; or that I saw fit to publish a series of foolish and slanderous statements about men who were no longer able to defend themselves—statements which I could not possibly substantiate, and which anyone better acquainted with the facts could most easily refute.

Believe me Sir, with equal truth and respect, very faithfully your obedient servant, RTGould

Unsurprisingly, nothing more was heard from Professor Hobbs.[369]

Wartime work

As a postscript to a letter of 10 October 1941 to Hinks at the RGS, enclosing the Hobbs correspondence, Gould noted: 'I'm at present working, in such time as I can spare from picking and storing apples, on a very "hush hush" mechanical job for a Government Department—a queer job to tackle in a tiny, rural spot like this'.

True to their word, the Government had remembered Gould's offer and had found him something to contribute, though not in Hydrography, but for the Ministry of Supply.[370] In writing to Geoffrey Callender the following year, on 7 September 1942, he revealed that the job had been '. . . a long investigation of German time-fuses . . . they seem pleased with the results'.[371]

The Harrison Timekeepers

The reason Gould was writing to Callender was to bring him up do date about his circumstances and to ask a favour. Gould had visited the museum briefly at the beginning of June and had seen 'what was left' of the old workshop in the Queens' House. In the Spring of 1941 an incendiary bomb had landed on the House and this area had sustained damage. 'As you probably know, most of my tools have been looted. It's rather a pity, but luckily all the special gadgets which I had made myself for dealing with the Harrison machines were left, as being of no use'. The museum's Establishment Officer, Reg Lowen, had posted Gould his remaining tools and Gould noted: 'I have a little workshop here and I could, at a pinch, restart work on the Harrisons in it at any time'. Gould was primarily thinking of H3 of course, which was still in pieces and boxed up, currently in store at one of the museum's outstations in Minehead where it had been sent soon after the outbreak of war.

As for the favour, Gould had been asked to give one of a series of lectures on the history of science at Cambridge, his paper to be called '*The Quest for The Longitude*', and knowing that the timekeepers were in store at the Observatory there, he wondered if he may be allowed to show one of the watches at his lecture, on 30 October. Callender suggested he ask the Astronomer Royal, whose decision it would be, and evidently agreed that work could continue on H3 at Upper Hurdcott. In his answer to Callender[372] Gould adds: 'I should be very glad to have no.3 here, and will devote all my spare time to reassembling it . . . all that remains is the assembly and adjustment. This is, I should think, as safe a place for it as one could find. We are a mile up the Nadder valley from Barford, the nearest village, four miles from Wilton, and seven from Salisbury. Within a half-mile radius, there are not more than a dozen houses'.

Gould duly wrote to Spencer Jones, and in his reply,[373] the Astronomer Royal willingly gave his consent to the loan of both watches for Gould's lecture 'They have been three years in Cambridge and it would provide an opportunity to verify that they are dry and in good condition'. Following the lecture and Gould's inspection of the Harrisons in the cellars of Cambridge Observatory he submitted a comprehensive report to Spencer Jones, copied to Callender. The report included a little plan showing the exact placing of the timekeepers in the cellars (those particular basement rooms were subsequently filled in,

during works in the 1960s, owing to problems of damp) and noted that
H1 had indeed suffered from damp with all steel parts badly rusted.

In a covering note Gould adds 'It was certainly a little disheartening,
at first sight, to find that the same sort of thing, on a smaller scale, had
happened to No.1 as happened to it long ago. But this was probably
inevitable; and in the past I've had a lot of disheartening experiences with
the Harrison machines. Things have always come out right with them in
the end—and I shall look forward to overhauling No.1 once more when
better days arrive. Harrison 4 and Kendall were a *success fou* at my lecture
and I am very grateful for being allowed to exhibit them'.

Thankfully, H2 and the watches were in sound condition, though the
silver cases of the watches were now heavily tarnished, and the report
states 'I would recommend, when opportunity offers, that the cases
should be chromium plated; and, at the same time, that the plates of the
movement should be gilt'. The former advice was eventually adopted,
though, in his reply to Gould, Spencer Jones remarks: 'I would much
prefer rhodium plating, because a rhodium plated silver article looks
like silver and does not have the metallic look of chromium'[374] and
Gould fully concurs in his reply, admitting that he had not heard of
rhodium plating before.

Arrangements were now made for Gould's tool rack (rescued from
the looted workshop) to be sent, though Reg Lowen objected, claiming
it was too bulky. Evidently this news caught Gould in a bad mood and he
wrote to Callender sharply 'I take it that Lowen possesses intelligence
and a screwdriver. If he uses both, he will find that the tool-rack—which
by the way is quite light, though bulky—takes to pieces . . .'. The case
containing H3 in pieces was also sent by rail from Minehead and collected from Salisbury station by a local carrier, Mr Jarvis, who delivered
it to Upper Hurdcott. Gould wrote to Lowen and Callender on
14 December to let them know everything (except one glass jar of no
value) had arrived safely, and work now started again on H3. Gould again
had his much-desired source of horological interest, but progress, as
ever, was slow.

The Brains Trust

1942 saw a steady stream of broadcasts on *Children's Hour* and other
programmes, but the BBC soon had a more celebrated role for Gould. In
a letter to Callender of 29 September, 1942, Gould mentions 'I have been

invited to take part in a "Brains Trust" broadcast on Oct.13—I don't know if you ever listen in'.

Less well known today, *The Brains Trust*, first broadcast in January 1941, was at the time one of the most popular and well known programmes on British radio. Originally to be titled *Any Questions?* it was similar to the programme of that name which is still current on the BBC today (2005), both on radio and TV.[375] Questions from listeners were put by the Question Master, usually Donald McCullough (not to be confused with 'Uncle Mac', Derek McCulloch of *Children's Hour*) to a panel consisting of a variety of public figures, each notable for some achievement or some specialization, and whose opinion the listeners would value hearing. The significant difference between today's programmes and *The Brains Trust* is that with the latter, specific party political questions were avoided—in that sense it was wholly different from the current programme—and a much broader kind of subject was considered. Questions ranged from philosophy to scientific fact and even fiction. An element of *The Brains Trust* programme was re-invented briefly in another BBC radio programme broadcast during the 1980s entitled *Enquire Within*, in which all kinds of abstruse and unusual quandaries, outside 'general knowledge' were put to the programme for answer. The difference with the Brain's Trust was that a small panel of minds were expected to have an answer, and were given no prior warning, before the day of recording, of the questions (except for one 'Open Question' which all members were notified of some days before). Researchers on programmes such as *Enquire Within* spent weeks finding specialists from outside, before the programme went out.

So the remit was broad, even including questions intended for comic relief, much needed during the 1940s. The original panel, which was to include resident panel members and guest speakers, included as principal resident speakers, Cyril E. M. Joad, Professor of Philosophy at Birkbeck College, University of London, famous today for his coining the cliché 'it all depends on what you mean by . . .'; the biologist, writer and general science expert Julian Huxley; and Commander Campbell, the resident comic, famous for beginning his stories 'when I was in Patagonia . . .' Over the years there was a great variety in the kind of guest speaker chosen to join the Trust, including many famous and distinguished names. By the time Gould was invited to join, as a resident expert, the programme was nearly 2 years old and was hugely popular.

By 1944 it had a regular audience of 12,000,000 listeners and a weekly mail of 3,000 questions.

According to a newspaper article published when Gould died, the idea for inviting him onto the Brains Trust had come from one of his 'Stargazer' fans: 'The little girl used to sit entranced by the radio when "Stargazer" talked in *Children's Hour*. One day she could not resist writing the letter that made the B.B.C. big wigs sit up. "Why don't you have him on the Brains Trust?" ' she wrote, 'He knows everything'. The article continued: 'So, on to the Brains Trust went the tall, handsome, greying ex-Naval officer, and every session was enlivened and stimulated by his charm and dazzling erudition'. Waxing lyrical, the article, which was headed 'The man who was never wrong' (Gould would have been amused) praised his ability to 'delight, intrigue and flabbergast millions'.[376]

In a book about the programme the Producer, Howard Thomas, tells us:

'To these "residents" we . . . added . . . a man who almost chose himself, or rather was elected by sheer public esteem. This was Lieutenant–Commander Rupert Gould R.N. Gould was no newcomer to broadcasting, but apart from his appearances in the Children's Hour he was not well known to listeners. I heard him giving one of these talks, and as he poured out facts I wondered whether he was as good "without the book". He was. I invited him to take part in our programme and his first appearance was sensational. The newspapers hailed him as a 'discovery', and for days afterwards paragraphs about him kept appearing. A big, heavy man, with an owl-like face, Gould had very little to say at the preliminary luncheons. In the studio he talked so softly that I had to place him dead in front of the microphone in order that his words might be picked up. When the time came to deal with the questions, his answers were such a flow of knowledge that everyone else was left breathless . . . At one session, Joad mentioned one obscure book, and Mrs Hamilton mentioned another. Gould was able to clear up their vagueness by giving the titles of the books, a brief outline of the plots and the names of the characters. Time after time Gould trotted out dates and details with astonishing speed and accuracy. Once I asked him how he organised this mental card index of his. He told me that his was a photographic mind: 'I can visualise the actual page of the book where I had read the information'. The first time Gould appeared with our original three Brains Trusters I think they were all slightly sceptical of the depth of his knowledge, but Gould soon had them as impressed as were ordinary listeners. Once I invited him to give 'Second Thoughts' but he told me that up to then he had made no mistakes, and there was nothing to correct. I have never seen one letter from a listener correcting anything Gould has said; that was indeed a rarity in such a programme. Gould was the eternal small boy . . . out of his pocket I have seen him fish queer pipes, heavy old copper coins, a watch which showed the date and shape of the moon, a glass frame in which was a threepenny piece covered with a circle of paper the same size, on which he had written the Lord's Prayer in four languages. He even supplied a magnifier to read the writing.[377]

Naturally, in the chapter in Thomas's book titled *The Brains Trust Stumped* he notes: 'All this was in the days before we had Commander Gould in the programme . . . The question that closed the first Brains Trust season and was left unanswered was certainly one Gould would have enjoyed: "Why do even the most reliable wristwatches refuse to work on

58. The Brains Trust in action. Gould, on the right, has the art critic James Agate on his right and Jenny Lee M.P. on his left. Professor Joad is in the foreground on the left, sitting next to Donald McCullough, the Question Master. Next to him is J.G. Crowther, Head of the Science Department at the British Council. (Copyright © BBC)

some people's wrists?" Anything to do with watches or typewriters could be answered by Gould without a second's hesitation . . .'.

A contemporary account of meeting Rupert at this time was put on record by Reginald Pound, the Editor of *The Strand* magazine (to which Gould was a contributor), who wrote, in jocular style:

Commander R T Gould, whose astonishing memory feats are part of the entertainment provided by the Brains Trust broadcast by the B.B.C., sitting in my visitors' chair: face and figure of an eighteenth century scholar who, with a wig, would be a typical Kneller subject. Two rows of safety pins, large ones, affixed above a waistcoat pocket for no immediately obvious purpose. He talked in a quick recitative about the first typewriters, old watches, some mysteries of history including the inevitable Man in the Iron Mask, longevity, strange disappearances, and the Pyramids. Of the extraordinary mental facility one knew in advance from his frequent radio demonstrations of it. Of the safety pins one didn't. They seem to have no special relevance and were apparently not holding anything up, not even his reputation.

Then, commenting on the phenomenon of the Brains Trust itself, Pound noted, a touch cynically:

'*Memo to a future generation:* Kindly note that the recordings of the B.B.C.Brains Trust which you have been listening to in a reminiscent programme called "Scrapbook of the Forties" are not to be taken as evidence of the highest level of contemporary intellectual power among your great-great-grandfathers, but rather as souvenirs of a cerebral circus performing at this period to very large and uncritically appreciative radio audiences. What we call the Brains Trust is essentially an entertainment, characteristic of this era, in which great numbers of people desire information, knowledge, and enlightenment preferably, and in most instances necessarily, without any counter-contribution of effort, mental or physical. Just as the superficially informative weekly papers developed out of popular education half a century ago, educational advances since then, particularly in what we call the secondary schools, have created an undignified appetite for other people's opinions which can be passed off as one's own.'[378]

There was perhaps a little jealousy among his colleagues on the Brains Trust . Commander Campbell, the resident jester, remembered Gould in his autobiography (1953), but while complimenting his 'good friend Commander R.C.Gould [sic]' on his 'phenomenal memory', Campbell suggested that this was not always as clever as it seemed. A question was asked about the meaning and value of the mathematical term Pi (the ratio of a circle's diameter to its circumference) and (Campbell tells us) Gould answered that 'as far as it could be calculated it was three point- and then he added thirty-five more figures. McCullough took these down on a piece of paper as he spoke them, and at the finish said 'Gould, can you repeat those figures?' 'Certainly', said Gould, and proceeded to do so. I was not so impressed afterwards, however, when I worked it out myself. 'Pi' is represented by 3.142857 with the last six figures repeated to infinity!'. In fact this statement of Campbell's is just completely incorrect, as the value of Pi is (to thirty five decimal places, as Gould gave it) 3.1415926535 89793238462643383279950288. Gould would no doubt have been highly amused by Campbell's claim to have 'worked it out myself'![379]

It was not all dry facts with which Gould regaled the Trust and listeners though. If a member of the panel was affecting a particularly pretentious or pompous line on a question, Gould took great pleasure in bringing them down to earth. His favourite *bête noir* was Professor Joad, who would often come up with a particularly fanciful answer, after a considerable digression, only to be 'put in his place' with a few well chosen words by Gould. Happily, a few of the recordings of Brains Trust programmes have survived in the BBC sound archives and a number of episodes with Gould are still extant[380] enabling us to hear his dulcet tones across the years.

Gould often had answers to the comic questions too, sometimes with fantastically rapid wit. When the tricky question was sprung on the panel 'What is the difference between fresh air, and a draught?', it was Gould who, without a moments hesitation replied: 'It is fresh air when you pull the railway carriage window down yourself, and it's a draught when the person opposite you does it!'.

The programmes were usually recorded for broadcast some days later, and were always preceded by a lunch at the Dorchester for the panel, to warm them up. Gould's first recording for the Brains Trust took place on 8 October 1942. He noted to the Producer's assistant afterwards that it had not been quite the ordeal he had expected, though when he listened to it go out, he felt he could do better.[381] The fan mail, which he had already experienced in a small way as 'The Stargazer', now poured through the letter box at Upper Hurdcott, forwarded by the BBC.

On 18 March 1943, the Brains Trust team, including Gould, made a very special recording, in front of the film cameras, and produced by Strand Films, at Elstree Studios. Howard Thomas's assistant, Reita Hendry, wrote to Gould the following day to say that she hoped he was none the worse for 'the rather strenuous morning at Elstree'.[381a]

59. The Brains Trust on Film on 18 March 1943. The panel are (l-r) Prof.Joad, Julian Huxley, Sir William Beveridge, Donald McCullough, Gould and Commander Campbell. (London News Agency)

Broadcasting other than on *The Brains Trust* continued during the war, *The Times* on 19 October 1943 announcing, for example, 'Please Listen to Lt.Cdr R.T.Gould's appeal on behalf of the National work of the Royal Alfred Seaman's Institution . . . for British Merchant Seamen, their families and dependents, BBC Home Service Programme, 8.40 p.m. Sunday next . . .' General talks, as well as talks on the *Children's Hour* kept him busy too.

Occasional Odd Volumes dinners were held during the war, in spite of rationing, and Gould managed to attend the meeting at the Savoy on 13 April 1943, and the one in October, to see some of his OV brothers again. In general though, one has the impression he had, for the most part, lost touch with the majority of his old friends.

Cecil and Jocelyne

Both the Goulds' children were by now on active service. Cecil was Pilot Officer Gould, serving in Intelligence, at this time in Egypt but soon to be in Normandy. Jocelyne was an officer at Bletchley Park, the secret, code-breaking establishment in Buckinghamshire, where she met her future husband, Flight Lieutenant Frederick Stacey R.A.F.V.R. As a married couple, they visited Rupert and Grace at Upper Hurdcott. Jocelyne recalled not taking much of a liking to Grace during the two days they stayed, and Cecil was no more impressed with her when he visited on a separate occasion.

The Stargazer Talks

Work continued to progress slowly on H3. Gould wrote to Professor Callender on 25 July 1943 to apologize, saying that in addition to his doing jobs on the house at Upper Hurdcott, and his broadcasting duties, he had been working against the clock to finish a new book, with ten drawings to illustrate it. Against the clock or not, as Callender pointed out to Gould in his reply: 'I suppose that many months must elapse before the book appears under present conditions of the publishing trade' and sure enough, it was a whole year before it saw the light of day. It was titled *The Stargazer Talks*, and contained the text of fifteen talks he had given in previous years on the *Children's Hour*. Some, like the 'Devil's Hoofmarks', 'Sea Serpents' and the 'Canals of Mars', were subjects he had covered elsewhere, but a number were quite new. Included among these were such diverse subjects as 'The Man in the Iron Mask', 'The Beginning of Mechanical Flight' and the 'Mystery of the *Mary Celeste*'. Gould was one of the first to point out, in this story, that the mystery was not quite what it seemed as the ships papers and the ship's boat was missing, so it was clear *how* the crew had left, the real mystery was *why*.

The book had been offered to Methuen's back in late 1941, but they had declined and it was now published by his old friend J. G. Lockhart (who had himself published two books which included the mystery of the *Mary Celeste*) at Geoffrey Bles.[382] The book, priced five shillings, was a great success. The press enthused: 'My only complaint about this book is that it contains only fifteen talks. I could have done with more' (*The Sunday Times*), and 'They are just "talks" but oh! So interesting and strange, and written as though he spoke' (*The Daily Sketch*). Although the first edition states that it was 'First published 1943', it seems this was a misprint, as the reprint (necessary owing to the first selling out quickly) states 'First published July 1944, Reprinted September 1944'. The book was then reprinted again in April 1946.[383]

H3

Gould's letter to Callender also included a long report on the work now needed to H3: 'I cannot say how long the work will take—but I hope and intend to get the machine re-assembled and going before the autumn, so that I can move it into the house for the winter. My workshop here is a disused loose-box in a stable—one of the out buildings'. The repolishing necessitated by the poor storage at Minehead would have to be done by Gould himself as Buck's Gold and Silver Plating Company had now gone into receivership.

One or two new parts needed would also have to be outsourced somewhere else, as Gould discovered that Hopgood of Blackheath had retired from business and gone away, leaving no address. The two remontoir springs would also now need to be replaced, owing to severe rusting, but here Gould was in luck. By a rather extraordinary coincidence, Ganeval & Callard, the firm of springmakers who had made the present rusty ones, had moved from London for the duration, to Shaftesbury, and were just a few miles away. Gould also asked for the old display case for H3 as a dust cover once work was complete, as 'my workshop here is very dusty, and spiders are numerous'.

Writing in early 1944 included preparing the text of *Oddities* for a second edition, followed in May by the same treatment on *Enigmas* (Methuen's having agreed to release the rights to *A Book of Marvels*), and it was at this time that his old correspondent R.J. Cyriax got in touch again, and was able to make many useful comments on both books. The two authors would now stay in touch for rest of Gould's life.

Big Ben

In August 1944 Gould's friend, the clockmaker Alfred Gillgrass, knowing of Gould's interest in the Big Ben story, sent him a copy of a guide to Westminster which contained inaccurate information about the great clock and its famous bell. This was probably the beginning of a project which resulted in Gillgrass writing *The Book of Big Ben*.[384] Gould, having failed to do something for Gardner, and not having time to do anything himself now, consented to write a Foreword for Gillgrass. The clock's main protagonist was Edmund Becket Dennison, Lord Grimthorpe,

whom Dodo had met some years before when visiting friends at
St Albans. Grimthorpe was another of those historical figures who interested Gould greatly, as is pretty clear from his Foreword:

'Grimthorpe was aptly titled. He was one of the grimmest men who
ever lived. Yes, what he knew he knew thoroughly. His work on horology is first-class, and has stood the test of time. He was also, I believe, a
very sound lawyer—certainly he amassed a fortune of something like
two millions at the Parliamentary bar, disposing of it (as is the custom of
eminent lawyers) in a self-drawn will which it took the courts several
years to interpret. It had I believe, seventeen codicils, some of them
written in pencil on the backs of envelopes. As for his being of a quarrel-
some disposition, "you'd be surprised!". Controversy was the breath of
his nostrils—and he was merciless and entirely fearless, not only
verbally but in writing. Some of his published letters to Government
Departments make one's hair stand on end; mine did when I turned up
a photograph of him in his last years. You never saw such a face—it
would stop a clock'.

The Heart Attack

In October 1944, a Yorkshireman by the name of Eric Whittle wrote to
Gould, via the BBC, with questions about John Harrison, as he was
beginning some local research on him. Mr Whittle continued his
research for many years, corresponding with Gould on a number of
other occasions and ultimately publishing a small book *The Inventor of the
Marine Chronometer*.[385] Gould's reply in October 1944, typically packed with
useful 'Harrisonian' information for the young historian, apologized for
the delay, stating that he had been rather seriously ill, with pneumonia,
a later reply revealing that this illness was still with him.

Then, in December, a terrible and unexpected blow struck. Cecil said
he had always worried that his father, a lifelong smoker, considerably
over-weight in middle age and unfit, yet still at times very energetic with
tennis and practical things, was asking for trouble. And so it proved, as
that month Gould had what was, by all accounts, a massive heart attack,
quickly followed by a series of small strokes (often the result of a heart
attack). Telling R. J. Cyriax about it in correspondence some time later,[386]
he recalled:

'I woke one night in the throes of what I later realised was a violent attack of
cardiac asthma. The doctor who then attended me expected, as I learned

afterwards, that the end might be a matter of days—however, I rallied and as soon as I was able consulted a specialist at Winchester. He gave me the once over with great thoroughness . . . the heart trouble must have persisted, unnoticed for a long time'.

The Times reported on 28 December 1944 that Commander R. T. Gould was 'now able to go for short walks. He has been ordered to rest for some time'.

Canterbury and a Gold Medal
1945–1948

With the war finally coming to an end in the spring of 1945, the Astronomer Royal wrote to Gould to let him know that H4 and K1 were being taken to the Chronometer Workshops, which, since the early move from Bristol during the war, were based at a house named Lynchetts in Woolley Street, Bradford-on-Avon. In fact, Spencer Jones had taken them to exhibit at his Royal Institution lectures that January, and had them away from Cambridge already. Gould may well have known this because the lecture by Spencer Jones was reviewed in *The Times*, the reporter noting, of H4, that at the end of the lecture 'the children were allowed to handle this precious piece of ingenious workmanship', presumably closely supervised!

It transpired that Gould was a little offended that the watches were being sent to Bradford-on-Avon, and that he had not being asked to deal with them. He remarked in a letter to Callender at the NMM: 'This was the first I had heard of his intention and I was, I confess, just a shade surprised at being relieved, a little summarily after a good many years work, of their maintenance. But naturally, I accepted the position . . .'.[387] Nevertheless, he promised to attend when the timekeepers were sent to Bradford-on-Avon to give the staff in the Section a preliminary introduction to them.

David Evans

Gould, still having injections for his heart trouble, was unable to make dates in March but felt that in April 1945 he would be much fitter. Spencer Jones informed him that George W Rickett was still the Officer in Charge, the head of the repair shop being Holborn Warden. He also informed Gould that: 'The work on the two timepieces will be done by Evans, a young Scot, very keen and intelligent, and any tips you can give him about the construction and design will be much appreciated'. The only place to stay was The Swan, in the town, Spencer Jones remarking 'there is a shortage of staff, but I have stayed at many worse places during the war'. The 'Evans' that Spencer Jones mentions was David Evans FBHI (1919–1984), who would in due course become one of the world of horology's most respected and revered craftsmen. After heading up the Chronometer Workshops he went on to teach watchmaking practice to many grateful students during the 1960s at the Birmingham School of Jewellery and Watchmaking.

Gould visited Bradford-on-Avon on 16 May 1945, and oversaw David Evans dismantling K1, noting afterwards in a letter to Spencer Jones on 20 May: 'I was much impressed by Evans skill and his knowledge of horology and I am sincerely glad that the upkeep of no.4 and K1 will in future, no doubt, be in the hands of such a man—much younger than myself, more versed in minute work and having much better eyesight'. He still wished to have responsibility for the larger machines however, those being 'not at present beyond my powers, and I hope to continue with it for some time longer'.

Discussing the watches, he reiterated the recommendation that the cases should be rhodium plated, and that the movements of both watches might be electroplated. This suggestion was put by the Astronomer Royal to the staff at the workshops, who approved the plating of the cases but, concurring with Spencer Jones's own feelings, recommended against the gilding of the movements. At the same time Spencer Jones wrote to Gould asking if he wished to have the other two larger timekeepers sent to him at Upper Hurdcott for cleaning there, something even the enthusiastic Rupert would surely have found too much to take on at the time.

Curator of Navigation

On 29 May, Gould wrote (marked *Private*) to Callender at the Museum saying that owing to his poor health, H3's progress had been slow, and that as he will almost certainly be leaving Upper Hurdcott in September, and his movements would then be uncertain, it would be best if H1 and H2 (and H3 if he couldn't finish it in time), were all sent to Bradford-on-Avon.

In the longer term however, he believed he would be the right person to maintain them, and that the logical place for him to do this would be at the museum, where they were all destined to return in due course. So, getting to the real point of his letter, he went on: 'since, in that case I should, in any event be working at the Museum, I am emboldened to enquire whether you see any likelihood of my proving acceptable to yourself, and/or any other authorities concerned, as a successor to Maxwell [who had recently retired] in the charge of the Navigation Room?'

Money was evidently still a significant issue for Gould, but he chose to minimize this, in case it helped: 'May I say that whatever salary may attach to the post is not my first consideration? Since my Mother's death in 1937 I have enjoyed a quite sufficient income—previously, since leaving the Admiralty, I had practically nothing beyond my microscopic pension. My chief motive is that I feel I am qualified for the post and since with care of my heart, I still may expect a reasonable span of life (I shall be 55 in November), I should like to fill it to the best advantage with interesting and congenial work. If you have other views, or if you think that I am still *persona non grata* in other quarters, please regard my suggestion as withdrawn. You will not, I think mind my saying that my unhappy difference with my wife has now been made-up, and that, except to her, I have not mentioned my suggestion (and shall not, without your approval) to a living soul—I have pulled no strings'.

Callender replied the following day, the letter not surviving unfortunately, but it appears from Gould's response that he was told no permanent posts were being filled at the time. Gould responded by asking if he could be accepted on a temporary basis, perhaps for 12 months, reviewable every 3 months. He would accept Maxwell's salary of a flat £400 a year and would be prepared to sacrifice his Navy pension (£128,15s, gross) while receiving it, if necessary. It seems Callender felt this would be a possibility, and he was going to propose it to the Trustees at their meeting on 25 June. As it turned out, the decision appears to have been 'no' on the assumption that the Treasury were unlikely to allow it, but Callender let Gould know straight away that he was pursuing the matter, and that for the time being he should remain silent.

The museum's staff file then reveals a letter sent to Callender from the Treasury (responsible for agreeing all posts of middle ranking and above) on 7 July, informing him that actually they had no objection to Gould being appointed in Maxwell's place. So, after all, Gould was offered the post, on a 1-year contract, on 13 July, at £400 a year plus £60 war bonus.

Needless to say, Gould was overjoyed, and wrote to Callender 'I am writing within an hour of receiving it, to acknowledge your very kind and welcome letter of July 13th. May I say that I am fully convinced that the result can only have been brought about through your advocacy; and that it shall be my business to show that I am not altogether undeserving of it'. He hoped to travel to the museum in the next couple of weeks and hoped, if it was permitted, to bring H3 with him. A day was agreed but at short notice Gould was obliged, owing to pains in his stomach, to have to postpone the meeting. Writing to apologise he said 'I am so sorry this should have happened just now, as it must give a bad impression of my general health. Actually this is quite good; and as I say I have not had any trouble of this kind—and very few of any kind—for ten years'. Clearly he was anxious to present the picture of a reliable and fully fit new starter, but sadly the truth was very different.

It was at this time that Gould's last publication (during his lifetime), *Communications Old and New*, saw the light of day. A simple summary of ancient and modern methods of communications, from the Tom-Tom to Wireless Telegraphy, the book had been in the pipeline for some time and the 2nd edition of *Oddities* (1944) records it as a title 'by same author'. In fact it did not appear until 1945 (the book is undated). Gould sent a copy of the book to Callender on 12 July 1945, describing it as 'a pot-boiler which Cable & Wireless commissioned me to write some time ago and which has just appeared'. The book has characterful illustrations by C.F. Tunnicliffe and is one of Gould's rarer, more sought-after titles today.

Meanwhile, carrying on with help at Bradford-on-Avon, on 16 August 1945, Gould oversaw David Evans's overhaul of H4 and saw the newly rhodium plated case of K1 (H4's had not been unpacked but was also done) which he considered an excellent job. Writing to Spencer Jones on 17 August, he offered to escort the watches to the museum on about the 15 September when, he said, he was leaving Upper Hurdcott for good. He also hoped that the three large machines could all be brought to Greenwich for him to tackle, and H3 was taken there on 10 September, still in pieces, from Upper Hurdcott, with the other two large timekeepers following on from Cambridge soon after. Gould and Grace themselves had moved from Upper Hurdcott at the end of the month, the lease on Upper Hurdcott having expired, and had settled in Kent.

Gerry Salmon

Back in May 1943, the couple had been visited in Wiltshire by Geraldine Salmon, a writer of historical novels, well known in her day under her pseudonym of J G Sarasin. Salmon, who was staying in Dorset for the duration, quickly became a firm friend of Rupert and Grace and, according to Cecil, as the war's end approached, Gerry Salmon offered the couple accommodation at her home. This was a large house called *The Grange* which she shared with her mother, at Harbledown, a village near Canterbury in Kent. The Grange had been requisitioned as a billet for soldiers during the war, but was now released to Geraldine and Mrs Salmon and they returned to Harbledown, along with Geraldine's aunt and cousin.

Initially, Rupert and Grace moved to the Grange too, but this did not last long. Gould's daughter Jocelyne has explained why.

Kew

As noted in his letter to Callender, right at the end of the war, Gould was getting on better with his wife Muriel. Although she had refused to take him back after Miss Gurney left in 1938, she and Rupert were generally on better terms. But now he was with another woman, Grace Ingram, and Muriel was apparently jealous. According to Jocelyne 'She didn't want him back, but she didn't want Grace to have him'. On balance then, she made her decision to try and have him back, and Jocelyne tried to help by being the intermediary.

Arranging to meet her father at Waterloo Station, Jocelyne eventually persuaded her father to return to his wife. She recalled:[388]

I went to Waterloo and he didn't recognise me . . . He was just staring ahead of him and I could see that he'd have passed me. I said "Daddy it's me". Then we went to Kensington Gardens and I told him that mother wanted him back. So we sat there talking about this. He said "well its so difficult because you have persuaded me but of course when I go back Grace will persuade me" '.

But as Rupert had always said, Muriel was his first love and he decided to give the marriage another try. Leaving Grace with Geraldine Salmon at Harbledown in September 1945, he moved in with Muriel at Kew and she reported to her daughter a couple of days later, reassuringly, that her parents 'were husband and wife again'.

Alas, the optimistic outlook did not last long. On 17 September, Gould started work at Greenwich, travelling every day by train, but for one week only. The minutes of the next NMM Trustees Meeting succinctly explain that the new Curator of Navigation 'being in an obviously weak state of health, had been persuaded to take medical advice and had by Doctor's orders been forbidden to undertake regular employment.[389] He accordingly on 24th September handed in his resignation with an expression of his profound disappointment at the enforced severance of an association thus happily begun. He hoped, however, to be able to continue work on the renovation of Harrison's Chronometers. The Trustees expressed their regret at losing so valuable an addition to the staff'.

This was a bitter blow, but it didn't stop Gould from active involvement in horology of course, and on 27 September he was at the Royal Astronomical Society putting the Harrison clock back after its return from Oxford. The RAS notebook records: 'Remounted clock. 2pm left it going, fully wound . . . Clock showing correct time by TIM Arc 6¼ degrees', and over the next few months he would visit several times to check on its going and timekeeping. Owing to his deteriorating health though, it was not be long before he was obliged to hand over the care of this clock to others as well, and Charles Frodshams would soon be looking after it for the Society instead.

Things were no better in the newly reunited marital home than they were for permanent employment. Jocelyne recalled 'after a week of him being there, Mama I think realized it was all a big mistake . . . and she said to me "I've got to go away, I've got to go away" '. Feeling trapped, Muriel took off for a couple of weeks, leaving Jocelyne to stay with Rupert. 'She claimed she didn't want him there in the end. She'd got him away from Grace [but] she didn't want him there and she wasn't really very nice to him. I mean it wasn't in her nature to be really nasty, really venomous, but she wasn't nice to him, she found him a nuisance'. And so this brief reunion ended sadly. A brave face was put on things when required; a small celebration was held by Cecil's parents at the house for the announcement of his new posting as Assistant Keeper at the National Gallery early in the new year, and the couple stayed together, though effectively living apart, for over 7 months.

From the beginning of 1946, Gould's health was never again good enough for much activity, and the remaining years—there would be less than three of them—were spent reading and doing a little writing, though no more books were on the cards. According to *The Times* on 28 January that year, he did agree to accept the role of President of The Johnson Society, based in Lichfield.

Interest in Harrison's work was always present of course, Gould's fascination for this subject never deserted him. In February 1946, a local Yorkshire paper reported: 'Few, if any of the company enjoying their glass and a chat in the Angel Hotel, Brigg, about a fortnight ago, were aware of the identity of the two strangers sitting quietly in a corner talking with a local resident. One was Lt.Cdr GouldR.N.(Retd), well known broadcaster and member of the BBC Brains Trust; the other was Mr Alfred Gillgrass of Leeds . . . Earlier in the day they had travelled from Leeds to Brocklesby Hall where, by permission of the Earl of Yarborough, they had gone . . . to see a stable clock . . .'. This was the exceedingly fine and important turret clock by John Harrison at Brocklesby Park. It was the precursor to, and the 'test bed' for Harrison's precision pendulum clocks. Brocklesby's clock is well documented today, but was hardly known in the horological world at that time. It is a measure of Gould's determination to keep his horological interests going that, weak as he was, he made his way to North Lincolnshire and climbed the stairs at Brocklesby to see a work by Harrison.

60. Gould's close friend Geraldine Salmon. A successful writer of historical novels, she published under the pseudonym of J.G.Sarasin. (Lander, Canterbury).

61. The Grange at Harbledown, near Canterbury in about 1945. Gould spent the last 2½ years of his life here at Geraldine Salmon's house. (Anthony Salmon)

On 13 May 1946, Rupert made the final break with his wife and left London. Happily for him, Grace Ingram was still prepared to have him back and he now returned to live with her permanently at Gerry Salmon's at The Grange, Harbledown. But his health was still deteriorating, and even that very day, he told George Rickett of the Chronometer Section,: 'as soon as I got here I had to go to bed as I was coughing up a lot of blood . . . the haemorrhages were simply a bye-product of my dilated heart. Still, they weren't pleasant'.[390] In fact Gould's health was very poor throughout this last chapter in his life. The strokes had partially paralysed the whole of the right hand side of his body, and when out walking he would always have Grace on his arm. He was unable to get into the bath unaided, and a neighbour, Mr Burgess, would visit three times a week to help him bathe.

The Harrisons

By the 8 May 1946, the Harrison timekeepers were still not returned to the NMM from Bradford-on-Avon. Callender and his Trustees were becoming impatient to discover what progress was being made, as visitors were constantly enquiring when they were to return. Rickett was soon able to report to the Astronomer Royal that, duly spurred on by this prompt, the Section had arranged for the completion of H4 and K1, and they were returned and set up at the museum by David Evans on 24 May 1946. Rickett continued: 'Regarding Harrison 1,2 & 3, I gathered from a telephone conversation with the museum Secretary on Thursday that these machines would be forwarded to Bradford-on-Avon by van at an early date, together with many odds and ends of tools, etc. used by Gould in the maintenance of the machines. It is practically certain that Gould will at no time in the future be fit to undertake the reassembly of No.3'. He went on to say that the way he felt best to pacify the Museum was to get H2, which was in reasonable condition, back to them as soon as practicable (about 4 weeks), but H3 and H1 would take much longer.

Gould was consulted and, writing to Lowen at the museum on 3 June 1946, concurred with all this. Later in the month he wrote to Rickett at Bradford on Avon to say that he now felt: 'I should turn over practically all of my collected material to the Observatory, or to you, and resign my Harrison work to Evans, including No.3. I feel my day is done. In my present state of health—which I cannot expect to improve—with chronic asthma, a half disabled right hand and failing eyesight, I am no longer fit for intricate work on the machines. At best, I can only expect to act as a consultant which, I need hardly tell you, I shall be glad to do as long as I live . . .'

In June, H4 had developed a fault and was taken back to Bradford on Avon on the 13th, then being returned to Greenwich in mid July when all three large machines were taken back to Bradford on Avon for restoration. Replying to a note from Gould, the Astronomer Royal summarized everyone's view at the time: 'I am sorry that you are unable to do any further work on the Harrison timepieces, but I think that Evans, with the aid of your notebooks and other materials, will be able to get them into satisfactory going order. He may need to consult you from time to time. He is the right sort of person for this work as he has the enquiring type of mind, gives a great deal of thought to what he is doing, and is a highly skilled craftsman . . .'

Gould replied on 24 August, sending his Harrison notebooks (Nos. 2–12) as a gift for their use, 'But I need hardly say that so long as my knowledge and experience are available, in consultation, they will always be at the service of the Harrison machines'. H2 was by now running reliably at Bradford on Avon and was returned to the NMM in October, Gould visiting in November to find that it had stopped owing to problems with the machine's 'Achilles heels', the balance connecting wires.

New books, written single-handedly, were out of the question for Gould by this date, but a second edition of *The Marine Chronometer* was still something he dearly wanted to see completed before he died. In a letter to Professor David Torrens on 11 September 1946, he asked for help. He thanked Torrens for the kind things he said in the September issue of *The Horological Journal* about the book, and went on: 'I have a favour to ask. If medical opinion is any guide, I have not very long to live; in fact, X-Ray photographs of my dilated heart suggest, I understand, that it is remarkable that I should still be alive. On the other hand, I much want, if I can, to get out a second edition of the book—my interleaved copy is bulging with notes made to that end . . . Could you find time to jot down a few suggestions as to eliminating outstanding defects? Some of these I know already . . .'

He admitted to what he believed to be technical weaknesses, such as too heavy an emphasis on Harrison and too little on Le Roy and Arnold, and said he felt the part on Huygens needed completely rewriting. He stated that he thought the style was stilted 'but then, it was my first book; and I am still teaching myself to write more or less clear English', all really rather unfair on himself. Meanwhile, Malcolm Gardner bought the book ('Copy B') from Gould on 19 June 1947, (for £5) with the agreement that Gould could hold on to it until the 2nd edition had gone to press, or Gould had died, 'whichever was sooner'. This must have been another occasion on which Gould needed money, Gardner's receipt shows him paying Gould a total of £68/10/- (68 pounds, 10 shillings), for an 8 day marine chronometer, two lots of books and a small manuscript.

Gould attended his very last meeting of the Odd Volumes on 27 July 1947, taking Cecil as a guest. It was at this dinner, it will be recalled, that Cecil was introduced as not having his father's good looks, inheriting, as he does, his mother's features. A rather cruel comment, and quite unlike Rupert, but perhaps understandable as far as his attitude to Muriel was concerned. The attendance register contains the usual signature, but now in a very shaky hand.

The Third Sex

Writing was naturally much slower at this period in Gould's life, but it had not totally ceased at Harbledown. It was at this late stage that Gould began a project which had been on his mind for some years, a study on the subject of bi-sexuality and homosexuality. The resulting book was to be written in conjunction with a local Doctor, Henry Treble, and was to be called *The Third Sex*, but the plan failed when Gould's health declined further. Treble was a partner in the practice headed by Gould's own G.P., Dr Wacher. In later years Dr Treble, himself homosexual, told the whole story of the book to one of his patients, who subsequently became a close friend. This was Mr John Burgess (the son of the local man who helped Gould at The Grange) and it is Mr Burgess who has kindly provided this insight into Gould's activities in his last years. Treble himself told Mr Burgess that he and Gould were collaborating on the book, and that Gould felt quite strongly that the subject needed a compassionate and objective study. Gould had remarked to Dr Treble that he had 'seen too many lives ruined' and that a better understanding of different sexual orientation was long overdue. It is not clear however where the data for such a study would have come from, and what Gould's part in the book would have been, but it provides an interesting footnote in the story of Gould's many interests.

Typewriters

Some smaller academic projects did see completion at this time. Writing to Mr Garratt at the Science Museum on 26 July 1947, to exchange typewriter information, Gould mentioned that 'I have just done the article TYPEWRITER for the new edition of Chambers Encyclopaedia. They proposed to allow 500 words to cover the whole subject!'. This work no doubt formed the basis of a series of articles, called *The Story of the Typewriter*, which Gould wrote for the journal *Office Control and Management*, published between January and September 1948. It was this work, referred to in Chapter 19, which was then published, just after his death, as a booklet. The editor, Dudley Hooper, noted in his Introduction (which was a tribute to Gould) 'When I saw him last, at the Business Efficiency Exhibition in 1947, he was a sick man; I think he knew he would never recover, but looking round the exhibition (he was thinking of writing a comparable work on the adding machine) he remarked that the less time one had to live, the more one found to arouse one's curiosity . . .'

Commander Campbell, his colleague on *The Brain's Trust*, remembered Gould's last broadcast at this time. It was his fiftieth. Campbell recalled: 'He was then a very sick man and even I who was sitting next to him had difficulty in hearing him. It was not surprising that the BBC had dozens of telephone calls in the first five minutes asking Commander Gould to 'speak up'. Campbell recalled 'Once he said "Campbell, I have the frame of an ox and the constitution of a cat" '.[391]

Gould did try to maintain some of the less strenuous activities, including his membership of the Council of the British Horological Institute, to which he had been asked to return a few years after the judicial separation. Also he kept his membership of the SNR's Council to which he had recently been appointed. In October he combined a Council meeting of the SNR at the museum with a visit to oversee the securing of H2, which was being temporarily moved during redecoration of the gallery, though he frankly admitted that with Evans present too, he found his advice was almost unnecessary.

The BHI

Gould's colleagues in the world of Horology were now beginning to realize that one of their most revered figures was not long for this world, and the decision was made by the BHI Council that he should be awarded the Institute's Gold Medal for his contribution to Horology. Gould attended that Council meeting, and was asked to temporarily leave the Council Chamber while the matter was discussed, the young Assistant Secretary, Charles Allix, escorting him into an adjoining room. Mr Allix still (2005) remembers the conversation they had that afternoon, while the deliberations took place, and recalls that, after a few minutes Gould suggested the two of them should slip over the road to the pub while the Council made their mind up! Needless to say Mr Allix advised against it.

The presentation of the Gold Medal took place at the BHI's Annual General Meeting, held at Goldsmiths' Hall on Wednesday, 29 October 1947. The medal was presented by the BHI President, Gould's old friend the Astronomer Royal, Sir Harold Spencer Jones, who had himself just received the same medal that evening for his contribution to the nation's timekeeping.

These awards were recorded for posterity in photographs published in the *Horological Journal*, and looking at the photograph of Gould's

62. R.T.Gould receiving the BHI Gold Medal from the Astronomer Royal, Harold Spencer Jones, at Goldsmiths' Hall on 29 October 1947. Although only 57 years old, Gould was a very sick man by this time. (*Horological Journal*)

acceptance, one is inclined to look twice: the man receiving the medal appears so frail it is difficult to recognize the large and imposing figure that was Cdr Gould. The HJ reported 'Commander Gould said that the award filled him with great gratitude and a deep humility. In fact, this was the crowning hour of his life'.

Harbledown

A little writing occupied his time into the year 1948, but for that part of the year that remained for him, he was too weak to do much. One of his last efforts was to write a long letter to the *Horological Journal* about Grimthorpe, the Dents and the Westminster clock, which resulted in a spate of letters, Gould's last being published in March.

The last few months appear to have been very quiet at Harbledown. The young neighbour, John Burgess, recalls that Rupert and Grace would sometimes take trips down to Folkestone or Dover in a hired chauffeur-driven Rolls Royce, and Rupert would sit for the afternoon watching the ships in the channel, using a pair of powerful binoculars lent to him by Mr Burgess. But apart from such gentle, stress-free activities, there was apparently very little work of any kind at this stage; there is virtually no correspondence of Gould's surviving beyond the Spring of 1948 and he probably spent the Summer in total rest. In late September, developing severe pneumonia, his health deteriorated for the last time and Dr Wacher admitted him to Canterbury Hospital. Cecil went to visit him there and remembered him being sad and a little bitter, Cecil recalling that 'he wasn't very nice to me'.

Rupert Gould died at Canterbury Hospital, in the late afternoon of 5 October—on his father's and his brother's birthdays. Coincidentally, his death in Canterbury connected him to a much earlier part of his family history—it will be recalled that Rupert's great-great-grandfather, Thomas Gould, had been born in that city in 1806. Gould's death certificate records the cause of death as 'Parkinsonism' and 'Paroxysmal Sachyondia'. It was always said that Gould had, for the latter years of his life, suffered from Parkinson's disease. According to current specialists however, it is apparently just as likely that his heart attack, and the consequent strokes, had produced 'Parkinson-like' symptoms, a common result of a stroke, and that he did not actually suffer from the disease itself.

A few of his old friends in the Sette of Odd Volumes had heard rumours that he was unwell, and were beginning to worry about him. In mid-October Vyvyan Holland commented in a letter to Ralph Straus that he had heard that 'Brother Hydrographer' was seriously ill and that his death was probably imminent. In fact Gould had already been dead for several days.

Obituaries

Obituaries appeared in *The Times* as well as many other national and local papers, including the *Illustrated London News*, and, of course, the *Horological Journal*, which included three separate appreciations, from the historian H. Alan Lloyd, the electric-clock pioneer Frank Hope-Jones, and his recent colleague and 'Harrison successor' David Evans. His old acquaintance and admirer, Heinrich Otto, took great pains to produce a large and somewhat rambling tribute to Gould, which he tried hard to get published, even sending copies to Cecil and to Muriel. But, being his own worst enemy, and fussing too much over the matter put people's backs up and the piece was not published in the UK, only in the US, in the journal of the Horological Institute of America.

The degree to which Rupert had been short of money was then revealed once his administration was completed by Cecil. The value of his whole estate, published on 3 March 1949, was £428/10/ 10 (net).

The funeral was held at Ashtead Church at 3.00pm on Monday 11 October, and Rupert was buried in the same grave as his mother, the monument marked in Latin (of course) *'Non Omnis Moriar Multaque Pars Mei Vitabit Libitinam'*: 'I will not wholly die and the greater part of me will escape the grave'. Perhaps this was an expression of hope that his memory and that of Dodo, will live on (it would not have had overtly religious overtones as far as Gould was concerned). Coincidentally, Gould's young neighbour at Harbledown, John Burgess, worked at the funeral directors who undertook the arrangements, and he remembered that many (about 20) wreaths were sent, Grace's a cushion of red carnations, marked: 'to my dear Rupert from your beloved Grace'. The many wreaths marked the considerable number of people or institutions wishing to pay their respects but, while Rupert's son Cecil, and Grace were present at the funeral, almost no one else attended. Neither Muriel nor Jocelyne were there. Cecil believed that both subsequently regretted this, the picnic for the children on Rupert's grave the following year, described in the Introduction to this book, perhaps being a small indicator that Muriel felt she had been pretty tough on her husband at times. She almost never again would refer to their differences though, and never again spoke of her relationship with Miss Gurney, to a large extent preferring to pretend those episodes in her life simply hadn't happened.

Of his father's death, Cecil wrote:

'In our tormented family life, his death affected me less at the time than in later, more reflective years. His life, as he himself realized, had been a sad waste of great and varied talents. I think that his failure to make a success of his adult life was partly due, as in other cases which I have known, to his having been too successful as a boy. His brilliant school record had been achieved at the cost of overwork which had sapped much of his energy and exhausted his ambition. His entering the Navy had been a mistake, and the shattering of his career when my mother took him to court had left him with no sense of direction and largely dependent financially on Dodo . . .'

It must be said that this is a rather pessimistic view, by a son who, by his own admission, never much liked his father, but Cecil's continuing remarks are certainly exactly right: 'The most lasting monument to his life and work is his standard book on the marine chronometer and his restoration of the Harrison time-pieces'. For his contribution to the world of horology, this achievement alone, in one man's life, should surely be enough to justify a feeling of contentment, but when one considers the many other fields in which R.T. Gould also made ground-breaking contributions, there should be no doubt.

His enthusiastic championship of his many interests, made so accessible by his incomparable broadcasts on the BBC, inspired listeners to take an interest and to realize their potential. Not everybody has the enormous mental capacity that the celebrated *Stargazer* did, but by demonstrating such effortless erudition, he showed that knowledge can be empowering, encouraging a generation of young people to strive to learn.

Of his research and publications, hardly a day can go by in this, the twenty-first century, when someone in the worlds of horology, crypto-zoology, tennis, typewriter technology, or broadcasting, doesn't refer to what Rupert T. Gould has said about a particular matter of their concern. He may not have believed it as, in pessimistic mood, he lay dying in Canterbury Hospital, but his name would live on across the world for the great achievements of his remarkable life.

Gould summed up his own pessimistic view of his life, expressed when he was apparently well and in his 50th year, in his personal copy of *Oddities*. In the text, he commented on the nature of the eccentric seventeenth century inventor Orffyreus, saying of him that:

'He passes from our sight ... an exasperating and yet pathetic figure—morose, self-centred, childishly passionate, vacillating and yet tenacious, his own worst enemy, forgetting the duties of ordinary human intercourse in his passion for mechanism and wrecking his life as a result. *Non Defecit Alter*'.

Alongside, as an annotation, dated 27 July 1940, Gould has remarked: 'The other, of course, being myself'.

Epilogue

John Burgess reports that on Rupert's death Grace went to live with her brother 'in a large house in the country' and died there a few years later during a severe epileptic fit. Geraldine Salmon lived on into old age, eventually going into a home in Herne Bay in the early 1970s.

Muriel Gould lived alone for the rest of her days, moving several times from one flat to another, restless as always, and died in London in 1978.

Cecil remained as a professional art historian at the National Gallery for the rest of his working life, contributing many published works on his speciality, the Renaissance, eventually being appointed Deputy Director and retiring in 1978. He died in 1994, was unmarried and had no children.

Jocelyne and Frederick Stacey settled in Reading in Berkshire and had two children, Sarah, unmarried and today a successful journalist and writer, and Simon, a Chartered Surveyor, married with two children, Rupert's great-grandchildren, Ross and Alexandra. Latterly, Jocelyne and Frederick were divorced, and Jocelyne then shared a house with her brother Cecil in Thorncombe in Somerset (see plate 16). She died in 2001.

The Harrison Timekeepers

The extraordinary story of the restoration of the Harrison timekeepers does not end with the death of Rupert Gould, of course, and Appendix 3 includes a note of those responsible for their continuing conservation over subsequent years.

Postscript: A ghost story

The house at Upper Hurdcott still stands—just. It is now (2005) in a ruinous state and completely engulfed in brambles and ivy from top to bottom, the roof partly open to the sky. The apple trees are still there, buried in the jungle that was the garden and the remains of Gould's workshop, without roof, survives on the side of the house.

After Rupert and Grace left, the house was let to a succession of families, but a few years after Gould's death it was said that the house had become haunted and locals claimed it was the spirit of Commander Gould! A couple visiting the house in the early 1950s found that the last occupants had left in a hurry: furniture was still there, food was still in the kitchen and some washing was still on the line outside. But the house had been deserted and was never lived in again. Rupert would have been amused.

References and Notes

CHAPTER 1: CHILDHOOD 1890–1905

1. Cecil Gould and Jocelyne Stacey (née Gould) kindly provided many reminiscences and much family information in extended audio interviews. The tapes are held by the author.
2. Much useful biographical information has also been taken from Cecil Gould's unpublished autobiography, *Tosca's Creed* (revised edition 1992), a copy of which Cecil Gould kindly presented to this author for use in the biography.
3. Apparently Rupert's brother Harry, when a baby, had uttered this sound, instead of the more usual 'mama', when seeing his mother, and the name stuck.
4. According to Cecil, Rupert's son, the Monks were descended from the famous General George Monk, 1st Duke of Albermarle (1608–1670) who, after switching sides during the Civil War, was instrumental in bringing about the restoration of Charles II to the British throne. There is unfortunately no definite proof of this connection.
5. Gertrude S. Monk (24, Dressmaker), Mary A. T. Monk (20, Dressmaker), Eliza Worth Monk (16, Pupil Teacher at the Church National School), Harry E. Monk (13, Scholar), Gertrude J. Monk (10, Scholar), with Ambrose Crebor, a 20-year-old carpenter from Whitchurch, staying as a lodger.
6. John (12), Emma (6), Thomas John (4), Richard (3), and Sarah (1).
7. Cecil claimed the Goulds also had distinguished forebears, being descended apparently from a bastard son of Edward IV. Apparently, a wing of the Gould family also had a Devonian connection, the most notable from that line being a contemporary of William's, the Rev. Sabine Baring-Gould (1834–1924), the popular hymn-writer (works including 'Onward Christian Soldiers'), though what the exact relationship was is uncertain. As will be seen, William's immediate paternal line came from Kent.
8. In 1851, the Goulds were still in Dover but had moved to George Street. In the census that year Thomas and Ruth are shown with their son Thomas, now 14 years old, as the oldest (John and Emma may have left home or may have died in the interim), Richard, now thirteen, but no Sarah, who probably died in infancy. Instead, two more children had arrived: Ruth (7) and Steven (2).

9. The children were William Monk Gould, Minnie Eliza Gould, Harry L Gould, Gertrude Monk Gould and Amy Jane Gould . The 1861 census finds the family living in Prison Road, Lidford, with one 18-year-old servant. In 1871 they are at 15 Mermaid Street Rye, and the 1881 census reveals that they have moved to 26 Watchbell St in Rye. William Monk Gould, their son, had left home, but Gertrude (16, and born in Malvern in Worcester) was there with her parents, plus one servant, Jane Locke (Scottish and also 16).

10. L.J. De Bekker, *Black's Dictionary of Music and Musicians*, Black, London, 1924, p. 243.

11. Bank of England *Equivalent Contemporary Values of the Pound: A Historical Series 1270–2004*, London, 2004.

12. British Library, *The Catalogue of Printed Music in the British Library (up to 1980)* Vol. 24 London, 1983.

13. The source is an obituary for Mary Badcock in a clipping from a newspaper, of uncertain title and date, in the family's records. Mary Badcock died in 1948.

14. Thomas Skinner studied at St Andrews and the Royal College of Surgeons, graduating MD in 1853. He took the Simpson Gold Medal in Gynaecology and Obstetrics in 1855–1856 and was then private assistant to Sir James Young Simpson himself (1811–1870), who was Professor of Midwifery at Edinburgh, Physician to Queen Victoria in Scotland and the famous specialist in anaesthetics in childbirth. Simpson and his protégé Skinner pioneered the use of chloroform in that application. On moving to Liverpool in 1859, Skinner established a busy consulting practice in orthodox (allopathic) medicine, also joining the Liverpool Medical Institute. In 1876, once a convert to Homeopathy, he visited the United States, and during that time he developed a 'centesimal fluxion machine' for making high potency homeopathic remedies, the device thereafter being known as the Skinner Machine, an example of which is preserved at the London Homeopathic Hospital. He oversaw the publication of *The Organon*, the hugely influential international journal of homeopathy, described as his 'chief contribution to medical science'. He was thus a highly significant, pioneering figure in US and UK homeopathy, being best known in the history of medicine today for this work. After moving to London in 1881, Skinner's practice was first at 25 Somerset St., Portman Square; then at 6 York Place, Baker St., and lastly—for only a few months— at 115 Inverness Terrace. Skinner's wife Hannah died in 1897 (14 July), at Folkestone on the South East coast (recuperating from illness perhaps?) and it seems Thomas now sold the home in Beckenham and moved up to London. Apparently his son (Dodo's brother) Hilton Skinner, and his wife Emily and son Douglas Hilton had also been living with the parents as

they too now moved to a house of their own in Hayes, a few miles from Beckenham. Thomas Skinner soon married again and Thomas and Lillian—43 years younger than her husband—are shown in the 1901 census as living at Flat A, 4 Montague Mansions. They then lived briefly at the Inverness Terrace address but Thomas Skinner died in 1906, following internal injuries as a result of a fall caused, of all things, by slipping on a banana skin. Although (after his homeopathic cure) generally of very sound health, Skinner did suffer from diabetes (it is said many of the family had died of it over the years) and he was prone to occasional attacks of gout, an affliction his grandson Rupert would inherit from him. See John H Clarke M.D., *Thomas Skinner M.D. A Biographical Sketch*, London Homeopathic Publishing Co., London 1907.

15. The Liverpool address at that time was 1 St James's Road.

16. The original letter survives in the family records.

17. Patrick Hanks and Flavia Hodges, *A Dictionary of First Names*, OUP, 2003.

18. Jessie Miller, 22 years old, Kate Mussell, 19 years old, and a 40 year old visitor, Jane Salmon, a dressmaker from Ireland.

19. Sarah Quail, *Southsea Past*, Phillimore, Chichester, 2000.

20. According to Cecil Gould, Harry did not begin to speak until he was 2½ years old, though this fact alone is not necessarily a pointer to lack of intelligence.

21. Obituary to W. M. Gould, *The Times*,. 10 April, 1923, p. 15.

22. Noted in his correspondence with Professor A. W. Stewart (photocopies held by the author).

23. Ibid., Ref. 22.

24. A housemaid, Beatrice Gilbert, 23 years old; a cook, Minnie Gauntlett, 21 years old; and a 25 year old parlour maid Amelia Dunn.

25. Hugh Owen, 'Eastman's Royal Naval Academy, Southsea, in the 1870s', *The Mariner's Mirror*, Vol.77, No.4 (Nov.1991), pp. 379–387.

26. Howard Thomas, *Britain's Brains Trust*, Chapman and Hall, London, 1944, p. 112.

27. *Oddities*. 2nd Ed. P. 99.

28. Article by RTG, 'The Navigator at Sea', in the booklet *The Kings Navy*. ?1938.

29. Ibid., Ref. 1.

30. The year before Rupert's birth, 1889, saw Queen Victoria's Review of the Fleet as a highly diplomatic exercise in honour of her nephew, Germany's Kaiser Wilhelm (the Kaiser was actively building his own Navy at the time) and during Rupert's childhood in the mid-1890s, the Kaiser and his cousin the Prince of Wales would engage in friendly pseudo naval 'battles'. Both were skilled yachtsmen and in 1895 the Prince was 'triumphant' in Britannia, but was 'defeated' the following year by Wilhelm. At the time few from Britain or Germany could have guessed the very real cataclysm that would descend on Europe eighteen years later.

31. The period during Gould's childhood was quite unprecedented for technological advance. Over the previous half a century, the railways had made the entire country readily accessible to its population; without them the Goulds would have taken several days to reach Scotland each summer. With the electric telegraph, almost instant communication across the entire British Empire (by far the largest global community, and now at the height of its power) was also possible. The year 1879, just eleven years before Rupert was born, saw the introduction in Britain of the electric telephone (the first exchange), and in the mid-1890s the developed 'talking machine', in the form of Phonograph and Gramophone were both just commercially available. Even the tape recorder, in its very early stages of development as the *Telegraphone*, had appeared just before 1900. Some years later, in the mid-1930s, Gould himself would have his voice recorded at the BBC on the *Blattnerphone*, a German version of this machine using magnetic wire recording. And it was of course during the 1890s that Guglielmo Marconi was building and developing the first radio transmitters and receivers, his early experiments conducted on the Solent (Isle of Wight), just mile or so from the Goulds' home in 1897. It's even possible young Rupert witnessed the birth of this new medium, one which would some years later make R.T. Gould a household name. In 1890, practical electric lighting had only recently been introduced in Great Britain with Edison's and Swann's electric lamp bulbs and its use, and that of mains electricity in ordinary households in the following century was well understood and eagerly awaited, but not yet a reality. Like almost every Victorian-built town house in the land, 11 Yarborough Road was gaslit by. The modern photographic camera, though first developed 50 years before was, by 1890, only just becoming an instrument readily available to the man in the street. And that exciting medium, the cinematograph would also develop during Rupert's childhood in the 1890s, the French inventors Auguste and Louis Lumiere patenting their design in France and England in 1895. The first public films ('The Baby's Meal' and 'The Arrival of a Train'), were shown at the London Polytechnic the following year. Gould was later to be a great fan of movies, both silent and later 'talkies', and throughout his life was a regular cinema-goer.

32. The first in the world was the motorcycle made by Hildebrand & Wolfmuller of Munich in 1894.

33. *The Loch Ness Monster and Others*, pp. 11–12.

34. Membership No. P269.

35. During the review, Parsons' revolutionary steam launch *Turbinia* astonished and enthralled its audience (and much perplexed Admiralty staff), by zigzagging its way through the Fleet at a speed unheard of for any ship before.

The steam turbine was very soon adopted in ocean liners and larger warships, greatly increasing their maximum speeds and flexibility.

36. Ibid., Ref. 19.
37. R.T.G. *The Stargazer Talks*, p. 121.
38. Chanute, Octave, *Progress in Flying Machines*, 1894.
39. *Conways All the Worlds Fighting Ships 1906–1921*. Conway Maritime Press Ltd, London 1981.
40. Ibid., Ref. 39.

CHAPTER 2: NAVY TRAINING 1906–1913

41. Dr Jane Harrold and Dr Richard Porter, *Britannia Royal Naval College 1905–2005*. Webb, Dartmouth, 2005.
42. Michael Partridge, *The Royal Naval College Osborne, A History 1903–21*. Sutton/RN Museum Dartmouth 1999.
43. R.T. Gould personal file. The National Archives: ADM196/96, p. 4. N.B. many of the notes on this file are abbreviated, but are given in full when quoted here.
44. 'The Isis Class and Prize List: . . . the Easter examination of the cadets of his Majesty's ship Isis,. *The Times*, 18 April, 1907, p. 8.
45. A Navy List error gives this new officer the name of 'Robert' Gould.
46. HMS *Formidable* was a 'Formidable' class battleship of 14,500 tons.
47. HMS *Queen*, a 'London' class battleship of 14,150 tons.
48. *The Times*, 11 December 1908, p. 6.
49. 'Coon', the wife of Gould's colleague Archibald Colquhoun Bell, see page 71.
50. Such an increasing military presence doesn't have to be spoken about often to be constantly in the subconscious of the community either. Those of us (especially those children with a more vivid imagination) who were brought up in the 1960s in the shadow of American airbases (armed with nuclear bombers) in the United Kingdom, will surely know the insidious effect of a presence like this. With such a constant and active military force in the community, many of us in the 1960s were convinced (wrongly in our case, happily) that, sooner or later, a third world war was inevitable and suffered regular depression as a result.
51. The dedication reads: 'To Admiral of the Fleet Earl Beatty, G.C.B., O.M., under whom it was once my privilege to serve'. Strangely, Gould's dedication to Beatty was dropped from the Holland Press reprints of the book in the 1960s and 1970s; Gould would have been furious. It was added again in the Antique Collectors Club edition in 1989.
52. W.S. Chalmers, *The Life and Letters of David, Earl Beatty*, Hodder & Stoughton, London, 1951, pp. 90–106.
53. 'The Sketch' Supplement. *The Sketch*, 16 March 1910.

54. This strange little device (British patent No.25383/08) was used by Drapers and their customers for matching colours by artificial light. An example is preserved in the Science Museum, London (Inv.1953–386). The accompanying booklet explains: 'The Colourphone consists of two iridescent screens bound in the form of a pocket book. By placing the material about which you are doubtful on one of these leaves and reflecting artificial light upon them with the other, their similarity or dissimilarity will be instantly seen'.

55. This promotion was announced (as was usual practice) in the *London Gazette*, on 7 April 1911.

56. This note is in an annotation of Gould's in his personal copy of *Oddities*. He notes of M.R. James: 'I first heard of the Berbalangs from him—now dead alas! He read Skertchley's account to Harry & myself after we had dined with him at Cambridge, *c*.1911. RTG 16-IX-41'.

57. The drawing is now in a private collection.

58. HMS *Hawke* was a 7,350 ton 'Edgar' class cruiser, Commander Hugh Marryat, Gould acting as Sub Lieutenant alongside Lieutenant Basil Taylor who joined the ship on the same day.

59. Notified in *The Times* on 6 January 1912, p. 11.

60. HMS *Kinsha* was a 616 ton 'Kinsha' class river gunboat.

61. HMS *Bramble* was a 710 ton 'Bramble' class steel gunboat.

62. Anthony Preston and John Major, *Send a Gunboat!* Longmans, 1967.

63. The 1913 Diary of R. T. Gould is now in a private collection.

64. Stewart Correspondence, Ibid., Ref. 22.

65. Arthur Ransome's book *Oscar Wilde, A Critical Study*, (Secker, 1912), had made reference to Douglas's affair with Wilde. It seems Gould felt Wilde had been badly treated and that Douglas was hypocritical in his actions. In later life Gould would become a close friend of Wilde's son Vyvyan Holland.

66. The flimsy was a personal certificate given to an officer on leaving an appointment. It was written on thin ('flimsy') paper, and was a statement of that officer's conduct, for the officer's own use should he need it as evidence of previous good character. His official personal file would of course normally have remained closed and unavailable to him, and Pritchard's showing him part of it was very unusual, a measure in fact of his extremely good performance.

67. HMS *Kent*, was a 9,800 ton 'Monmouth' class 1st class armoured cruiser.

68. The great prima ballerina Anna Pavlova (1885–1931) had her own, hugely successful company touring Europe at this time and, with the famous impresario Diagalev's productions, was taking Europe's capitals by storm.

69. S.W. most probably stands for Sopwith-Wright, the company which soon after became the Sopwith Aviation Company. The most likely model of biplane was that known as the 'School Biplane' or 'Hybrid'. By coincidence, this happened to be the model the Royal Navy had just chosen to purchase, though Gould was probably unaware of this.

70. Rupert's Aunt Jane lived at Woodleigh, a house in Brougton Park in the north of Manchester.

CHAPTER 3: THE WAR, A BREAKDOWN AND MARRIAGE 1914–1920

71. The Bat was a fine motorcycle; the company, founded by S.R. Batson in 1902, but sold to Theodore Tessier the following year, was one of the first to exploit motorcycle racing and record breaking for its publicity, something which would have appealed very much to Rupert. The Bat No.2, which had been brought out that year, was the sports touring machine with a 770cc JAP side-valve V-twin engine capable of taking the bike up to 75 miles an hour. This was one of the fastest production motorcycles at the time and one of the most technically sophisticated, though with reliability as something of a recognised weakness. The exact model bought by Rupert can be seen in *The Handbook of Classic British Bikes* (Anon), Bookmart Ltd., (Abbeydale Press) 1999, p. 26.

72. This involved depositing a measured quantity of acetylene crystals into the chamber of the lamp and adding water to the canister which was then closed, the crystals then producing acetylene gas which was ignited in the burner above to produce a bright white light. In 1913 the expression 'lighting up time' meant exactly that for most cycles and automobiles.

73. Margaret Campbell, *Dolmetsch: the man and his work*, Hamish Hamilton (1975).

74. See the exhibition catalogue: *The Illustrators: The British Art of Illustration 1780–1996*, Chris Beetles, 1996. Three of the five Gould drawings in this show were of Paganini. Beetles notes: 'Reconciling Romantic and decadent preoccupations, he depicted composers and performers in nocturnal settings, possessed by their art and wasted by neurasthenia . . . Most paradigmatic is the emaciated yet diabolic figure of Niccolo Paganini playing his own composition with self-obsessed virtuosity. Its allusion to Beardsley is far more than stylistic and thematic, and appears to associate the mythic qualities of Paganini with the character of Gould's own model'.

75. HMS *King George V* was a 23,000 ton, 'King George V' class battleship.

76. Ship's Log of *HMS King George V*, 8/1/13–27/10/13: ADM 53/45747.

77. HMS *Achates*, a 935 ton 'Acasta' class destroyer, was a tender to (i.e. came under the administrative and accounting umbrella of) HMS *Hecla*, in the 4th destroyer flotilla. Commander Walter L. Allen had four officers under him: Lieutenant R.T. Gould as Navigating Officer, Lieutenant Hubert C. Oliver, Engineer Lieutenant Commander Leonard Walker and Gunnery Officer Edward J Jackman.

78. Ibid., Ref. 63.

79. *The Case for the Sea Serpent*, p. 212.

80. Sir Andrew Macphail, *History of the Canadian Forces*, The Medical Services, Hospital Ships and Enemy Action, (www.gwpda.org/naval/rcnmedoo.htm). *The Maine* had been fitted out by a group of American women for service in the South African (Boer) war and subsequently acquired by the Admiralty. She was the Royal Navy's only hospital ship at the time, but was too badly damaged to be re-floated and was broken up at the wreck site after the war.

81. The log of *Achates* then duly records Lieutenant Willis as joining the ship at 7.15 p.m. on Monday 3 August, and Willis's signature appears regularly in the ship's log thereafter.

82. E. Jones and N. Greenberg (2006), 'Royal Naval Psychiatry: Organisation, Methods and Outcomes 1900–1945', due for publication in *Mariner's Mirror* in 2006.

83. Professor Edgar Jones, private correspondence with author.

84. Julian Putowski and Julian Sykes, *Shot at Dawn*, Leo Cooper, London (revised edition) 1992.

85. E.T. Meagher, 'Nervous disorders in the fighting forces', *Journal of the Royal Naval Medical Service*, Vol. 10, 1924, pp. 1–15.

86. Gould made these assertions in correspondence with Professor Stewart, on 21 December 1935 and 12 April 1936.

87. Personal Communication with Professor Edgar Jones. It must be stated however, that there is a definite link with Gould's breakdowns and national crises. Gould himself admitted so, and three out of the four mental breakdowns he experienced in his life occurred at such times (The First World War, the General Strike and The Second World War).

88. Ibid., Ref. 1. The name he cited is actually indistinct. It might have been 'Moore', but enquiries in all likely quarters have failed to produce details of a likely contender with either name.

89. Letter from Straus to L de G Sieveking, 27 December 1931. Straus stated that he had 'practical knowledge of medical psychology: I was busy with the shell-shocks in the war'. MSS Dept, Lilly Library, Indiana University, Bloomington, Indiana.

90. A week or two after his holiday in Harrogate with the Goulds, Basil Hallam joined the Royal Flying Corps and trained to join the Kite Balloon Section, being posted to France the following year. He too would die there, less than a year later, in a somewhat mysterious accident when it seems his parachute (at the time these were very primitive) failed to open on his bailing out from a balloon: another death for Rupert to have to come to terms with.

91. George Cowper Hugh Matthey (1883–1972). Short biographical piece in *A Brief History of the Partners & Directors of Johnson, Matthey & Company*, (probably published in-house), London, (no date).

92. This remark was made in an annotation in Gould's own copy of *Oddities*, p. 235.

93. Archives of the Hydrographic Office, Taunton, Parry papers I, Misc File No. 4, Item 29.

94. Ibid., Ref. 93 Parry papers I, Misc File No. 4, Items 1 & 2.

95. 'Timekeeping at Sea', Ibid., Ref. 93 Misc File No. 50, Folder No.1.

96. Ibid., Ref. 93 File: H9057/1919 dated 24 November 1919.

97. Ibid., Ref. 93 File: 'Report on The Antarctic Charts Published by the Hydrographic Department, February, 1920', dated 21 February 1920.

98. Ibid., Ref. 93 File: H5514/1921, dated 12 September 1921.

99. Revealed in a letter to R. J. Cyriax, dated 13 May 1933. Archives of the Scott Polar Research Institute, ref:MS397/2.

100. Ibid., Ref. 93 File: H6948/1921.

101. Described by R.T.G. in the Stewart correspondence, Ibid., Ref. 22.

102. *The Times*, 15 November 1916 and the 21 May 1917 respectively.

103. Reported in *The Times* on 13 June 1917.

104. The vicar of the church, the Rev W.S. Swayne officiated, the entry in the Register witnessed by Muriel's parents, Rupert's father, one W.W. Maw and the Lord Mayor, W.H. Dunn. St Peter's is now used by the Armenian church.

105. Aunt Jane's will was a predictably complicated and long one recognising family, servants and many charities and church funds. Legacies included £1,000 and a number of articles of family jewellery and plate to her niece, Dodo, and the interest from the capital of the residuary Trust fund to her twelve nephews, nieces (including Dodo), great nephews (including Rupert) and grand nieces.

106. Ibid., Ref. 2.

107. The car was probably second hand, being a right hand drive, 1916 Model C (though registered in England 1918), but it would not have been inexpensive at the time. Although a relatively small company, and little known today, Scripps Booth (who had been bought out by General Motors in 1918), had a name for quality and style in the light car market. Famous owners included the Danish Royal family, who had several of them, the King of Spain, The Queen of Holland, the tenor John McCormick, Mrs Jay Gould, Mrs R.C.Vanderbilt and Winston Churchill. See: *The Scripps Booth Register*, (Newsletters) and S. Medway, 'Artist's Conception, The Novel Cars of James Scripps-Booth,' *Automobile Quarterly*, Vol. 13, No. 3, Third Quarter, 1975.

108. Ibid., Ref. 2.

109. Letter, RTG—Frank Dyson, 19 September 1921 RGO Archives RGO7/89.

110. The house was demolished in the 1930s for a new block of flats.

111. Letter from RTG to G.C. Williamson dated 3 August 1920, now in the archives of the Manuscript Department, Lilly Library, Indiana University, Bloomington, Indiana.

CHAPTER 4: JOHN HARRISON AND THE MARINE
CHRONOMETER

112. *The Marine Chronometer, its History and Development*, Potter, London, 1923.

113. John Harrison, *The Principles of Mr Harrison's Timekeeper*, London, 1767 (Facsimile reprint, British Horological Institute, Upton, Newark, 1984).

114. Douglas W. Fletcher, 'Restoration of John Harrison's First Marine Timekeeper', *Horological Journal*, Vol. 94, No. 1125, June 1952, pp. 366–369.

115. *Four Steps to Longitude*, National Maritime Museum, Greenwich, The exhibition was from 19 January to 30 September 1962.

116. Jonathan Betts, 'Arnold & Earnshaw, The Practicable Solution', *The Quest For Longitude*, (W.Andrewes, ed.), Harvard, 1996, pp. 312–328.

117. A guinea was £1.1.0. (one pound, one shilling and no pence). There were twenty shillings in a pound (and twelve pence in a shilling), so 100 guineas was £105. H5 was cleaned by Samuel Atkins and his son Charles Edward Atkins in 1892, before it was put on display at the Guildhall.

118. Only one of the drawings is actually signed by Bradley and some of the others are believed to be by students of Bradley's at Kings College. Bradley was promoted to professor at the college c.1848. The five extant drawings have the references MS. RGO 6/586 f213r–f217r. When Gould inspected them, the drawing of H1 was accompanied by 'A list of the principal parts with numerical references to the accompanying drawings'. Gould made a copy of this list of 26 parts, and noted: 'The figures underlined (Nos.13–26) are to be found in the extant drawings—the remainder are not. Presumably many of the drawings have been lost'.

119. Poole, rather ill-advisedly, published a description of H4 in the *Horological Journal* (HJ) at the time, without fully understanding the intricacies of the watch. In later years, Gould acquired back-copies of the HJ and, having read Poole's description, was absolutely scathing in his annotations in the margins of his copy of the article. Summing up with typical Gould irritation, he noted.: 'Poole, why didn't you *learn* something about No. 4 before writing about it?', and after Poole had made a particularly fatuous remark, Gould simply noted: 'Ass!'. Gould's annotated copies of the Horological Journal are now preserved in the BHI Library at Upton Hall.

CHAPTER 5: RESEARCH AND THE FIRST
RESTORATIONS 1920–1922

120. The correspondence of the Astronomers Royal, Frank Dyson and Harold Spencer Jones and Assistant William Bowyer, is preserved at the RGO Archives, Cambridge University Library. The relevant files are RGO7/89 and RGO8/119, the letters filed in chronological order. Where Dyson/

Spencer Jones/Bowyer correspondence is quoted, these are the sources unless other references are given.

121. R.T.G., 'John Harrison and his Timekeepers', *Mariner's Mirror*, Vol. XXI, No. 2, April 1935.

122. Gould wrote up a study of this very fine chronometer (No. 37 by Ferdinand Berthoud) in a paper dated 27 March 1920 (now preserved in the BHI Library). Edgell sold the chronometer to Gould in November that year.

123. For example, Heinrich Otto, who was something of a fan of Gould's, expressed real concern about the safety of the timekeepers in his hands (see Appendix 5 for example). In the 1950s the noted antiquarian horologist, David Torrens, was quite critical of Gould's over-cleaning of H1 (Torrens' notebook with his comments, is in a private collection) and in more recent years, members of the Royal Observatory's (MoD) chronometer workshops have suggested, with some justification, that Gould's restorations were rather less professional than is generally believed.

124. The inscription reads:

John Harrison's first timekeeper.
This machine, the first successful chronometer ever made was invented and constructed by John Harrison, between the years 1728 and 1735. He sailed with it to Lisbon, on board H.M.S. 'Centurion', in 1736, and returned in H.M.S. 'Orford'. During both voyages it enabled the ship's longitude to be obtained with great accuracy.
It embodied the following devices, invented by Harrison: the gridiron compensation for the effects of heat and cold, the helical balance spring (in tension), the 'grasshopper' escapement, and the maintaining spring, or going fusee. This time-keeper was deposited in the Royal Observatory in the year 1765. (NMM inventory Ref: ZAA0856)

125. The lecture was announced in *The Times* Court Circular on the day, 13 December 1920.

126. 'The History of the Chronometer', *The Geographical Journal*, Vol. LVII, No. 4, April 1921, pp. 253–270.

127. Ibid., Ref. 120. Gould wrote to Greenwich on 3 January 1921 to make arrangements for an Admiralty lorry to send it there from the RGS.

128. Miss Cayley took most of those required for the book; she had also been at the Science Museum taking photographs for Gould's book during the previous month.

129. The drawing is now in a private collection. Interestingly, even Gould's keen eye was not infallible. In this drawing, the set-up ratchet wheel is depicted the wrong way round!

130. Ibid., Ref. 120. Dyson to Gould, 25 May 1921.

131. Letter from Gould to Mill on 9 August 1921. Archives of the Scott Polar Research Institute, Ref: MS100/38/1–12.

132. Ibid., Ref. 120. Gould to Dyson 8 September 1921. It seems that Gould also had the help of others in working on H4. Watchmaker J.F.Cole (1875–1963) described Gould as 'A time waster', saying 'when he was repairing Harrison's No 4 Timekeeper he spent hours with us in St John's Street'. *Horological Journal*, March 1963, p.96.

133. The NMM pieces are Ref. Nos: ZAA0129; ZAA0130; ZAA0131 (returned from the Science Museum, their inventory ref. 1922–718) and ZAA0854.

134. Gould promised Dyson he would keep the archives securely locked up when not in use, in a cupboard a few feet from the end of his bed where, he said, he secured H4 at night.

135. F. Von Osterhausen, *Paul Ditisheim Chronometrier*, Simonin, Neuchatel, 2003.

136. Ibid., Ref. 93, H5514/1921. This turned out to be Shackleton's last voyage; he died of a heart attack on *Quest* on 5 January 1922.

137. 'Notes on the History of the Date or Calendar Line', *The New Zealand Journal of Science and Technology*, Vol. XI, No. 6, April 1930, pp. 385–388. Copy provided via the New Zealand Government Astronomer's office, by courtesy of the UK Hydrographer.

CHAPTER 6: THE MAGNUM OPUS 1921–1923

138. Letter, Gould to the Director of the Science Museum, 10 July 1922. SM file ref: 01346. The chronometer had been Inv. No.1921–277. It was eventually sold by Chamberlain's widow to the Mariners Museum, Newport News, VA., USA, and is now in their permanent collections (Inv No. NA.0095).

139. This volume is now in a private collection.

140. These volumes and notes are now in a private collection.

140a. All the original drawings for the book were sold to Sir David Salomon who left them to NMM where they are now preserved (Ref:__)

141. For further explanation, see Ref. 116.

142. Ibid., Ref. 22, Gould to Stewart, 10 May 1936.

143. Gould's pre-publication copy, received on 13 March 1923, now referred to as annotated Copy A, is now in a private collection.

144. Gould's interleaved copy, now referred to as annotated Copy B, is now in a private collection.

145. Note in the Preface in annotated Copy A (see Ref. 143)

146. Note in the Preface in annotated Copy A (see Ref. 143)

147. Annotated copy Q is now preserved in the Clockmakers' Company Library at the Guildhall, London. The ACC edition, due out in 2006, is said to contain many of these annotations.

148. Letter, Maurice Aimer to Andrew King, dated 30 January 1976.

149. John Stacey & Sons, Leigh Auction Rooms, Essex on 29 July 1986. This book, which may or may not be 'Copy C', is also now in a private collection.

150. The surviving typescript of the book was sold in Malcolm Gardner's catalogue No. XII, (1958).

151. Potter's published an advanced notice in *The Horological Journal*, Vol. 64, No. 766, June 1922, p. xi.

152. *The Nautical Magazine*, Vol. 109, No. 5, May 1923, pp. 457–458.

153. *The Mariner's Mirror*, Vol. 9, No. 6, June 1923, pp. 191–192.

154. *Times Literary Supplement*, April 12, 1923, p. 241.

155. Thos. D. Wright, *The Horological Journal*, Vol. 65, No. 777, May 1923, p. 178, 181.

156. Kipling's letters to Gould are now preserved in the British Horological Institute library, at Upton Hall.

157. Letter, Gould to Hinks, 15 October, 1923. Archives of the Royal Geographical Society, ref: CB9.

158. *Geographical Journal*, Vol. LXII, No. 5, November 1923, pp. 378–380.

159. *Nature*, Vol. 113, No. 2838, 22 March 1924, pp. 416–417.

160. *Nature*, Vol. 113, No. 2842, 19 April 1924, p. 570.

161. *Nature*, Vol. 113, No. 2850, 14 June 1924, p. 857.

CHAPTER 7: HOROLOGY: THE OBSESSION

162. Letter, Gould to Ditisheim, 7 January, 1923. Private collection.

163. Letter, Gould to Admiral Parry, 11 January, 1923. Ibid., ref. 94.

164. Letter, Gould to the Secretary of the Royal Society of Arts, 23 February 1923. Private collection.

165. Letter, Gould to Heawood, 18 April 1923. Ibid., Ref. 157. William Monk Gould's death, on 7 April, was duly notified in *The Times* on 10 April 1923, p. 15, col.D.

166. Letter, Gould to Ditisheim, 23 August 1923. Private collection.

167. Ibid., Ref. 135.

168. Letter, Gould to Admiral Parry, 22 December 1923, Ibid. 94.

169. The Gould notebooks, Nos. 2 to 18 are (somewhat confusingly) given the NMM Manuscript Department reference nos: GOU/1 to GOU/17. Gould's Book No.1, which only came to the NMM many years later, has the NMM No. MS88/027. See Appendix 3 for a brief summary of the contents of all the notebooks.

170. Gould himself even comments on this in later years. In a note in book 16, against an entry of 19 January 1929, he remarks: 'If I didn't know this *must* be my own writing I would cheerfully swear that it wasn't. RTG 21/x/1936'.

171. A few pages of book 13 also contain notes on H1's first cleaning but chiefly only a list of parts.

CHAPTER 8: H2 IS RESTORED 1923–1925

172. Twenty seven pounds, ten shillings and no pence (see Ref. 117). The cost was equivalent to about £800 in today's money, Ibid., Ref. 11.

173. This pocket book of Gould's is now in a private collection.

174. The text of the lecture was published in the *Horological Journal* in five parts, from December 1923, et seq.

175. An 'orrery' (named after the 4th Earl of Orrery who commissioned one of the first about 1712), is a scientific instrument intended as a three dimensional representation of the Solar System, with small globes, representing the known planets, orbiting a central sun globe, all travelling at the correct relative rates and positions. Gould's completed R.U.S.I. orrery restoration was notified in *The Times*, 5 March 1924. The orrery, which dates from c.1770 and is signed by James Simonds, is now in the collections of the National Maritime Museum (NMM Ref: ZAA0651). In 1937 Gould would have a chance to restore the original orrery itself, see p. 304.

176. Letter, Gould to Heawood, 2 May 1924, Ibid., Ref. 157.

177. Letter, Gould to Ditisheim, 8 May 1924, NMM Archives, ref:C1420.

178. These, in summary, were: Connecting wires for linking balances; connecting wires for connecting balances to balance springs; and connecting wires for axially controlling the balances; repairing the pallets, repairing the remontoir detent and repairing the centring springs for the balance anti-friction arms.

179. Letter, Gould to Dyson, 22 September 1924. Ibid., Ref. 120.

180. *The Daily Telegraph* carried a notice on 18 October 1924.

181. 'Pioneer Chronometers', *Nautical Magazine*, Vol. 113, No.1, (1925) pp. 22–27.

182. Letter, Gould to Bowyer, 25 November 1924. Ibid., Ref. 120, RGO8/119. The expenses were:

> Two new remontoire springs £1/15/0
> Set of 12 phosphor-bronze wires (including spares) 15/0
> New pivots for compensation gear 10/0
> Replacements in motion work 7/6
> New Key (Made to pattern of No.4's key) £1/0/0
> Total: £4/7/6.

183. Heinrich Otto, 'In Memory of Lieutenant-Commander R.T. Gould', *H.I.A. Journal*, Horological Institute of America, January 1950, pp. 17–18, 33, 47, and 50.

184. Letter, Dyson to Lyons, 5 October 1925, Ibid., Ref. 120.

185. These notes were received by the curator, G.L. Overton, on 5 February 1926, though they are apparently not now in the Science Museum's files. A copy of the notes is however preserved in the BHI Library, and Gould's own notebook No.11 is another copy.

CHAPTER 9: THE SETTE OF ODD VOLUMES

186. His membership was approved at the July 'business meeting' of the Sette, Gould noting in his acceptance letter to the antiquarian, G.C. Williamson,

on 26 July 1920, his feelings of 'pride and unworthiness'. Letter, Gould to Williamson, in Williamson MSS, MS Department, Lilly Library, Indiana University, Bloomington, Indiana.

187. Ralph Straus, *Pengard Awake!*, Chapman and Hall, London, 1920.

188. Rule No.1 from the Sette's rules, published in each year's annual.

189. Noted in his letter to Williamson. Ibid. Ref. 186.

190. Sette names would sometimes be 'recycled' after a member died or resigned and would be used by a new member, where appropriate.

191. Letter Gould to Professor Stewart, 12 April 1936, Ibid. Ref. 22.

192. Ibid., Ref. 1.

193. J. Lewis May, *John Lane and the Nineties*, J. Lane Bodley Head, London, 1936. John Lane was mentor/surrogate-father to his relative Allen Lane, who went on to found the Penguin books empire.

194. Waugh and his elder brother Evelyn were sons of the publisher and literary critic Arthur Waugh, another contributor to Lane's *The Yellow Book*.

195. A. P. Herbert, *In the Dark, The Summer Time Story & the Painless Plan*, The Bodley Head, London, 1970.

196. It was Lane who had first introduced the use of metal plates to repair broken limbs and pioneered the treatment of the cleft palate. Lane was, controversially, also a noted liberal on contraception and, for his sins, was the specialist who first championed the practice of prescribing liquid paraffin for constipation!

197. Bernard Quaritch had begun using meetings of the Sette for commercial purposes, which many members found unacceptable.

198. Hall chose his Sette name, *Inquisitor*, advisedly. His entry in the D.N.B. notes: 'Officer prisoners from German submarines who had stubbornly refused to respond to ordinary interrogation often became as putty in his hands. There was a hypnotic power about his glance which broke their resistance . . .'.

199. The watch had centre seconds and a verge escapement. Gould published his creation in a note to the *Horological Journal*, Vol. 66, No. 782, October 1923, pp. 30–32. Unfortunately, the whereabouts of this watch is now unknown.

200. 'The Blick' was manufactured by George C.Blicksenderfer, Connecticut, U.S.A. Andrew Simpson of Leicester City Museum Service (who looks after the bulk of Gould's collection of typewriters today) says: 'The machine shown is an early model, possibly a model No. 5 or No. 6 of 1893–1896, or even the "featherweight" model of 1896. The machine was a typewheel design having instantly interchangeable typefaces. It also had the advantage of visible writing, something we take for granted now . . . The keyboard layout on the machine in Gould's picture is an "Ideal" arrangement rather than the "Qwerty" we are familiar with today. I can understand why he liked the machine, they are still rather

special, with many features well ahead of their time'. Letter from Mr Simpson, 23 February 2005.

201. The other guest was Leonard Carr–Laughton, naval correspondent to the *Morning Post*, one of the founders of the Society for Nautical Research and first editor of the *Mariner's Mirror*.

202. The dinner was preceded by a short programme of violin and piano solos and songs, mostly by English composers, one by Keel himself.

203. Evan Baillie Noel had, with J. O. M. Clark, just published his monumental, two volume work *A History of Tennis*. Gould's guests that evening were Jack F. Marshall, C.D. Skinner and R.H. Hill.

204. *The Times*, 15 January 1924, p.1, col. G. A notice under Domestic Situations Vacant, reported that Mrs Rupert Gould was offering a 'moderate salary' for a Nursery Governess & needlewoman.

205. Ibid., Ref.2, p. 9.

206. The redrawn title page of the *Mariner's Mirror* began in January 1924.

207. This was eventually published in *The Geographical Journal*, Vol. LXV, No. 3, March 1925, pp. 220–225.

208. The Latin term for the legendary 'Sea Serpent', *Megophias Megophias*, had not been introduced by Gould, but was a classification taken from the work of the pioneering Dutch researcher, Dr A.C.Oudemans. See Chapter 12.

209. Opusculum No. LXXX, on *Sea Serpents*, was published in November 1926. The editor G.C. Williamson evidently had difficulty understanding Gould's insistence on the print run: 333 with only odd numbers used, totalling 167 copies. Having explained it all carefully once, Gould was obliged to write him a long and very irritated letter explaining the logic of it all again. They evidently took their 'oddness' in the Sette very seriously!

210. It is however, a standard 'fill-in' drawing by Gould (of a lady dusting a shelf of odd volumes), done in 1923 for the ladies night in June that year, which illustrates the cover of all the menus during the period of his absence. The Sette members knew he was unwell and perhaps this was a way of acknowledging that he was with them in their thoughts.

CHAPTER 10: SEPARATION 1925–1927

211. Kay Redfield Jamison, *Touched with Fire: Manic-Depressive Illness and the Artistic Temperament*, Simon & Schuster Ltd, New York, 1996.

212. Private correspondence with Professor Edgar Jones.

213. Letter, Muriel Gould to Arthur Hinks, 5 June 1925. Archives of the Royal Geographical Society, ref: CB9 (GO-GP).

214. The Ditisheim copy of the Thacker transcription is extant in a private collection. *David Penney Postal Auction, Catalogue No. 7*, 31 March 2003, Lot. 133, p. 40.

215. Letter from Hinks to Cockerell, 4 December, 1923. Archives of the Fitzwilliam Museum.

216. Letter, Gould to Cockerell, 3 June 1926. Ibid. Ref. 215.

217. Even a group as exclusive as the Sette of Odd Volumes could not fail to have been aware of the social tension at this period. On one occasion the staff at the Imperial Restaurant, Oddenino's itself, went on strike for better pay, a cause they won within a couple of hours.

218. Letter, Gould to Cockerell, 6 December 1926. Ibid. Ref. 215.

219. Ibid. Ref. 2

220. Victoria Glendenning, *Vita, the Life of Vita Sackville-West*, Penguin, 1983.

221. Radclyffe Hall, *The Well of Loneliness*, Jonathan Cape, London, 1928.

222. Paul Ferris, *Sex and The British, a Twentieth Century History*, Michael Joseph, London, 1993, p. 45.

223. Ibid. Ref. 1.

224. Letter, Gould to Bowyer 2 September 1926, Ibid. ref. 120.

225. Gould was announced as a member of Council in the *Horological Journal*, Vol. 69, No. 819, November 1926, p. 45.

226. Gould's three guests on 23 November, to hear A. J. A. Symons on Frederick Baron Corvo, were the marine artist and councillor of the Society for Nautical Research, Cecil King (1881–1942), Leonard Carr–Laughton of the SNR and the Naval publisher and agent F.C. Bowen.

227. It was one of the last major pieces of work done by Gould for the Hydrographic Department.

228. Collyer–Bristow & Co., in Bedford Row, London.

229. Ibid., Ref. 222, p. 95.

230. Public Record Office (National Archives) Ref: XC6129 J77/2421/5570.

The notes read:

'11.6.17—suggested gagging her during intercourse;
Summer / 18 & on—drunken. Progressively worse til she left him;
Sept/17—May/18 & on—first paroxysms, terrifying her;
April/25—culminated: threatened suicide; cried; said he had murdered her;
May/25—demonstrated to her the vice of self-abuse and said he was victim to it;
Apl/27—burst into her room drunk;
16. 5.27—told her to go and he would take children;
21. 5.27—drunk and incapable;
Jany/27—21.5.27—constantly drunk, abused, threatened etc;
Mental Strain: Unsafe Etc'.

231. For administrative convenience, these three separate branches of the High Court were lumped together into this Division, without any particular significance as a group. The 'Admiralty' branch has no formal connection with the Royal Navy's Admiralty organisation. This rather confusingly named branch in the High Court deals mostly with merchant service cases and enquiries into shipwrecks. Because this Division has sections which deal with deaths (probate), broken marriages (divorce) and shipwrecks (included under Admiralty) it is known jocularly by some in the legal profession as 'the court of wrecks'!

232. John Juxon, *Lewis & Lewis, The Life and Times of a Victorian Solicitor*, Ticknor & Fields, New York, 1984.

233. Ibid. Ref. 230. The allegations read:

1. On June 11 1917 at Newquay in the County of Devon the respondent suggested to the Petitioner that he should gag her during sexual intercourse.

2. During the Summer of 1918 and onwards throughout the married life and particularly on or about the last Tuesday of each month from June 1918 the respondent frequently drank to excess and returned home in a drunken condition. The respondent's bouts of drunkeness became progressively more frequent until the 21 May 1927 when the Petitioner, terrified thereby, left him.

3. From September 1917 to May 1918 inclusive and onward throughout the married life the Respondent frequently suffered from paroxysms terrifying the Petitioner thereby.

4. The said conduct culminated in April 1925 at the Cottage, Lynwood Avenue, Epsom, when Respondent threatened to commit suicide, cried in the night and lay on the floor and frequently leaped from his bed and came to the Petitioner saying he had murdered her and by the said conduct greatly terrified her.

5. On an occasion in May 1925 at the Cottage aforesaid the Respondent demonstrated the vice of self-abuse and told her that he was a victim of this habit and to obtain satisfaction used to take with him into the lavatory pictures of girls gagged and women in high boots.

6. On an occasion in April 1927 at Glenside, Kingston Road, Leatherhead, the Respondent burst into the petitioner's room in a drunken temper and terrified her.

7. On the 16 May 1927 at Glenside aforementioned the Respondent was violently enraged with the Petitioner told her to go and that she must do exactly as he told her and that he would take the children from her.

8. On the 21 May at Glenside aforesaid the Respondent was drunk and incapable.

9. From January 1927 until the 21 May 1927 at Glenside aforesaid the respondent was constantly drunk had frequent fits of uncontrolled temper often told the petitioner to go to Hell and that he did not care if she went on the streets and he intended to get the petitioner down off her perch and did not care if he saw her in the gutter and frequently threatened to burst into her room.

10. From the aforesaid conduct of the Respondent the Petitioner has suffered from severe mental strain and it is unsafe for her to continue to live with him.

234. Ibid. Ref. 1.

235. Gould received an order of 28 September from the Court, to pay £36/4/2 interim costs to Muriel, along with a further £200 to be lodged in Court as estimated cover of the final costs. A reply from Gould 's solicitors, submitted on 8 October, evidently objected to this, but 17th saw Murial filing costs and then, on 20th an affidavit concerning Ruperts objection (not on file). On 8 November, the Court then ordered Rupert to pay alimony, pending suit, at £140 per annum, to include the maintenance of the children, with arrears to be paid at £10 per month.

236. *The Times*, 22 November 1927, *Law Notices* announced the impending judgement.

237. *The Times*, 24 November 1927, Report: *Judicial Separation for a Wife*, p. 5.

238. Ibid. Ref. 11.

239. Ibid. Ref. 222.

240. *The Daily Mail*, Thursday, 24 November 1927.

241. The day after the judgement however, it would not be unreasonable to expect Gould to be feeling a little less than charitable towards the female of the species. By a strange coincidence the *Daily Mail's* cartoon that day, just a page away from his own headline, must have struck a peculiar chord with Gould. The paper covered a bizarre decision by the Post Office that in future women should be *banned from winding* Post Office clocks, being considered too ham-fisted to do the job safely! Parliament had recently legislated for younger women also to have the vote, and the cartoon also takes a swipe at the newly emancipated young lady, busily winding (and hence damaging) *Big Ben* while wearing a dress with the legend 'Flapper Vote'!

242. A caricature of Emin Pasha, The Governor of Equatoria, (a province in southern sindan), who was the subject of the talk by the author, bibliophile and collector of musical boxes, A.J.A. Symons, 'Speculator' to the Sette.

243. Ibid., Ref. 163.

244. Lockhart was admitted to the Sette on 25 October 1927.

245. Letter, Gould to Professor Stewart, 14 September 1936, *Ibid. Ref. 22*. On Hirst's premature death in 1936, Gould wrote to Stewart:

> I have been much upset during the last two days. Do you remember my sending you, some time ago, a couple of letters about my wife, one of which began 'Dear Val', and was addressed to a man who had been my greatest friend, but who had come to side with her? Well, he was Lt. Cdr Valentine D. Hirst R.N. and he died very suddenly on the 22nd. He was a week younger than I am; and—as I told you—we had been friends at Dartmouth and ever since, until I left the Admiralty. I have never met anyone quite like him. Even at Dartmouth, he was an outstanding character, with a most likeable personality—pleasant faced, bright, intelligent, always cheery, with a delightful voice and a charming manner, good at games and his work, with all the makings of a first-class officer. I admired him immensely—and in a way I think he admired me, until the war parted us—and when we met again (which we often did—he was Cecil's godfather, for instance) we were both a bit different. I was out of the Service, and married; he was still afloat, and finding life complicated—as Bossut said of Clairaut, he has *'entraine par un gout vif pour les femmes'*. Finally, he met an affinity, who proved to be a married woman. She got a divorce, and he married her. I believe they were happy, but I'm not sure. But his service career, which had opened so promisingly, ended in failure. He had a fine chance—he was 1st. and (G) of the 'Malaya' under Roger Backhouse—but he made nothing of it. He had a violent row, I believe, with Backhouse, and that was his finish. He was passed over for promotion to Commander (he got the usual 'step' [to Lt. Cdr] on retirement) and after two years' spell at Bermuda he retired at his own request in 1932 and settled down in Sussex with his wife and daughter. And now he is gone. I shall always regret that after our interchange of letters on the subject of my wife we never met or communicated again. I used to say to myself that if I met him I wouldn't cross the road to speak to him. But all the time I knew that it wasn't true. If we had met again I should have hailed him—and I think he would have hailed me—as his friend, whom it was a delight to see again. Still, at least our friendship ended in silence—not, as perhaps it might have done, in bitter recrimination and mutual insult'.

CHAPTER 11: ODDITIES AND ENIGMAS 1928–1929

246. Letter, Gould to Stewart, Ibid. Ref. 22.
247. Letter, Gould to Ms Drake, 3 December 1928. SPRI archives: UDC.92.
248. Letter, Gould to Mill, 9 February 1934. SPRI archives MS100/38/1–12.
249. See also mention of them in his book *Oddities*, in Appendix 4.
250. Reported in *The Times* and *The Star* the following day.
251. Ibid. Ref. 22.
252. Robert Graves and Alan Hodge, *The Long Weekend, a Social History of Great Britain 1918–1939*, Faber and Faber, 1941, p. 289.
253. George Orwell, on Charles Reade, *New Statesman and Nation*, 17 August 1940.

254. R. Austin Freeman, *Mr Polton Explains*. Hodder and Stoughton, London, 1940.
255. The expression 'Fortean' relates to the work of Gould's American contemporary Charles Fort (1874–1932), another pioneer in the field of paranormal research, whose work Gould had read closely. The *Fortean Times* defines 'forteanism' as: 'open minded exploration of all kinds of mysteries'.
256. Dr Mike Dash, correspondence with the author.
257. John Gilbert Lockhart, *Mysteries of the Sea*, 1924, and *A Great Sea Mystery, the True Story of the 'Mary Celeste'*, Philip Allen, London, 1927.
258. Ibid. Ref. 2.

CHAPTER 12: THE CASE FOR THE SEA SERPENT 1930

259. A. C. Oudemans, *The Great Sea-Serpent*, n.p., Leiden, 1892.
260. Bernard Heuvelmans, translated by Richard Garnett, *In the Wake of Sea Serpents*, Hill& Wang, N.Y. 1968.
261. The ship is a careful drawing of a 'Nelson' class battleship, the most up-to-date at the time and the pride of the Royal Navy.
262. In the early 1930s road fatalities were proportionally *15 times* higher than today. The heavy death toll on the roads at the time prompted the 1934 Road Traffic Bill, introducing speed limits and driving tests.
263. Times Literary Supplement F.A. Bather, 15 January 1931.
264. Times Literary Supplement Gould, 22 January 1931.
265. Times Literary Supplement Bather, 5 February 1931.
266. Times Literary Supplement Gould and Bather, 26 February 1931.
267. *The Discovery of the South Shetlands*. See Appendix 1.

CHAPTER 13: THE RAS REGULATOR 1927–1929

268. Letter, Gould to RAS 24 January, 1927, in Gould's Notebook.
269. Letter, Cottingham to RAS, 28 February, 1827. Ibid. Ref 268.
270. Letter, Gould to RAS, 20 March, 1927. Ibid. Ref. 268.
271. Letter, Gould to RAS, 8 June, 1927. Ibid. Ref. 268.
272. Letter, Gould to Phillips 5 January, 1928. Ibid. Ref. 269. Phillips lived near the Goulds in Ashtead.
273. Alfred Thomas, James Harold Agar Baugh: Pendulum Maker, *Antiquarian Horology*, Vol. 26, No. 3, September 2001, pp. 257–262.
274. Letter, Agar Baugh to Gould, 21 July 1928. Ibid. Ref. 269.
275. Arnold chronometer No.323, now in the NMM collection, ref: ZAA0117.
276. This was probably the clock signed 'Harrison 1725' which at that time belonged to one Captain Harrison (no relation). It had been sold to him by Alfred Fryer of Ulseby, who had discovered it in a bank in Hull. The

clock, which is probably by Sutton of Barton and has been substantially altered since new, is now (2005) in the care of the Hornsea Folk Museum in Yorkshire.

277. This was the sale of watches belonging to Messrs Asprey & Co., Ltd 'collected for exhibition in their show rooms as examples of 17th and 18th Century workmanship'. There were several watches by A.L. Breguet, amongst other fine makers.

278. Letter, Agar Baugh to Gould, 7 September, 1928. Ibid. Ref. 269.

279. Buck's £5/9/6, not £3 and Hopgood's £12/10 not £4.

280. *The Practical Watch and Clockmaker* was published between March 1928 and April 1939, when it was incorporated into the HJ

281. Jules Sottas, The Corvette L'Aurore and its Model, *Mariner's Mirror*, Vol.16, 1930, p. 132.

282. Claus Bergen, *U-Boat Stories*, Constable & Co. Ltd., London, 1931.

CHAPTER 14: H3 IS COMPLETED 1929–1931

283. Letter Douglas to Dyson, 19 December 1927. Ibid. Ref. 120.

284. Note, Gould to Douglas, 20 December 1927. Ibid. Ref. 120.

285. Letter, Dyson to Gould, 22 December 1927. See also a letter from Gould to Hinks at the Royal Geographical Society (Ibid. Ref. 157), dated 9 January 1928, explaining what was happening with H3.

286. Note dated 9 October 1929. The identification of Douglas was added as a later footnote.

287. Jennet Conant, *Tuxedo Park: A Wall Street Tycoon and the Secret Palace of Science That Changed the Course of World War II*, Simon & Schuster, New York, 2002.

288. Copy A.

289. In 1960, when the horological historian Col. Humphrey Quill enquired of Loomis about the timekeeper he replied to Quill: 'I remember Mr Gould very well and the historic Marine Timekeeper that he gave me. I am chagrined to report that this Timekeeper was taken from my laboratory sometime during the war when the whole laboratory was being moved from Tuxedo Park to Cambridge, Mass.' See: Vaudrey Mercer. *John Arnold & Son*, Antiquarian Horological Society, London 1972, p. 18.

290. F.H. Green, *Old English Clocks*, Ditchling Press, 1931.

291. Gould's use of the Greek expression *tetelestai* here may also have a slightly deeper meaning, in that the word was often used classically in the context of a debt being fully paid (i.e. on a written receipt), and may represent Rupert's feeling of achievement and, subliminally, of guilt assuaged.

292. Gould to Miss Drake, 19 January 1930.

293. Science Museum files ref: 775/16/8.

CHAPTER 15: H1: THE FULL RESTORATION 1931–1933

293. The Navigators Equipment, now in NMM archives (as far as is known) an unpublished, illustrated typescript, written by Gould and dated 30 August 1932, for Geoffrey Callender. It was part of a volume to be titled *The Coming of Age*, presumably originating from an idea conceived 1931, the 21st year of the SNR's existence, though the book does not seem ever to have been published. The typescript is all marked up for the printers by Callender and is associated with two drawings by Gould, one of H1 and one of Le Roy's timekeeper done in 1933.

294. Ibid., Ref. 120, The summary reads: 'Total not to exceed £30; Insurance paid Sept 1931, £1:10:0 & Dec.1932, 10:0; 12/III/32 paid to Gould £5; 27/X/32 paid to Gould £5, Total expenditure up to 31/XII/32 = £12; Balance in hand = £18. Liabilities to be met: Work by Hopgood and Plating Co (Buck)'.

295. Tony Mercer, *Mercer Chronometers*, Brant Wright Associates, Ashford, 1978; *Chronometer Makers of the World*, NAG Press, Colchester 1991 (2nd Ed. 2004); and *Mercer Chronometers*, Mayfield Books, Ashbourne, 2003. Tony Mercer even claims that the restoration of H3 was undertaken by Mercers, but in fact Mercers never had any part whatever in H3's restoration. There are several other remarks made about Gould in these books which are decidedly misleading, and a large pinch of salt is recommended when reading them.

296. Gould to Cyriax, archives of the S.P.R.I. MS937/2.

CHAPTER 16: THE LOCH NESS MONSTER 1933–1934

297. *The Candidates Book* in The Sette of Odd Volumes Archives at Cambridge University Library.

298. *The Observer*, 29 October 1933.

299. Its rather extraordinary to consider that as this chapter is being written (2004), Katherine Hepburn, who took the role of Cynthia in that film in 1933, and who was already an established, famous film star at the time, died less than a year ago.

300. Gould to Cyriax, 15 January 1934. Ibid. Ref. 296.

301. Quoted from *The Loch Ness Monster and Others*.

302. Gould took Stanley C. Edgar as a guest that night.

303. The society known as *The Anchorites* was founded in 1919, *inter alia*, to foster relations between the Royal and Merchant Navies. It is probable, though not certain, that Gould was a member.

304. *Daily Mail*, 3 January 1934.

305. A Zest for Life, the story of Alexander Keiller, *Lynda J Murray*, Morven Books, Swindon, 1999.

306. Ibid., Ref. 22.

307. *The Observer*, 1 July 1934.

308. Peter Costello, *In Search of Lake Monsters*, St Albans, Panther, 1975, pp. 359–360.

309. *Halliwell's Film & Video Guide*, 2003 18th Edition, Ed. John Walker, Harper Collins, London.

CHAPTER 17: THE HARRISON TIMEKEEPERS AND THE NMM 1934–1935

310. Gould visited the Science Museum on 12 December 1932.

311. Letter, Mackintosh to Spencer Jones, 18 January 1934. S.M.file: 775/19/1.

312. Letter, Spencer Jones to Mackintosh, 22 January 1934. S.M.file: 775/19/2.

313. *A Short History of John Harrison's Timekeepers* (now in the BHI Library).

314. Letters from Callender to Gould. NMM file ref: NMM5 G34–37.

315. The lecture was given by Gould to the Anchorites on 17 April 1934.

316. One of these very rare leaflets is now in the NMM collection.

317. Gould noted in a letter to the Harrison scholar Eric Whittle, that, by the time of the lecture, H1 had still not settled down in its running. He noted: 'I was in mortal fear that it would stop before I had finished speaking'. Letter to Whittle, 16 January 1945 (private collection).

318. Science Museum file ref: 116/48/6.

CHAPTER 18: PROFESSOR STEWART, THE BBC AND TENNIS 1936

319. Ibid., Ref. 22. In fact the whereabouts of all the volumes is now uncertain, though photocopies of volumes 2,4,5, & 6 are retained by this author.

320. Ibid., Ref. 1.

321. *B.B.C. Children's Hour: A Celebration of Those Magical Years*, Wallace Grevatt, Book Guild, 1988.

322. *The Stargazer Talks*, 1943, p. 3

323. Private correspondence with the author, 20 June 1993.

324. Ibid., Ref. 321.

325. Ibid., Ref. 22.

326. BBC Archives, Caversham, ref: 910 GOU I.

327. Private collection.

328. Private collection.

329. SPRI Archives ref: MS100/38/1–12.

330. 1 July 1938, BBC Script (copy held by author).

331. SPRI Archives ref: MS397/2, 4 July 1933.
332. Ibid. Ref. 22.

CHAPTER 19: MANY PROJECTS 1936–1937

333. Ibid. Ref. 22, 7 March 1936.
334. Ibid. Ref. 22, 25 April 1936.
335. Gould's literary agent at the time was James B. Pinker & Son, Arundell St, The Strand.
336. Ibid. Ref. 22, 7 March 1936. Methuen's files on the matter still survive, and note: 'MS was to be delivered in Aug' 38. Up to Sept 13 it had not come in.Half the advance of £75 is due on del.y and approval of MS. Agent is Pinker'. A further note, in resigned tone, simply remarks: 'Let this sleeping dog lie'.
337 Ibid. Ref. 22, 5 December 1936.
338. For a further explanation of this puzzle, see Martin Gardner, *Mathematical Puzzles and Diversions from Scientific American*, G Bell & Sons Ltd., London, 1961.
339. The huge glass and iron building called the Crystal Palace, originally constructed in Hyde Park for the Great Exhibition in 1851, had been moved immediately after the end of the exhibition to an area in South London which was then itself given the name of 'Crystal Palace'. The structure caught fire and was completely destroyed that night in December 1936.
340. S.M. Ref: 116/48/8.
341. S.M. Ref: 116/48/9.
342. Gould to Ward, 5 October 1937. Private collection.
343. Ibid., Ref 1.
344. Gould to Maxwell, 22 May 1937. NMM archives ref: NMM 5.
345. Plate VIII in *John Harrison and His Timekeepers* depicts the timekeepers in these first showcases.
346. Ibid., Ref. 326.
347. Kevin Littlewood and Beverley Butler, *Of Ships and Stars, Maritime heritage and the founding of the National Maritime Museum, Greenwich.* The Athlone Press & NMM, London, 1998.
348. Ibid., Ref. 314, 10 May 1937.
349. A copy is preserved in the NMM library, Ref: D1190.
350. *Office Control and Management*, January–September, 1948.
351. The Orrery restoration notebook is at the end of the Science Museum file 'Royal Observatory' Ref: 775.
352. *Illustrated London News*, 18 December 1937.

CHAPTER 20: LEAVING DOWNSIDE,
LEAVING LONDON 1937–1939

353. The gross value of the Estate amounted to £55,762, about £1.7m today (2005).

354. The Queen's House is the extraordinarily fine and important building by Inigo Jones, built at Greenwich in the 1630s and the first example of Palladian (neo-classical style) architecture in Britain. It is the central building between the two wings of the National Maritime Museum, joined to them by colonnades. The exact location of Gould's workshop in the Queen's House is uncertain, but it is believed to have been somewhere in the basement.

355. Ibid., Ref. 2.

356. Discussed in depth in correspondence with Professor Stewart Ibid., Ref. 22.

357. Note made in the annotated copy of *Oddities*, (private collection).

358. Paul Tabori, *Harry Price, Ghost Hunter*, Athenaeum Press, 1950, 2nd Ed. Sphere Books, 1974, p. 262.

359. Undated, but almost certainly published in 1938, the book may well have been produced to coincide with the British Empire Exhibition in Glasgow that year.

360. Ibid., Ref. 331.

361. Geoffrey Callender's Diary, ref: NMM 5.

362. Ibid., Ref. 305, p. 101.

363. Ibid., Ref. 2.

CHAPTER 21: UPPER HURDCOTT AND
THE BRAINS TRUST 1940–1945

364. Correspondence with the author.

365. Ibid. Ref. 2.

366. Ibid. Ref. 157.

367. Ibid. Ref. 1.

368. The Gardner correspondence is in a private collection.

369. Archives of the S.P.R.I.: Hobbs—Discoveries in the Antarctic 1819–1823, ref: (*7): 91(091) [1939] The correspondence is attached within the bound volume.

370. Ibid. Ref. 157.

371. Ibid. Ref. 314.

372. Ibid. Ref. 314, 13 September 1942.

373. Ibid. Ref. 120, 22 September 1942.

374. Ibid. Ref. 120, 6 November 1942.

375. Indeed, the roots of the current radio programme can be traced back to the Brains Trust itself.
376. Newspaper obituary 'The Man who was never wrong'.
377. Howard Thomas, *Britain's Brains Trust*, Chapman &Hall, London, 1944.
378. Reginald Pound, *A Maypole in the Strand* (London: Ernest Benn, 1948), p. 26 and pp. 56–7.
379. Commander A.B. Campbell R.D., *When I was in Patagonia*, Christopher Johnson, London, 1953.
380. BBC Sound Archives: LP 25885; LP 25887; LP 25888.
381. Ibid. Ref. 326.
381a. A recording of this episode has survived in the BFI Archive.
382. Ibid. Ref. 257.
383. In 1973 the book was again reprinted in the US under the new title of *More Oddities and Enigmas*, with another Foreword by Leslie Shepherd, who had done them for the earlier US editions of *Oddities* and *Enigmas*.
384. Alfred Gillgrass, *The Book of Big Ben*, Herbert Joseph, 1946.
385. Eric Whittle, *The Inventor of the Marine Chronometer: John Harrison of Foulby 1693–1776*, Wakefield Historical Publications, 1984.
386. Ibid. Ref. 331, 21 October 1947.

CHAPTER 22: CANTERBURY AND A GOLD MEDAL 1945–1948

387. Ibid. Ref. 314, 15 June 1945.
388. Ibid. Ref. 1.
389. NMM Trustees meeting minutes, 8 October 1945.
390. Gould to Rickett, 5 June 1946, in MoD chronometer file, preserved along with the Gould notebooks at the NMM, as Gou/18.
391. Ibid. Ref. 379, p. 62–63.

APPENDIX 4

Oddities

392. Mike Dash (ed.), *Fortean Studies*, Vol. 1, John Brown, London, 1994, pp. 71–150.
393. Alfred Leutscher, 'The Devil's Hoof-marks', *Animals*, Vol. 6 No. 8, 1965, pp. 297–299.
394. David Sealy, 'Trailing the Devil', *The Skeptical Intelligencer*, Vol. 6, 2003, pp. 11–18
395. Correspondence between Doyle and Gould on this subject can be found in the archives of the Scott Polar Research Institute, ref MS 397/4/1–8.

396. Joe, Nickell, with John F Fischer, *Restless Coffins, Secrets of the Supernatural: Investigating the World's Occult Mysteries*, Prometheus Books, Buffalo, N.Y. 1991, pp. 129–188.

397. Lionel and Patricia Fanthorpe, '*Tales from the Crypt*', *Fortean Times*, April 2000.

398. Ann Savours, *The Search for the North West Passage*, Chatham, London, 2003

399. Richard J Cyriax, *Sir John Franklin's Last Arctic Expedition*, Methuen, London, 1939, reprinted by The Arctic Press, Plaistow and Sutton Coldfield, 1997.

400. *The Times*, 20 September 1930, p. 6.

401. Yvan Martial, 'Bottineau et sa nauscopie', *L'Express* (Mauritius), 6 September 2004.

402. Gould records one further appearance, reported by an eccentric American weather-prophet named Tice, who Gould describes jokingly, in the first edition of *Oddities*, as 'the Lord Dunboyne of his day'. In his copy of the book, annotated for the second edition, Gould notes here 'This might have landed me in a libel action . . . substitute 'John Partridge'.

403. http://seds.lpl.arizona.edu/nineplanets/nineplanets/hypo.html which, in Appendix 7, discusses Hypothetical planets.

404. Ian Wilson, *Nostradamus, the Evidence*, Orion, London, 2003.

405. Ibid. Ref. 120, Gould to Dyson on 14 December 1929.

406. Ibid. Ref. 120, Gould to Bowyer on 3 October 1929.

APPENDIX 5

407. The details of this affair have in part been taken from surviving correspondence, now in private hands.

408. Paul Chamberlain, *It's About Time*, Richard R. Smith, New York, 1941, p. 22.

409. Paul Chamberlain, 'Notes on the Lever Escapement', in *Horology, The National Magazine for the Advancement of Horology*, (US), a long series of articles appearing sporadically in this journal during the mid-1930s, including: March and April 1936.

410. *Horological Journal*, Vol. 77, No.924, August 1935, p. 375.

411. Charles Allix, 'Thomas Mudge', *Apollo Miscellany*, June 1950, pp. 1–8.

412. Malcolm Gardner, *Catalogue VI, Part Three*, Autumn 1950, Lot 1267.

Bibliography of Works by R.T. Gould

1. BOOKS

Jeremiah Horrox, Astronomer

pp. 37 (133 copies only)
Opusculum No. 75 of the Sette of Odd Volumes, London, 1923
Note. The year given on the title page for the lecture, 1923, was a misprint for 1922. Most copies of the book seen by this author have had the tiny print neatly scratched out '3' and a '2' permed in a—almost certainly by Gould.
(Dedicated to the memory of Harry Hilton Monk Gould)

The Marine Chronometer, its history and development

pp. xvi.,287, plates xxxix, Figs 85.
J.D. Potter, London, 1923
(Dedicated to Earl Beatty)

Reprinted (smaller format, & without dedication), Holland Press, London, 1960, 1971, 1973, 1976, 1978
Reprinted (cheap, paper back edition): Irvine, Texas, Albert Saifer, New York, 1987.
Reprinted with additional illustrations and alterations (not all of them improvements) and with the dedication restored, Antique Collectors Club Ltd., Woodbridge, 1989

The Sea Serpent

pp. 55, plates 15, (167 copies only—numbered 1–333)
Opusculum No.80 of the Sette of Odd Volumes, London, 1926
(Dedicated to Megophias Megophias)

Oddities. A Book of Unexplained Facts

pp. 336, Plates vii, Figs 27. (Jacket by RTG)
Philip Allan, London, 1928
(Dedicated to the Sette of Odd Volumes by RTG & publishers)

Reprinted (some in blue covers instead of orange) November 1928
American edition, Frederick A Stokes, New York, 1928

Second Edition.
pp. 213, Plates iii, Figs 20. (Jacket by Geo Mansell)
Geoffrey Bles, London, 1944

(Now only dedicated to the Sette by RTG)
Reprinted 1945
Reissue of Second Edition.
pp. viii, 228, Plates iii, Figs 20 (with an introduction by Leslie Shepard, an index and new dust jacket).
University Books, New York, 1965
(Dedication now omitted)
Reprinted 1966

Paperback Edition
pp. 301, 'Plates' iii, Figs 20 (with an additional Intro: Oddities, A Book of Occult Events)
Paperback Library, New York, June 1969

Second Reprinting of Reissued Edition
(As for University Books edition of 1965. Jacket by D.Critchlow)
Bell Publishing Co, New York, c.1980.

Enigmas. Another Book of unexplained facts

pp. 320, Plates vii, Figs 17. (Jacket by RTG)
Philip Allan, London, 1929
(Dedicated to 'My Mother')

Second Edition
pp. 231, Plates v, Figs 18 (2 chapters replaced. Jacket by Geo Mansell)
London, Geofrey Bles, 1946
(Dedication now includes mother's name and dates)

Reissue of Second Edition,
pp. 248, Plates v, Figs 18. (with an index and a new dust jacket)
University Books, New York, 1965
(Dedication now omitted)
Reprinted 1966

Reprinted with new title: Unexplained Facts: Enigmas and Curiosities (with new dustjacket)
Bell Publishing Co., New York, 1980

The Case for the Sea Serpent

pp., xii, 291, Plates vii, Figs 30 (End-paper maps and Jacket by RTG)
Philip Allan, London, 1930
(Dedicated originally to RTG's cat Bill, but changed to Armorel Daphne Heron-Allen)
Reprinted in slightly smaller format, 1934.
American Edition, G P Putnam's Sons, New York, 1934

Reissued
(printed cloth covers, otherwise straight reprint)
Singing Tree Press, Detroit, 1969

The Loch Ness Monster and Others

pp. xii, 228, Plates vii, Figs 46 (Jacket with map by RTG)
Geoffrey Bles, London, June 1934
(Dedicated to Veronica Keiller)

Reissued
pp. xiv, 228, Plates vii, Figs 46 (With a Foreword by Leslie Shepard and new
dust jacket by N.Frank) University Books, New York, 1969
(Dedication now omitted)

Paperback Edition of Reissue (cover by N.Frank)
Citadel Press, Secaucus, 1976

Paperback Edition of Original (cover by J.Drake)
Black Books 1996
(Dedication retained)

Begegnungen Mit Seeungeheuern (by RTG with G-G F von Forstner)
pp. 179, Figs 42. (Partly a combination of CftSS & LNM&O)
Grethlein & Co., Leipzig, 1935

Captain Cook

pp. 144 (Great Lives no.40)
Duckworth, London, 1935.
(No dedication)

Second Edition
pp. 128, Figs 6. (with a new introduction by Gavin Kennedy, typographical
corrections and redrawn maps but without the bibliography. Cover the
Dance portrait of Cook)
London, Duckworth, 1981

A Book of Marvels

pp. ix,180, Figs 14 (The Fountain Library: Essays taken from Oddities and
Enigmas)
Methuen, London, 1937
(Dedicated to Miss J Bower)

Canadian Edition, S J R Saunders, Toronto, 1937

The Stargazer Talks

pp. 128, Figs, 10 (Dust jacket by Geo Mansell)
Geoffrey Bles, London, July 1943

(Dedicated to V.G.I.: Grace Ingram)
Reprinted September 1944 & April 1946
Reissued with new title: More Oddities and Enigmas
pp. 128, Figs 10, (With new foreword by Leslie Shepard and new dust jacket)
University Books, Secaucus, 1973

Communications Old & New

pp. 54, Figs 28 (Illustration by C.F.Tunnicliffe, based on Cable & Wireless adverts done previously) R.A.Publishing (For Cable & Wireless), London, 1944 (released 1945).
(No dedication)

The Story of the Typewriter

pp. 48. Figs 19. (Posthumous. Originally articles in Office Control and Management). Edited by Dudley W. Hooper, with a foreword by Mancell Gutteridge. Office Control and Management, London, 1949
Reprinted 1950

2. PUBLISHED ARTICLES (INCLUDING OFFPRINTS) AND CHAPTERS IN BOOKS

'The History of the Chronometer', *The Geographical Journal*, Vol. 57, No. 4, April 1921, pp. 253–270, and reprinted in The Horological Journal, Vol. 63, Nos 756 and 757, August–September 1921, pp. 195–198 and 218–221; & Vol.64, Nos 759 and 760, November–December 1921. pp. 29–32 & 54–56.

'Lost Islands of the Southern Ocean', *Discovery*, 3, 195, July 1922.

'An Historic Mudge Chronometer' *The Horological Journal*, Vol.64 No.768, August 1922, pp.224–227 (Reprinted from Bazaar Exchange and Mart)

'Two Early Gravity Escapements' *The Horological Journal*, Vol. 65, No. 777, May 1923, pp. 182–184.

'A Winking Pendulum' *The Horological Journal*, Vol. 66, No. 781, September 1923, pp. 6–9.

'A Miniature Copy of Harrison's No. 4' *The Horological Journal*, Vol. 66, No. 782, October 1923, pp. 30–32.

'Early Marine Chronometers', *The Horological Journal*, Vol. 66 Nos 784–788. December 1923—April 1924, pp. 67–70, 91–96, 109–113, 128–131, and 143–147.

'The Ross Deep' *The Geographical Journal*, Vol. 63, March 1924, pp. 237–241.

'The First Sighting of the Antarctic Continent', *The Geographical Journal*, Vol. 65, No. 3, March 1925, pp. 220–225.

'The Landfall of Columbus: an old problem restated'. . . . .*The Geographical Journal*, Vol. 69, May 1927, pp. 403–425.

'The Modern Typewriter and its Probable Future Development', *Journal of the Royal Society of Arts*, Vol. 76, No. 3940, 25 May 1928, pp. 717–738.

'The Chronometer', *The Horological Journal*, Vol. 70, No. 837, May 1928, pp. 164–165.

'Some Unpublished Accounts of Cook's Death', *The Mariner's Mirror*, Vol. 14, No. 4, October 1928, pp. 301–319.

'Bligh's Notes on Cook's Last Voyage' *The Mariner's Mirror*, Vol. 14, No. 4, October 1928, pp. 371–385.

'The Society's Harrison Clock', RAS *Monthly Notices*, February 1929, pp. 398–401.

'The Legacy of the Old Masters', *The Practical Watch and Clockmaker*, Vol. 2 No. 13, 15 March 1929, pp. 50–53.

'Notes on the History of the date or Calendar Line', (written by RTG in Nov.1921) *New Zealand Journal of Science and Technology*, Vol. 11, No. 6, April 1930, pp. 385–388.

'John Harrison's third Timekeeper', *The Times*, 13 April, 1931 ('A Correspondent', but written by RTG).

'The Restoration of John Harrison's Third Timekeeper' *The Horological Journal*, Vol. 73, No. 873, May 1931, pp. 166–168, & Vol. 73, No. 874, June 1931, p. 175.

The Navigator's Equipment Typescript (at NMM) dated 30 August 1932 and marked up by Callender (SNR) for setting, and associated with two illustrations 'for the Navigator's Equipment' (H1 and Le Roy timekeeper), one dated 1933.

'A Restored Masterpiece', *The Times*, 28 February, 1933, p. 13 ('A Correspondent', but almost certainly written by RTG).

'The First Practical Marine Timekeeper', *The Horological Journal*, Vol. 75, No. 895 March 1933, pp. 294, 303.

'Double-Pendulum Clock by Quare and Horseman', *The Horological Journal*, Vol. 75, No. 896, April 1933, pp. 317–318.

'The Reconstruction of Harrison's First Timekeeper' *The Observatory*, Vol. 56, No. 709, June 1933, pp. 192–196.

'The Loch Ness Monster: A Survey of the Evidence', *The Times*, 9 December 1933, p. 13–14.

'The Bi-Centenary of Harrison's First Marine Timekeeper', *The Illustrated London News*, 2 March, 1935, pp. 342–343.

'John Harrison and his Timekeepers' *The Mariners Mirror*, Vol. 21, No. 2, April 1935, pp. 114–139.

Reprinted by NMM (Pink covers and minus list of those attending the lecture) c.1947.

Reprinted by NMM (Beige covers, photos arranged in centre pages) in 1958 and after, printed as required.

Reprinted by NMM (described as '4th edition') 1978.

Reprinted by NMM (Green covers with H4 and mostly retaken photo's) 1987 and 1991.

Translated into Japanese by T.Kotani, and introduced by R.Yamaguchi, for Japanese watch and jewellery magazine (identity unknown) in three parts, Vol. 17, 1976, pp. 137–145, 137–172, 222–234.

'The First Chronometer', *Radio Times*, 18 October, 1935.

'John Harrison and his Timekeepers', *The Horological Journal*, Vol. 78, No. December 1935, pp. 12–21, et seq.

'Paul M. Chamberlain—An Appreciation' Horology, December 1935, p. 19 et seq.

'Cheyne's Proposed Arctic Expedition, 1880', *Bulletin of the Geographical Society of Philadelphia*, Vol. 33, No. 4, 1935, pp. 99–105.

'In Quest of a Monster', *Children's Hour Annual*, 1935, pp. 121–130.

'Some of my Hobbies', *Children's Hour Annual*, 1936, pp. 94–103.

'Why Big Ben's Voice is Cracked', *Radio Times*, 14 August 1936, p. 15.

'Look Out for the Big Moon', *Radio Times*, 25 September 1936, p. 15.

'All About the Westminster Clock', *Radio Times*, 30 October, 1936, p. 15.

The Start of 'TIM', *Radio Times*, 13 November, 1936, p. 11.

'The Instruments of Navigation', *Shipping Wonders of the World*, The Amalgamated Press Ltd., London, 1936, Vol. 1, pp. 504–514.

'To the Uncharted South', *Shipping Wonders of the World*, The Amalgamated Press Ltd., London, 1936, Vol. 1, pp. 31–38.

'Scott's Gallant Failure', *Shipping Wonders of the World*, The Amalgamated Press Ltd., London, 1936, Vol. 1, pp. 169–176.

'The Franklin Mystery', *Shipping Wonders of the World*, The Amalgamated Press Ltd., London, 1936, Vol 1, pp. 350–359

'South with Shackleton', *Shipping Wonders of the World*, The Amalgamated Press Ltd., London, 1936, Vol. 1, pp. 419–426.

'Ross in the Antarctic', *Shipping Wonders of the World*, The Amalgamated Press Ltd., London, 1936, Vol. 2, pp. 929–937.

'Survey Ships at Work', *Shipping Wonders of the World*, The Amalgamated Press Ltd., London, 1936, Vol. 2, pp. 1449–1455.

'Mystery of the Mary Celeste', *Shipping Wonders of the World*, The Amalgamated Press Ltd., London, 1936, Vol. 2, pp. 1591–1594.

'Captain Bligh—The Navigator', *Shipping Wonders of the World*, The Amalgamated Press Ltd., London, 1936, Vol. 2, pp. 1633–1637.

'Romance of the Chronometer', *Shipping Wonders of the World*, The Amalgamated Press Ltd., London, 1936, Vol. 2, pp. 1641–1645.

'The Burlesden Ship', *The Listener*, 5 June 1937.

'Make your own Puzzles', *Children's Hour Annual*, 1937, pp. 132–144.

'My Sixty-Three Year-Old Typewriter', *Radio Times*, 27 August, 1937, p. 17.

'The Original Orrery Restored', *The Illustrated London News*, 18 December, 1937, pp. 1102 and 1126.

'The Navigator at Sea', *The King's Navy* (undated pamphlet) *c.*1938, pp. 51 & 54–58.

'Finding the Way at Sea', *The Daily Telegraph and Morning Post*, 7 November, 1938, p. 12, (Clocks and Watches Supplement).

'The Expert Talks', *Ack-Ack, Beer-Beer*, A Medley from the famous BBC Programme, (ed.Bill Maclurg) Hutchinson, 1943, pp. 144–147.

'The Charting of the South Shetlands', 1819–28', *The Mariner's Mirror*, Vol. 27, No. 3, July 1941, pp. 206–242.

'Note on the Society's Harrison Clock', RAS *Monthly Notices*, Vol. 106, No. 1, 1946, p. 12.

'John Harrison and his Timekeepers', *Argentor* (The Journal of the National Jewellers' Association, June 1947, pp. 105–136.

'It Looks Impossible', *The Children's Own Wonder Book*, Odhams, London, pp. 86–95 (Illustrated by W. Nickless) Reprinted 1947.

'The Man Who Discovered Longitude', *The Lasting Victories*, Lutterworth, London, 1948, pp. 112–115 and 120–121.

3. PUBLISHED LETTERS IN THE PRESS AND JOURNALS

'Dent's Factory', *The Horological Journal*, Vol. 63, No. 757, September 1921, p. 224.

'The Marine Chronometer', *The Horological Journal*, Vol. 65, No. 778, June 1923, pp. 210–211.

'Two Early Gravity Escapements', *The Horological Journal*, Vol. 65, No. 778, June 1923, pp. 211.

'The Compensated Balance', *The Horological Journal*, Vol. 66, No. 792, August 1924, pp. 239–240.

'Harrisons No. 2 & No. 3', *The Horological Journal*, Vol. 66, No. 792, August 1924, p. 240.

'The Compensated Balance', *The Horological Journal*, Vol. 67, No. 794, October 1924, p. 32.

'The Origin of Bridge', *The Times*, 9 May, 1927, p. 10.

'Bouvet and Thompson Islands', *The Times*, 23 February 1928.

'The Chronometer', *Lloyds List & Shipping Gazette*, No. 35418, 5 March 1928 p. 8, col. 5.

'New Antarctic Lands' . . . *The Times*, 2 January 1929, p. 11.

'Franklin Expedition', *The Times*, 20 September 1930, p. 6.

'The Making of Boswell's Johnson' *The Times Literary Supplement*, 27 February 1930, p. 166.

'Sea Serpents' . . . .*The Times Literary Supplement*, 1931, pp. 20, 44, 60, 99, 154.

'Re: Thomas Axe, Deceased', *The Horological Journal*, Vol. 73, No. 871, March 1931, pp. 127–128.

'Longitude Harrison', *The Times*, 20 April 1931, p. 10.

'John Harrison's third Timekeeper', *The Horological Journal*, Vol. 73, No. 873, May 1931, pp. 166–168. No. 874, June 1931, p. 175.

'Sea Stories of Today', *The Times Literary Supplement*, 1932, p. 943.

'Samuel Pepys'; . . . *The Times Literary Supplement*, 1934, p. 76.

Marine Chronometers, *The Horological Journal*, Vol. 78, No. 935, August 1936, pp. 22.

'Edward Bramsfeld', *The Times*, 5 March 1937, p. 15.

'Discovery of the Antarctic', *The Times*, 11 May 1937, p. 12.

'The Evidence of Eye-Witnesses', *The Times*, 31 May 1938, p. 12.

'Sea Serpents', *The Times*, 21 September 1938, p. 6.

'13 Bibles to the Square Inch', *The Times*, 20 January 1939, p. 8.

'Track Charts of Explorers', *The Geographical Journal*, Vol. 97, No. 2, February 1941, pp. 133–135.

'An Early Map of the South Shetlands', *The Geographical Journal*, Vol. 97, No. 5, May 1941, p. 332.

'Samuel Watson', *The Horological Journal*, Vol. 85, No. 1014, March 1943, p. 61.

'The Electronic Brain', *The Times*, 29 November 1946, p. 5.

'Grimthorpe and the Dents', *The Horological Journal*, Vol. 90, No. 1072, January 1948, pp. 28–30.

'Cause of Grandfathers Stopping', *The Horological Journal*, Vol. 90, No. 1073, February 1948, p. 92.

'Grimthorpe and the Dents', *The Horological Journal*, Vol. 90, No. 1074, March 1948, p. 157.

4. PUBLISHED ILLUSTRATION (OTHER THAN OWN BOOKS)

Ralph Straus, *Pengard Awake!* Chapman and Hall, London, 1920, (dust jacket by Gould)

A.C. Delacour De Brisay, *And Then Came War- An Outline of the European Tragedy*, Ivor Nicholson and Watson Ltd., London, 1937, (maps and drawings by Gould)

Richard .J Cyriax, *Sir John Franklin's Last Arctic Expedition*, Methuen & Co., London, 1939, (4 maps by Gould), (Reprint: Arctic Press, Plaistow, London, 1997.

5. REVIEWS AND OTHER LITERARY ASSOCIATIONS

'Jules Sottas, The Corvette L'Aurore and its Model', *Mariner's Mirror*, Vol. 16, pp. 116–133. (RTG for technical and historical interpretation and for notes on the *megametre*).

Claus Bergen, *U-Boat Stories*, (trans by Eric Sutton). Constable & Co., London, 1931. (RTG helped with technical translation/interpretation).

Austin Freeman, *Mr Polton Explains*, Hodder and Stoughton, London, 1940 (dedicated to RTG). Fourth Impression, 1950.

'The Four Voyages of Columbus, . . .' *The Geographical Journal*, Vol. 101, No. 5, May 1943, pp. 263–265.

Marine Sextants and Accessories, Marine Instruments Ltd., Glasgow and London, February 1946 (introduction by RTG).

The Book of Big Ben, Alfred Gillgrass, Herbert Joseph, London, April 1946. Foreword by RTG.

'A Russian Expedition to the Antarctic, . . . ' *The Geographical Journal*, Vol. 110, No. 3, September 1947, pp. 98–102.

6. BROADCASTS & SCRIPTS FOR BROADCASTS

(CH = *Children's Hour*. Rec = date of recording, broadcast uncertain).

'Umpiring at Wimbledon' (CH) 1 July 1938.
'Astronomy in the Black-Out' 12 April 1940.
'Cricket Stories & the Longest Day' 4 June 1940
'The Sky at Night' 8 August 1940.
'The Night-Sky', February 1941.
'Fifty Selected Puzzles' ('for occasional use'), January 1941.
'Some Unsolved Mysteries' (CH) 13 March 1942.
'Jules Verne' (CH) 12 June 1942.
'The Mystery of Kaspar Hauser' 7 July 1942 (Rec).
'Islands which Appear and Disappear' 7 July 1942 (Rec).
'The Devil Fish' 7 July 1942 (Rec).
'The Beginning of Mechanical Flight' (CH) 24 July 1942.
'I am an Expert' (*Ack-Ack, Beer-Beer*) 26 November 1942.
'Nova Puppis' 29 November 1942.
'Radio, and Weather-Forecasting' (Empire Newsreel), 4 January 1943 (Rec).
'Television in Modern Warfare' (Empire Newsreel), 4 January 1943 (Rec).
'The New 'Lifeboat' Charts' c.1943.
'The Ship's Lifeboat' c.1943.
'Isaac Newton' (CH) 6 January 1943.
'Appeal on Behalf of . . . Merchant Seamen's Institution' 24 October 1943.
'Treasure, in Fact and Fiction' 31 January 1944.
'Personal Choice' 23 February 1944 (Rec).
'Appeal on Behalf of the Royal Navy' early 1945.
'The Last Voyage of Sir John Franklin' 25 June 1945.
'The Great Waterfalls of the World late' 1945.
'The Story of The Crystal Palace' 24 November 1945.
'Marking Time (Watches) (*The World and His Wife*) 28 January 1946.

(Note: this list only includes those from scripts in the family records and from *The Stargazer Talks*. There are undoubtedly more in the BBC archives at Caversham, Bucks).

Recommended Reading

GOULD AND HIS TIMES (INCLUDING SOME OBITUARIES)

A.B. Campbell, *When I Was In Patagonia*, Christopher Johnson, London, 1953.

Robert, Graves, and Alan Hodge, *The Long Week-End*, Faber, London, 1940.

Paul, Tabori, *Harry Price, Ghosthunter*, Sphere, Books Ltd., London, 1974.

Howard Thomas, *Britain's Brains Trust*, Chapman and Hall, London, 1944.

[John Gilbert, Lockhart], 'Comment', *Blackwood's Magazine*, December, 1948, pp. 472–73.

The Times, 9 October 1948, p. 6.

Illustrated London News, Vol. 273, October 16, 1948, p. 442.

Horological Journal, Vol. 90, Nos.1081 & 1082, October & November 1948, p. 594 & pp. 655–656.

Mariner's Mirror, Vol. 35, No. 1, January 1949, p. 2.

Geographical Journal, Vol. 112, No. 6, December 1948, p. 258–259.

Heinrich Otto, 'In Memory of Lieutenant-Commander R.T. Gould,' *H.I.A. Journal*, Horological Institute of America, January 1950, pp. 17–18, 33, 47, 50.

Jens Thygessen Schmidt, 'John Harrison and his Timekeepers', *H.I.A. Journal*, Horological Institute of America, February 1950, pp. 15–17, 24, 34, 41, 43.

Albert L. Partridge, 'Lt.-Commander Rupert T Gould, R.N.' *NAWCC Bulletin*, The National Association of Watch and Clock Collectors (USA), Vol. 4, No. 4, June 1950, pp. 158–163.

Leslie Shepard, 'Introduction', *Oddities*, Paperback Library, New York, 1964, pp. 5–8.

Gavin Kennedy, 'Introduction', *Captain Cook*, Duckworth, London, 1978, pp. 9–13.

LONGITUDE/CHRONOMETERS (IN ADDITION TO GOULD'S OWN PUBLICATIONS)

Col. Humphrey Quill, 'John Harrison, Copley Medallist, and the £20,000 Longitude Prize', *Notes and Records of the Royal Society of London*, Vol. 18, No. 2, December 1963, pp. 146–160.

Richard Good, 'John Harrison's Last Timekeeper of 1770', *Horological Journal*, Vol. 97, No. 1167, December 1955, pp. 804–809 (reprinted in *Pioneers of Precision Timekeeping*, Autiquarian Horological Society, c.1964).

Peter Amis, 'The Bounty Timekeeper', *Horological Journal*, Vol. 99, No. 1191, December 1957, pp. 759–770 (reprinted in *Pioneers of Precision Timekeeping*, Autiquarian Horological Society, *c*.1964).

Col. Humphrey Quill, *John Harrison, the man who found Longitude*, John Baker, London, 1966.

Eric Whittle, *The Inventor of the Marine Chronometer: John Harrison of Foulby 1693–1776*, Wakefield Historical Publications, Wakefield, 1984.

Anthony Randall, 'The Technology of John Harrison's Portable Timekeepers', *Antiquarian Horology*, Vol. 18, Nos. 2&3, Summer & Autumn 1989, pp. 145–160, 261–277.

Jonathan Betts, *Harrison*, National Maritime Museum, London 1993.

Dava Sobel, *Longitude*, London, Fourth Estate 1995. Dava Sobel and William Andrews, *The Illustrated Longitude*, 1998.

William Andrewes (ed.), *The Quest for Longitude*, Harvard, 1996.

B.B. Schofield, *The Story of HMS Dryad*, Mason, Havant, 1977.

C.L. Lowis, *Fabulous Admirals and some Naval Fragments.* Putnam, London, 1957.

LOCH NESS AND SEA SERPENTS

A.C. Oudemans, *The Great Sea-Serpent*, n.p., Leiden, 1892

Constance Whyte. *More than a legend. The Story of the Loch Ness Monster*. Hamish Hamilton, London, 1957

Bernard Heuvelmans, (trans. Richard Garnett). *In the wake of Sea Serpents*, Hill & Wang, New York, 1968.

Peter Costello, *In Search of Lake Monsters*, Coward McCann & Geoghegan, New York, 1974.

Ronald Binns (with R.J. Bell), *The Loch Ness Mystery Solved*, Open Books, Shepton Mallet, 1983.

Loren Coleman and Patrick Huyghe, *The Field Guide to Lake Monsters, Sea Serpents and Other Mystery Denizens of the Deep*, Tarcher/Penguin, New York, 2003.

ODDITIES, ENIGMAS, ETC.

Alfred Leutscher, 'The Devil's Hoof-marks,' *Animals*, Vol. 6, No. 8, 1965, pp. 297–299.

David Sealy, 'Trailing the Devil,' *The Skeptical Intelligencer*, Vol. 6, 2003, pp. 11–18.

Simon Singh, *Fermat's Last Theorem*, Fourth Estate, London, 1997.

Eric Frank Russell, *Great World Mysteries*, Dobson, London, 1957.

BROADCASTING

Wallace Grevatt, *B.B.C. Children's Hour: A Celebration of Those Magical Years*, Book Guild, 1988.

Marilyn Lawrence Boemer, *The Children's Hour: Radio programmes for children 1929–1956*, Scarecrow Press, Metuchen, N. J., London, 1990. (The U.S. equivalent).

Howard Thomas, *Britain's Brains Trust*. Chapman & Hall, London, 1944.

POLAR RESEARCH HYDROGRAPHY

Richard J. Cyriax, *Sir John Franklin's Last Arctic Expedition*, Methuen & Co, London 1939. Reprinted: Arctic Press, Plaistow, 1997.

Ann Savours, *The Search for the North West Passage*, Chatham Books, London, 2003.

Henry Stommel, *Lost Islands – The Story of Islands That Have Vanished from Nautical Charts*, University of British Columbia, Vancover, 1984.

TYPEWRITERS, TENNIS & MISC

G. C. Mares, *The History of the Typewriter; Successor to the Pen*, (Reprint: Post Era Books, Arcadia, CA, 1985).

Michael H. Adler, *The Writing Machine*, Allan & Unwin Ltd, London, 1973.

Summary of Gould's Harrison Notebooks and the Later Work

These notebooks are now all preserved in the manuscripts section of the National Maritime Museum (NMM). They were not all acquired by the NMM at once however, and their listing needs a little explanation. The numbers used to describe them (i.e. 'Book 1', 'Book 2', etc.) represent Gould's own numbering system.

Book 1 is primarily a nineteeth century manuscript which happened to have a number of blank pages at the back which Gould used to make a few brief notes on the restoration of H1. For this reason, when he presented his Harrison notebooks to the Chronometer Section (1946, though he had bequeathed them in his will), he kept back Book 1 as being of little interest, and it remained in his family's possession for many years after his death.

In August and October 1946, he presented (in two batches) his books 2–17 to the Ministry of Defence's Chronometer Section. Book 18 appears to have been kept at the Royal Astronomical Society and only joined the others after Gould's death. Here they remained until 1979, when they were presented to the NMM by the Section. Between 1945 and 1947, there had been a little correspondence between Gould and the Section and these letters, and some prints of H4's mechanism, were all tied together by the Section, making a further 'notebook' presented to the museum, though not numbered in Gould's sequence.

On their arrival at the Museum, the Manuscript Section gave these eighteen books the NMM reference numbers GOU/1 to GOU/18, with Gould's Book 2 being GOU/1, Book 3 being GOU/2 etc., the additional file being given the reference GOU/18. Then, in 1988, Cecil Gould presented the museum with his father's notebook No.1 which now has the reference GOU/19.

Book 1. (GOU/19) Nineteenth century copy of the manuscript by Harrison entitled *An Explanation of my Watch . . .* (1763) preserved in the Clockmakers Company library (MS 3972/1), made for Bennett Woodcroft of the Patent Office. Gould has added two and a half pages of typescript commentary, and annotated the manuscript itself heavily, with a view to having it all put into typescript form. The end of the book contains several pages of notes by Gould on the first cleaning of H1, and a few notes on the second restoration in the 1930s.

Book 2. (GOU/1) 'Harrison No.2'. The restoration of H2, and a few notes on H3, during the period October 1923 to October 1924 at Epsom.

Book 3. (GOU/2) 'Harrison No.3'. The beginning of the restoration of H3, starting in September 1924 and then continuing from October 1929 to November 1930. Here the narrative of H3's assembly is continued in Book 4, but when this is nearly full, in January 1931, the notes begin again in Book 3, continuing until completion in March 1931. Finally there is a brief note about a repair to H3 in August 1937.

Book 4. (GOU/3) 'Harrison No.3'. The continuation of the restoration and assembly of H3, from November 1930 through to January 1931, when H3's completion is continued again in Book 3. The final pages of Book 4 contain a brief note, from December 1933, concerning the winding of H2 and H3 at the Science Museum, and two pages of notes, from December 1930, on the longcase clock with the 'winking pendulum' (see p. 131).

Book 5. (GOU/4) 'Harrison No.1'. The full restoration of H1, from September 1932 to October 1934, including its return to Greenwich and continuing repairs. One further note, from October 1937, concerns a stoppage of H1. The last six pages of the book consist of a diary kept by Gould during August 1932 when the children were staying at Downside.

Book 6. (GOU/5) 'Harrison No.4; Kendall 1; Harrison No.2'. The later restoration of H4, from May to June 1935 and K1, from June to July the same year. The book continues with the cleaning work on H2 from October 1936 to April 1937, then continued in Book 7.

Book 7. (GOU/6) 'Harrison No.2'. The continuation from Book 6 of the cleaning of H2, containing the completion of the job in April 1937, with a couple of notes of stoppages in July 1938.

Book 8. (GOU/7) 'Harrison 3'. The beginning of the second restoration of H3, starting in October 1937, the book finishing with the move of H3 to the Queen's House (NMM) in Greenwich in April 1938.

Book 9. (GOU/8) 'Harrison 3'. The continuing second restoration of H3, from May 1938 to August 1939 when H3 was packed up for protection, and continuing again in July 1943 at Upper Hurdcott, intermittently until August 1945 when, still fully dismantled, it was packed for transporting to Bradford-on-Avon.

Book 10. (GOU/9) 'Harrison No.1'. Three different sets of typescript instructions and explanations for the staff at Mercers on the work required on parts of H1. 'Notes on the Winding Mechanism' (October 1932); 'Notes on the Balance Springs' (November 1932); and 'Notes on the Compensation Mechanism' (December 1932).

Book 11. (GOU/10) 'Notes on John Harrison's Second Marine Timekeeper (and a few on his third)' A twenty-seven-page typescript monograph, written for the Science Museum in 1925, (see p. 145).

Book 12. (GOU/11) Offprint of 'The Restoration of John Harrison's third Timekeeper' *The Horological Journal*, (Vol. 73, No. 873, May 1931, pp 166–168 and Vol. 73, No. 874, June 1931, p175), with a number of Gould's annotations.

Book 13. (GOU/12) 'Various papers, in chronological order, re the cleaning & repair of Harrison 1'. Included are some photographs and sketches of H1, and correspondence and notes dating from '1920' to October 1934.

Book 14. (GOU/13) 'Various papers, in chronological order, re the cleaning & repair of Harrison 2'. Included are 'Harrison's second marine timekeeper— procedure for repairing the wires connecting the balances' (October 1927), and correspondence dating from October 1923 to May 1947.

Book 15. (GOU/14) 'Various papers, in chronological order, re the cleaning & repair of Harrison 3'. Included is lengthy correspondence dating from September 1929 to February 1932 and then from December 1937 to December 1945.

Book 16. (GOU/15) 'Rough note book, kept by me during the repair, cleaning and adjustment of the Royal Astronomical Society's Harrison Clock'. The entries in this notebook are arranged in a most haphazard order (see p. 211). The book begins (at the back) with a few notes on the going of the clock (running at the Society probably) dating from February 1927. The entries continue (at various places in the book and at various locations) from this point through until February 1929 on its return to the RAS.

Book 17. (GOU/16) 'RAS Harrison Clock'. Copies of correspondence dating from November 1924 to February 1929.

Book 18. (GOU/17) 'The Harrison Clock'. Record of the going of the RAS Harrison regulator from April 1929 to June 1949.

CHRONOMETER SECTION'S ADDITIONAL FILE (GOU/18)

A file of correspondence chiefly between Gould and the Chronometer Section, dating from April 1945 to December 1946. It includes Gould's own summary of what the first twelve books contain. At the front are a number of modern pulls from the copper printing plates of John Harrison's *The Principles of Mr Harrison's Timekeeper* (see p. 98), which were still in the possession of the Admiralty, and which were considered of historical interest.

H3 DRAWINGS (REF: HSR/Z/23)

As part of his work in restoring H3, Gould had full-size photostat copies made of the five drawings commissioned from Thomas Bradley (*c*.1840) of H3, and these were then annotated extensively by him during the course of the work at various times. These annotated copies, along with two torque graphs for H3, two eccentricity charts for the escape wheel of H3, and an eccentricity chart for the escape wheel of H2, were also presented to the Chronometer Section in May 1972 and are all now preserved in one folder at the NMM.

RESTORATION WORK ON THE HARRISON TIMEKEEPERS AFTER GOULD

H1, which had suffered from some corrosion while in store at Cambridge during the war, needed considerable work. It underwent restoration by D.W. Fletcher at the Chronometer Section (by this time established at Herstmonceux in Sussex) in 1952, in preparation for its display, along with H2 and H4 at the British Clockmakers, Heritage Exhibition at the Science Museum from May to September that year. In 1961 H1 was completely dismantled and cleaned again, by Roger Maber at the Chronometer Section, during which time Gould's key-winding was all removed, the somewhat corroded original compensation gridirons were replaced (the steel parts being replaced with stainless steel!) and the original type pull-winding was restored. Since then it has run virtually constantly without attention, its only significant stoppage being when all the timekeepers were moved from the NMM's Navigation Gallery to a new display in the Royal Observatory in 1985.

H2 was serviced by Roger Maber at the Chronometer Section in November 1960, during which time he improved one or two of Gould's less tidy repairs and made accurate dimensioned drawings of some of the parts. The timekeeper has run reliably at the museum ever since.

H3 was eventually put back together and returned to the NMM for display by David Evans. It was then dismantled and cleaned again by Roger Maber in the spring of 1961. After fifteen years of reliable running it was then restored again at the Chronometer Section by Bert West and Roger Stevenson in January and February 1976, in preparation for display at the NMM's exhibition to commemorate the bicentenary of Harrison's death. At the same time, the glazed case of H3, which had been rediscovered in the museum's stores in the early 1970s, was cleaned at the NMM and photographed, with H3 within it.

H4, needing lubrication, had to be cleaned several times at the Chronometer Section in the 1960s and 1970s. In February and October 1962 it was cleaned by Bill Roseman (Officer in Charge of the Section) and Bert West; in June 1964 by Roger Maber; in December 1967 by Bert West, Roy Shergold and Roger Stevenson; in December 1971 and April 1972 (+new mainspring) by Roger Stevenson; in February 1976 by Roger Stevenson and Robin Thatcher; and in October 1979, by Robin Thatcher. In September 1982, H4 was cleaned at the NMM (Old Royal Observatory) by Jonathan Betts (JB) and Bert West (with Anthony Randall as an observer) at which point the care of all the Harrison timekeepers was taken on by the staff at the NMM. In August 1984, H4 was cleaned again by JB; in February 1986 (+new hooking on mainspring) by JB, and in 2004 (for research on diamond pallets by Jonathan Hird at the Cavendish Laboratories, Cambridge) by JB.

After its return to the museum following *The Second World War*, K1 was not generally kept running when on display, and does not appear to have been cleaned again until 1983, when it was overhauled and photographed by JB (with Anthony Randall as an observer).

K2, the Bounty Timekeeper, which was until 1963 the property of the Royal United Service Institution, was overhauled and repaired by Peter Amis in 1957. On the closure of the Institution in 1963, ownership of the watch was transferred to the National Maritime Museum, who then asked the RGO Chronometer Section at Herstmonceux to take on its regular care along with the other timekeepers by Harrison and Kendall. In 1964 it was cleaned by R. Bowie, in 1972 by Roy Shergold, in 1976 by Roger Stevenson and Robin Thatcher, and in 1984 by JB.

K3, the other Kendall which James Cook took on his third voyage of discovery, remained in the possession of the Admiralty until it was transferred on loan to the museum in 1967. The timekeeper was not kept running on display, and does not appear to have been overhauled until 1986 when it was cleaned and photographed by JB.

For the sake of completeness, the fate of the two other watches in this pioneering story could be noted.

H5 was never in Admiralty possession and, after its acquisition by the Clockmakers Company in the nineteenth century (see p. 102) was not kept running in their museum, so has rarely been overhauled. It was cleaned and overhauled for the Company by Richard Good in 1955, again by George Daniels in the 1970s, and then in 1983 by JB (with Anthony Randall as an observer).

The Jefferys watch (see p. 96) had suffered near-destruction during *the Second World War*. It had been stored in a safe in a jewellers shop in Hull when the building was destroyed by bombing and the watch had been completely 'cooked' during the subsequent fire. After the war it was carefully cleaned by David Evans of the Chronometer Section, though the watch would never again be fit to run, all parts having been annealed during the fire. With the permission of the Corporation of Hull Trinity House and the Clockmakers Company, in 1993 the watch was again dismantled and studied at Greenwich by JB (with Anthony Randall as an observer and photographer).

Summary of the Content of Rupert Gould's first two books on Unsolved Scientific Mysteries

ODDITIES (1928)

The Devil's Hoof-marks

The first, and one of the most popular, of the mysteries Gould describes in *Oddities* is one he first read about as a boy in the Summer of 1900. On holiday with the family and staying at a large old house in Haslemere, he came across an account of the mystery in an old newspaper in the library of the house.

The story concerns the much-publicized appearance, across the county of Devon in February 1855, of a long series of hoof-shaped prints in the snow, said by the more melodramatic members of the community to be the tracks of the Devil himself. Many locals refused to go out at night and there was much local alarm. The tracks, which were very well documented, seemed to run in a single row continuously, as though made by a bi-ped placing one foot directly in front of the other as it proceeded, but apparently crossing high fences, over rooftops, rivers and enclosed courtyards without interruption.

The discussions of this case over the years have been as heated, and the explanations have been as many and various, as they have been bizarre: from the tracks of a *kangaroo* to fake hoof-marks made by a band of *gypsies*, intent on frightening a rival clan. It was certainly not one to which Gould was prepared to offer a definite solution, though he did point out that the explorer Sir James Ross had observed similar 'hoof-marks' on the icy and desolate island of Kerguelen in the South Atlantic, in 1840, marks they supposed were those of an abandoned pony, though none could be seen.

A theory put forward by one Thomas Fox in the Illustrated London News from 1855, published at the time of the sightings, and illustrated by Gould in *Oddities*, was that the footprints were in fact made by a rat, the front paw prints converging together at the front ('knock-kneed' in style), the straight back prints joining the rear end of the front prints to form the kind of horse-shoe shape '∩' that was seen. A very comprehensive summary of the many writings on this mystery, edited by Mike Dash and extending to no less than 80 pages, is to be found in Fortean Studies Volume 1 (1994).[392]

In the 1960s one of the more likely explanations was published by Alfred Leutscher,[393] and recently reiterated by David Sealy, a retired member of the Natural History Museum's staff (South Kensington).[394] Sealy incidentally, remarks of Gould's work in analyzing unsolved mysteries, that: 'he was one of the first to do so in a properly scientific, and sceptical, manner. He gave no support to any supernatural explanations . . . His books are classics of their kind'. Leutscher's 'complete' solution attributed the footprints to a kind of wood mouse.

According to the theory, the marks were actually made in the opposite sense from that proposed by Fox: the rear prints of the mouse being closer together and turned out, the prints in front joining the rear and effectively elongating them, forming a U shape, similar to the horse-shoe shape but formed with the animal proceeding with the open end of the marks forward. If this is indeed the solution to the mystery, then Gould was very close to the answer with the theory of the paw marks of the rat.

On 13 February 1935, a short summary of The Devil's Hoofmarks was given by Gould in one of his radio broadcasts on the BBC's Children's Hour as *The Stargazer*. This much-abbreviated version of the story was then published in his Stargazer Talks (1943) and an updated version appeared a year later, in 1944, in a second edition of Oddities. The chapter in this second edition, and the summary in the Stargazer Talks, included a note that it had been pointed out to him, that statistically it would have been impossible for any one creature to have made all the footprints in such a short space of time and that there must have been quite a number of such creatures about that night. This perhaps supports the theory that the animal could have been a rodent of some kind, which would also seem a very likely candidate for the footprints on the remote and barren Kerguelen Island.

The Vault at Barbados

This chapter details the story of a burial vault, cut out from solid rock within the churchyard of Christ Church, Barbados, which was said, between the years 1812 and 1820, to have been troubled with the coffins being moved about within the vault by 'unknown forces'. Five times this is reported to have occurred, in spite of the entrance to the vault being sealed by several independent and respectable people, the seals being intact before each inspection. After the fifth time, the bodies were moved to burial in the ground and the tomb left empty. Noting similar cases on record in England and one on an island in the Baltic, Gould considered all logical explanations based on the evidence he had, including earthquakes and flooding, and was unable to find any convincing answer attributable to causes either human or natural.

Touching on the possibility that the cause may have been supernatural, Gould regarded himself 'unqualified to comment' (being sceptical of such explanations, and a wholehearted disbeliever in Spiritualism), and instead cited, amongst others, the famous author Sir Arthur Conan Doyle, a confirmed spiritualist, who had himself written an account of the mystery and with whom Gould had corresponded on that subject. Conan Doyle was a wholehearted believer in the psychic nature of this phenomenon and was certain the cause had been threefold.

First the movement was partly due to spiritual 'forces desiring the more speedy decomposition of the bodies' (hence ensuring these were moved from the vault to the ground); second, there were physical forces causing the movement, which had been derived from the 'effluvia' emitted by the men who had carried the coffins into the vault, and whose 'effluvia' was then trapped in the sealed space 'as in the cabinet of a genuine medium'; and thirdly Conan Doyle declared that because two of the bodies were of people who had committed suicide 'there remains a store of unused vitality which may, where circumstances are favourable, work itself off in capricious and irregular ways . . .'.

Resisting the temptation to dismiss these explanations out of hand, Gould carefully considered each, before concluding their improbability, though pointing out that he is not a total disbeliever in 'supernatural phenomena': 'on that subject I neither affirm nor deny anything'.

Nevertheless, he had a very healthy scepticism about 'genuine mediums' and there was a polite tension between the two authors on the subject. Conan Doyle wrote to Gould in 1930, after having read his views on spiritualism as expressed in *Oddities* and its sequel Enigmas, stating, that 'I think every remark you make upon it is open to question.' This prompted Gould to answer sharply that he hadn't intended to attack spiritualism, but that 'if I ever do so, it will be in a book entirely devoted to the subject', though he doubted he was impartial enough on the matter, adding conclusively 'I have very few human sympathies and I find it difficult to understand the mentality of the average believer in spiritualism. To my mind they are mostly credulous, imperfectly educated and ill-balanced; the raw material from which, in all ages, religious and other bodies have been formed—sometimes by fanatics and sometimes by charlatans. So long as ordinary life is drab, and men fear death, such leaders, however fantastic their teachings, will never entirely lack a following.' Conan Doyle replied on a postcard, simply stating that the seances he had witnessed had been very carefully observed.[395]

Unable to make even tentative suggestions as to a solution to the mystery of the coffins, Gould left the question of the Vault at Barbados unanswered. However, in modern times, the mystery was studied again by Joe Nickell, a leading sceptical investigator of the paranormal, who came to a rather startling conclusion concerning the matter.[396] Nickell cites Gould extensively, but points out a significant flaw in his research, the fact that Gould made no attempt to research at Barbados himself, nor to find someone to do it for him, relying only on secondary sources. After considerable research carried out for him on Barbados, and following up research on the backgrounds of those involved in the narration of the mystery, Nickell concludes that in fact there *never were* any 'primary sources', and that the occurrences never actually happened! The theory suggests that the whole story was in fact created by those reporting it, mostly known Freemasons, as an allegory on the 'secret vault' of Freemasonry, though this solution too is treated with scepticism by some paranormal researchers.[397] The vault itself is real enough though, and remains empty to this day on Barbados.

The Ships Seen on the Ice

As noted in Chapter 3, a particular professional interest of Gould's was the history of the discovery and charting of the Polar Regions; it was something he made a great specialization of and his research and writings on that subject, which span his whole adult life, are widely respected to this day. Since the early days of marine exploration, one of the most important aims of Arctic voyaging was to try and discover a route through the semi-frozen seas over the top of the North American continent. This was the much-sought 'North-West Passage' which it was hoped would provide a quicker and safer route to the Pacific than having to sail south round the notoriously dangerous Cape Horn at the tip of South America.

Perhaps the most well-known of all the expeditions to search for the N.W. Passage was that embarked on by Sir John Franklin, in the two H.M. Ships *Erebus* and *Terror*, in 1845, the ill-fated expedition which famously ended in disaster, all men dying either by illness, starvation or freezing to death. After July that year nothing was heard of the expedition and from 1848 a long series of search parties, several of them funded by Lady Franklin, Sir John's wife, were sent to discover their fate. Finally in 1857, one of Lady Franklin's expeditions, commanded by Captain F.L. McClintock R.N. in the ship *Fox*, found conclusive evidence of the disaster in the form of a note left by the last survivors in April 1848. This stated that Franklin had died the previous year, to date 23 others had perished, and that the remaining men had abandoned the ships to chance their luck trekking south, as they must have known, to almost certain death.[398]

The question Gould poses in this chapter is 'What became of the abandoned ships, *Erebus* and *Terror*?'. Eskimos who had been interviewed by McClintock had stated the ships were both soon crushed by the ice and sank, but Gould questions the veracity of much of the Eskimos' evidence and the subject of this chapter in *Oddities*, the mystery of 'The Ships seen on the Ice' provides another possible answer to the ships' fate.

In 1851 the English brig *Renovation* was sailing westward in the north Atlantic when some of the crew, and a passenger on board, witnessed the passing of a large ice-floe upon which were two abandoned, three-masted ships, one heeled over, the other upright. They corresponded perfectly in size and description with Franklin's ships *Erebus* and *Terror* and Gould concludes they may well have been the lost vessels. Unfortunately, the sighting was not reported to the proper authorities (the Admiralty) for some time and the extensive enquiry that followed, studied in detail by Gould for clues, was inconclusive. One of the most thoroughly researched accounts of Franklin's last expedition was written by Richard J. Cyriax[399] in the 1930s, following a huge amount of research and much interesting correspondence with Gould.

When anyone presumed to write upon a subject Gould regarded as his own, he tended to consider them 'guilty until proven innocent'. Cyriax, who had read this chapter in *Oddities* and had found Gould's chart 5101 referred to in Chapter 3, wrote to him in 1932 and quickly gained his confidence and respect. In due course Gould would come to regard Cyriax as the pre-eminent authority on Franklin and would draw the four maps for Cyriax's book, eventually published in 1939. However, one of the few aspects of the Franklin story the two experts never quite agreed upon was the question of the 'Ships Seen on the Ice': Gould always maintained his belief in their being Franklin's ships. In 1930 he confirmed this view in a letter to the Times[400] and his second edition of *Oddities*, in 1944, reaffirms it, though recognizing the counter view of his friend Richard Cyriax. Cyriax simply didn't believe they could have been Franklin's and concluded that the Eskimo accounts should be believed and that both ships sank in Arctic waters. This is now generally accepted as the most likely fate of the two ships, though it leaves the question open as to the identity of the ships on the ice.

The Berbalangs of Cagayan Sulu

This strange little story covers just six pages in the book, and is in some ways the most peculiar among this collection of peculiarities: in a sense it is perhaps the 'joker' in the pack. The subject is entirely taken from one article, written in 1896 by one E.F. Skertchley, a resident of Hong Kong, and published in the Journal of the Asiatic Society of Bengal, a highly respected sponsor and Gould's sole reason for including what would otherwise be regarded as 'an old wives tale'. It is a story that Gould first heard narrated to him and his brother by M.R. James many years before at Kings College, Cambridge.

Skertchley's article focuses on the little island of Cagayan Sulu (actually the island of Cagayan, in the Sulu Sea) in the Philippines. After some reassuringly straightforward descriptions of the people and culture of the place, things suddenly take on a rather weird and wonderful direction. Continuing his matter-of-fact tale of ordinary folk, Skertchley describes a small village in the centre of the island where, in dracula-style, there exist a community of folk, the Berbalangs, who are able to transmutate into ghouls with illuminated eyes and ghostly sound effects, killing and eating their human victims. He himself witnessed the death of a resident which was apparently associated with a 'visitation' by the Berbelangs.

Needless to say, the mystery Gould discusses in this chapter is not the likelihood of the Berbelangs having supernatural incarnations, but how and why these people managed to stage the appearance of ghosts and convince an apparently sane and logical English observer that they were indeed what they appeared to be. He concludes that, whatever the deception, Sketchley was correct in supposing that the Berbelangs were 'a most unpleasant people' and that as 'American civilisation' has now reached the Philippines, a thorough scientific analysis should clear up to what extent they really did have unusual powers. No evidence can be found of any systematic research having been carried out however.

Orffyreus' Wheel

The chapters on the Arctic (ships seen on the ice) and Hydrography (doubtful islands) were naturally of professional interest to Gould, but only the subject of this chapter, on matters mechanical, can be said to have been a passion of his. Gould was also fascinated by—and very critical of—those who believed in things which the real scientific evidence proved were impossible or at best were highly unlikely to exist.

He notes: 'Such is the flat-earther, the circle-squarer, the Ten Tribes man, the Jacobite, and the man who, measuring the Pyramids with a foot rule (or more commonly, relying on similarly-accurate measurements made by other people) establishes to his own satisfaction that the early Egyptians were only a little lower than the angels . . .'. Gould's chapter continues: 'Among this happy band (one can hardly add "of brothers", for in general one crank hates another most whole-heartedly) an honoured place will, I think, always be found for the man who is convinced that he has discovered the secret of "perpetual motion" . . . That place is his of right, because, like the King, he never dies. He is always with us—and there are always a good many of him'.

'Orffyreus', the German pseudo-scientist J.E.E. Bessler (born 1680), the subject of this chapter, was one such. Bessler, working in the early eighteenth century, created a number of perpetual motion machines, but unlike the legions of other cranks who have, over the centuries, claimed such impossibilities, Bessler's machines appear to have undergone considerable scientific scrutiny and were never fully explained. The first machine, shown at Gera in 1712, was a wheel of about 3 ft in diameter and 4 in. thick, encased in canvas and which, without any external supply of power, was apparently capable of accelerating up to a certain limit, at which point it could lift a weight of several pounds.

The following year Bessler exhibited at Draschwitz a similarly constructed but larger wheel, 5 ft in diameter and 6 in. thick, which accelerated up to 50 revolutions per hour and could raise a weight of 40 pounds. A third, still larger machine, exhibited at Meresburg was 6 ft in diameter and 1 ft thick. A committee of eminent men were allowed to examine the machine (though not internally) and considered it a true perpetual motion, though the eccentric, fanatical Bessler had many opponents and satirists regularly published pamphlets denying his claims. In 1717 in Hesse Cassel, when under the patronage of the Landgrave of that principality, he built his largest wheel, which was then subjected to particularly close scrutiny by a Dutch scientist who described it to his friend Isaac Newton in terms of astonishment. The machine was also seen by many other learned gentlemen, including the London instrument maker John Rowley who, after close inspection and experimentation with it, was convinced he had seen a real perpetual motion machine.

Bessler himself described in principle the action of his machine in a publication of 1719, but this only further compounds the mystery as what he describes is the old fallacy of the 'over-balancing wheel', the absurdity of which is explained in detail by Gould. If this really was what was inside Bessler's wheels then, Gould concludes, it is only surprising they appeared as convincing as they did, but he also suggests Bessler may have deliberately given a false description to mislead those wishing to discover the real secret within. Over seventy years after Gould's careful summary, we are no further in our understanding of the curious matter of Orfyrreus's wheel.

Gould concludes the chapter by remarking of the eccentric Bessler: 'he passes from our sight . . . an exasperating and yet pathetic figure—morose, self-centred, childishly passionate, vacillating and yet tenacious, his own worst enemy, forgetting the duties of ordinary human intercourse in his passion for mechanism and wrecking his life as a result. Non Defecit Alter' (the Latin loosely translating as 'when one goes, another takes his place', or perhaps, in the modern vernacular, 'there's always one')

Crosse's Acari

The Englishman Andrew Crosse was a scientific amateur who appeared, in 1836, to have created life during carefully controlled electrical experiments. Working on the artificial formation of crystals, he discovered in his sealed and sterilized apparatus the gradual appearance of hairy, six or eight legged mite-like creatures (*Acari*), sometimes appearing in very caustic solutions, and definitely appearing to be living, moving about at will. The assumption was that these creatures had been created as a result of the electrical experiments and the matter became the subject of great controversy and religious outrage.

It had never been Crosse's intention to generate living creatures in this way and he was always perplexed both by their appearance and by the reaction of people to them. Crosse described the great care he took to keep the experimental equipment free from 'infection' by external agents and he was certain that the mites were a product of the electrical experiment itself. Gould naturally enough observes that however careful Crosse may have tried to be, it is impossible to discount the possibility that the Acari were mites which, as ova, had simply found there way into the apparatus, though he doesn't express an opinion either way. In a postscript in the second edition of *Oddities*, Gould quotes his friend Dr A.C. Oudemans as saying he was certain the acari were simply the common *Glycophagus domesticus* which is very tenacious of life and capable of getting into tins which appear to be hermetically sealed, though Gould remarks that, in his opinion the theory does not go quite far enough to cover all the reported facts.

The Auroras and other Doubtful Islands

On this subject (if not on the islands themselves!) Gould was on the firmest possible ground and, among the fine collection of stories that make up *Oddities*, this is perhaps one of the best. During his years (1916–1928) as an Assistant to the Hydrographer, he made a professional specialization of it and wrote on the subject several times during his life (it will be recalled his first presentation to The Sette of Odd Volumes was on this subject). The chapter tells, amongst other things, of the discovery and careful charting of a group of islands, the Auroras, in the South Atlantic (between the Falkland Isles and South Georgia) in the eighteenth century, 'islands' which were then seen again several times by other ships, their appearance described and whose co-ordinates were checked and confirmed for 'new and correct' charts of the area, but which *never actually existed*.

It was the celebrated Captain James Weddell, whose name is now immortalised in the Antarctic sea named after him, who, in 1820, first cast doubt on the existence of this group of islands. One explanation for the extraordinary charting of these non existent land masses is what Gould refers to as 'expectant attention', that is that having been told by one misguided observer of the existence of an island in a certain position, later navigators imagine they encounter it, mistaking icebergs, cloud formations, mirages etc for solid land, and confirm the existing authority. Conversely, examples are given of very real islands that, for various reasons, appear and/or disappear and are therefore difficult to chart, an example being Garefowls Rock, which disappeared, leaving the last surviving colony of the now-extinct sea bird, the Great Auk, without a home.

Then there are islands the existence of which has long been doubted but which have subsequently been found and accurately charted. The historic problem for navigators of establishing the longitude accurately, before the mid-eighteenth century, and the solution of the problem by John Harrison naturally also gets a mention here. An example of a doubted island which proved to exist is that of Bouvet Island, discovered first by the Frenchman J.B.C. Bouvet de Lozier in 1739 who guessed it to be a cape on the long imagined great southern continent.

In fact Bouvet Island has the distinction of being the remotest place on the earth, being over 1,000 miles from any other land and having an area of ocean surrounding it very nearly as large as the whole continent of Europe. So it is not surprising that subsequent searches for it failed, including one in 1843 by Captain James Ross, whose ships by coincidence were the Erebus and Terror, which soon afterwards would sail with Franklin to the Arctic, never to return. The story of how countless expeditions searched for, and failed to find the likes of Bouvet island remind one of the great numbers of years and lives spent charting the oceans and coastlines of the world. By the many non-maritime and non-technical among us today, these charts and maps seem to be taken utterly for granted, as though they simply occurred as a fact of nature, like the coastlines themselves.

Not that every detail of the world's coastlines are certain, even today. This chapter in the first edition of Gould's *Oddities* was itself partly responsible for inspiring the search for Thompson Island, said to have existed (and shown on Admiralty charts) about 45 miles from Bouvet island in an area which Gould observed had not yet been systematically searched. As a result of the chapter in *Oddities*, and just 4 months after the book appeared, in August 1928, the Norwegian ship *Norvegia* was dispatched to cover this area and conclusively determined the island was non-existent, after which its presence on Admiralty charts was removed. Then again, even in the second edition of *Oddities*, there remained the question of the existence of Dougherty Island in the South Pacific, which if it was real, would be even more remote than Bouvet Island. The evidence for it was provided by a number of nineteenth century sources, but Gould then cites many more recent, unsuccessful searches for it, supposing 'expectant attention' may again be the basis of its charting. However, in 1944 he could only leave the question of its existence open and it is only in more recent years that it has been established that no such island does in fact exist.

Mersenne's Numbers

Mathematics was another of the academic subjects which held great fascination for Gould, though, uncharacteristically, his summary of its usefulness rather dismisses the whole subject, stating that Pure Maths in general is of limited utility. He describes the theory of numbers in particular as 'so utterly devoid of practical value, and so remote from the normal interests of even the majority of mathematicians, that most of the latter have discreetly left it alone'. On the subject of numbers, the famous '*Fermat's Last Theorum*' is discussed. Those interested in maths will recall that in recent years this topic has been in the news and a book has appeared on the subject.

A theorem is a mathematical statement professing to be true but not necessarily having a mathematical proof that it is so. The 'Last Theorum' of the great French mathematician Pierre de Fermat (1601–1665) states that the equation $X^n + Y^n = Z^n$ only holds true if $n = 2$ and that any higher power than 2 can never be applied. In other words, as Gould summarizes it, 'the sum of the squares of two whole numbers can be a square, but the sum of two cubes can never be a cube, the sum of two fourth powers can never be a fourth power and so on, for ever'. Fermat claimed to have a proof of this theorem, adding the tantalizing remark in the margin of his copy of the theorum 'I have found a wonderful and valid proof. This narrow margin cannot contain it'. Gould believed him, 'the balance of evidence seems to be in his favour. It awaits rediscovery'. In fact it was only many years after Gould's death that Fermat was proved right, by the extraordinary travails of a young English mathematician, Andrew Wiles, in 1995. The book about this exciting rediscovery, written by Simon Singh, was published by fourth Estate in 1997, hot on the heels of their most celebrated best seller, *Longitude*, by Dava Sobel (1995).

Mersenne's numbers concerns the question of *prime* numbers, that is, numbers which cannot be divided by any other number into 'whole parts', and *composite* numbers, that is, those which have *factors*, or numbers which they can be divided by. *Perfect* numbers are composite numbers where all the factors added together happen to equal the number itself, and perfect numbers are very rare. For example, there are only eight perfect numbers in all of those up to 2,305,843,008,139,952,128.

The largest perfect number known at the time of the second edition of *Oddities*, in 1944, (a number which ran to no less than *77* figures!) was calculated independently by three different people: R.W. Hitchcock in 1934, Gould himself in 1935 and by E.T. (Teddy) Hall (1924–2001), an 18-year-old school boy from Eton College who corresponded with Gould, and invited him to lecture at the school in 1942. Teddy Hall, incidentally, went on to have a distinguished scientific career. Most notable was his analytical work on Piltdown

Man and the Turin Shroud which revealed both as fakes, and his service on the Board of three national museums, the National Gallery, the British Museum and the Science Museum. He was also an avid horologist and collector, so it is not surprising he and Gould were keen correspondents, notwithstanding Hall's young years.

It was the computation of perfect and prime numbers that concerned Mersenne and 'Mersenne's numbers' form a statement (similar to a theorem but with limitations) summarising the basis on which a certain mathematical expression will result in a prime number and when it will result in a composite number. The statement was described as: 'one of the unsolved riddles of higher arithmetic', but no one yet knows how Mersenne (or Fermat, as it is also believed by some experts that Mersenne was passed the statement by him in correspondence) calculated and proved the statement to be true. Gould finally lists three other theorems for which there was (1944) still no proof: Euler's Biquadrate Theorum, Goldbach's Theorum and Lagrange's Theorum.

The mathematician and 'metagrobologist', David Singmaster, and Jeremy Gray of the Open University have kindly looked at this chapter of *Oddities* and have made a number of interesting and useful comments. In general, Singmaster observes that 'Overall, we are still as mystified by Fermat's results as Gould was . . .'. He notes that 'a great deal has happened since Gould wrote in 1928 . . . [but his] . . . article is an excellent exposition of the material for its time. There is nothing cranky or quirky in it. With hindsight, we now know that the exact details of Mersenne's conjecture are not very important, so I might cavil at Gould's strong interest in it, but this is certainly not a major fault'. He also notes: 'Euler's conjecture was disproved in 1967 when Lander and Parkin used a computer to discover $27^5 + 84^5 + 110^5 + 133^5 = 144^5$', and that 'Goldbach's Theorum remains unsolved, but quite close results have been obtained'. Gray points out that Mersenne was a Minimite Friar and lived in Paris, not *Nevers*, but apart from 'one or two mistakes' is generally impressed and able to give the coverage a clean bill of health.

The Wizard of Mauritius

This chapter tells of the well-authenticated case of a native of Mauritius who was able to foretell the arrival of ships en route to the island long before they could possibly be seen by naked eye or telescope. Gould first encountered a reference to this story in David Brewster's *Letters on Natural Magic*, 'one of the most wonderful books ever written' he says, which, coming from Gould, is saying something. The subject of the chapter's title is a Frenchman by the name of Bottineau, who claimed he had, by 1764 on the island of Mauritius, mastered the art of '*Nauscopie*', the ability to detect the arrival of ships approaching from below the horizon, by observing closely the effects on the atmosphere. Twenty years later, by which time he had returned to

France, Bottineau was seeking to make his fortune by selling his services to the highest bidder and had found a notable, and ultimately notorious, supporter in Jean Paul Marat (later famously murdered in his bath after his excesses during the French Revolutionary regime known as the *Terror*). In recommending Bottineau to the British government, Marat cited the extraordinarily strong evidence supporting Bottineau's ability to detect the arrival of ships three or even four days before their arrival over the horizon. This was supported by certificates, quoted by Gould, from the Governor and Officers of the garrison at Mauritius. Of even more importance, was Bottineau's ability to detect land, several days before arrival, when on board a ship.

Gould concludes from the surviving evidence that he has no choice but to accept that Bottineau, somehow, was indeed able to do what he claimed. As to how he did it, and whether he himself even understood the process, Gould points out that there are many precedents even today of phenomena which we accept but cannot explain, giving as an example the curious fact 'that a freely-suspended piece of paper, exposed to strong sunlight, will always tend to turn so as to set itself north and south. This fact has been repeatedly verified; but, so far, it has not been explained'. He especially chides the scientific community for being critical of the ill-educated observer/discover of an undoubted phenomenon for his being unable to explain it. He cites the case of people with the ability to dowse for water (which Gould regarded as perfectly proven) using their innate sensitivity with great effect, but being quite unable to say how it works.

And so it was with Bottineau, who could only state that, on the approach of vessels some miles below the horizon, he was able to discern a very distinct kind of atmospheric effect. Unfortunately, unable to find a sponsor in France or Britain, Bottineau died in obscurity and the whole question of the Wizard of Mauritius remains unanswered to this day, though at least two others, one on Mauritius and one on Tristan da Cunha, are on record as also having the same skill. Gould ends the chapter by suggesting that the matter could be relatively easily resolved if a researcher would only go to the island with a powerful camera and record the effects on the horizon when the arrival of ships, determined by radio, is known, but this has not yet been accomplished.

In more recent times however, Gould had a notable admirer who took a particular interest in this chapter in *Oddities*. No less a figure than Lord Louis Mountbatten had read *Oddities* and had expressed a great interest in having research done to find an explanation for *Nauscopie*. Since Mountbatten's assassination in 1979 however, it does not appear that the matter has yet been taken any further.[401]

The Planet Vulcan

In 1859 the amateur astronomer M. Lescarbault observed what he believed was a planet, within the orbit of Mercury, passing across the face of the Sun.

He did not announce it at the time however. Then, a few months later, U.J.J. Leverrier, the Director of the Paris Observatory, reported that by studies on the anomalous motion of the planet Mercury, he had determined there may be matter—perhaps a small planet or maybe a belt of asteroids—close to the Sun. Now hearing of Lescarbault's observation, Leverrier visited him, satisfying himself that Lescarbault had indeed observed something in the nature of a celestial body within the orbit of Mercury. So impressed were the scientific community with this amateur's discovery of a new celestial body, which he dubbed the planet Vulcan, that he was immediately awarded the Legion d'honneur. However, the 'discovery' was immediately challenged and unfortunately for Lescarbault and Leverrier, the planet refused to make further appearances when it should have done.[402]

It is still uncertain what Lescarbault saw; Gould suspects it may after all have simply been a sun spot, but if it were a celestial body it could only have been a relatively small asteroid. Even by 1944, the date of the second edition of *Oddities*, it was uncertain exactly what there was orbiting the Sun within Mercury's orbit. Of Gould's coverage of this subject, Dr Robin Catchpole of the Institute of Astronomy at Cambridge, comments that it is 'an excellent and interesting assessment that compares well with modern discussion'.[403]

Nostradamus

It was inevitable that Gould should wish to include a discussion of this remarkable soothsayer in *Oddities*; He chose the engraving of the prophet as the frontispiece to the book and the subject is the epitome of what fascinated him; it would have been conspicuous by its absence.

The piece begins with the salutary reminder that, since the early nineteenth century when an Act of Parliament (still in force in 1944) was passed on the subject: 'every Person pretending or professing to tell fortunes . . . shall be deemed a Rogue and Vagabond . . .' after which Gould introduces a gallery of such rogues by way of introduction. Interestingly he refuses to include Mother Shipton, stating 'She is as much of a myth as Robin Hood', the two sources for her existence both being of more than doubtful provenance.

Pointing out that even scientific men such as the first Astronomer Royal, John Flamsteed, practised Astrology, Gould notes that the Astronomer cast the horoscope for the foundation of the Observatory in 1675, but reminds us that Flamsteed hardly took it seriously (see p. 312).

Gould then comes to Nostradamus, and with him the author's candid admission that he believes the phenomenon of prophecy can, in certain instances, be possible. 'For the rationale of the matter, I cannot do better than refer to Mr J.W. Dunne's remarkable book An Experiment with Time. He shows, in a manner which arouses my respectful envy, the possibility of (so to speak) remembering the future as we do the past'. The essence of Dunne's subject is the observation that, occasionally, aspects of dreams we experience when asleep can actually appear in life the following day or soon afterwards. Gould evidently experienced confirmation of this phenomenon, and the present writer has to admit to having noted similar occurrences following dreams, occurrences which it stretches the imagination to ascribe to coincidence.

The coverage in *Oddities* of the prophesies of the Frenchman Michael Notredame (1503–1566)—Nostradamus in the Latin form—is quite brief. Beginning at the age of about 50, he wrote his prophesies in the form of rhymed quatrains and arranged them in sets of one hundred that he termed centuries. The first three centuries were published in 1555 with seven more a few years later. They were written in a difficult style and Gould avoids the tendency of the credulous to fit every one of the statements to later events, preferring to cite only four examples. All four seem on the face of it to be very extraordinary in how they fit the events they appear to predict, the fourth, relating to activities of Louis XIII of France and Henri, second Duc de Montmorency, executed at Louis's instruction in 1632, is so completely described in the 18th quatrain of the nineth century that Gould is wholly convinced that it was indeed prescient: 'in my submission, anyone who believes that this prophecy was fulfilled by pure chance is a much more credulous person than the one who, after examining the evidence, forms the conviction that Nostradamus possesed, and exercised, the power of foretelling the future'.

Ian Wilson, author of the book *Nostradamus, the Evidence*,[404] was 'genuinely surprised how good and sensible it is, and I wish I had known of it when writing my book'. He observes that 'Gould's chapter was genuinely innovative . . . and almost certainly the first into the English language with the information he put forward'. He continues 'Gould is undoubtedly very well-informed for his time. For instance, knowing about the editions with faked Mazarin prophesies. Also the spuriousness of Mother Shipton, etc. But the real star is his information on the Montmorency execution. I have to confess that I had not previously given that particular quatrain much attention. And looking up Peter Lemesurier's book on Nostradamus I see that even Lemesurier, who . . . has lived and breathed Nostradamus for decades, was unaware of executioner Clerepeyne. He thought "clere peyne" was something to do with "bright fire." '

ENIGMAS (1929)

There were Giants in those Days

The opening chapter looks at historical reports of people of very large stature, a subject of personal interest to Gould who himself stood 6 ft 4½ in tall without his size 13½ shoes on. The main part of the chapter discusses a race of men discovered in sixteenth-century Patagonia (southern Argentina) of very large stature, said to average 7 or 8 ft tall but with some estimated at 10 or 12 ft, and seen by such notable explorers as Ferdinand Magellan and Francis Drake. However, travellers in the following century only occasionally reported seeing inhabitants of unusual size and, sifting the evidence, Gould concludes that while the Patagonians were undoubtedly of very large stature, perhaps averaging 7 ft or more, they probably were never quite the size people had imagined they saw.

Three Strange Sounds

An odd collection of unexplained reports, the first of which tells of the peculiar sound made by the ancient Egyptian statue of Memnon on many recorded occasions in ancient times, always at dawn and 'a thin strident sound, like the breaking of a harp string'. Gould considers the likelihood of natural, and deceitful, causes, and wonders, if deception was at the root of it, how the sound had been created, and why. He concludes that the sound was more likely to have been a natural occurrence, perhaps owing to differential expansion in the rock.

The second strange sound concerns a phenomenon observed in the 1820s by Captain Parry during experiments to measure the velocity of sound in low temperatures, while on one of his voyages of Arctic exploration. In a particular experiment with cannon shooting blanks, the observers at some distance from the cannon consistently heard the report of the cannon, *after which* they heard the command 'fire'. There has never been a clear explanation for this observation, but one suggestion (made by Gould's friend Professor A.W. Stewart) noted in the second edition, was that as sound travels faster in warmer air, the report of the gun (which would be carried at a higher stratum as its projection was up into the atmosphere) would arrive sooner than the shout, which would have carried closer to the ice and hence through colder air.

The third strange sound is that of the 'Barisal Guns', not guns at all but sounds like the report of a gun heard in the Indian village of Barisal and surrounding area, though sometimes heard in many parts of the world, including in England, and with no logical explanation. Gould lists many of the suggested explanations, including fireworks, actual gun fire, bamboos bursting in jungle fires, thunder-claps, the collapsing of riverbanks, globular lightning, landslips, submarine eruptions and so on, However, all fail to fit the circumstances, though he admits it may be possible the sounds are caused by a variety of these causes at different times. He concludes (1946 second edition) that there is sufficient evidence of an unexplained phenomenon to warrant an investigation by the newly formed Ministry of Defence.

Old Parr and Others

Here Gould looks at the subject of great old age (something, sadly, he never experienced personally), and considers the likely veracity of records of reputedly very old people, some pretty well documented. Some can clearly be discounted as absurd, such as seventeenth- and eighteenth-century deaths recorded, without further comment, of English people aged 120, 133, 138 etc., and one record of a man in Hungary who is said to have died in 1724 at the age of 185. In the nineteenth century, however, the general view then formed that it was virtually impossible to reach an age exceeding 100 years, though the increasingly reliable records of births and deaths began to prove otherwise. Gould then focuses on three rather extraordinary cases, though it must be said that in this chapter the evidence for real 'enigmas' seems weak. The cases all occurred in the sixteenth and seventeenth centuries and the difficulty of verification of evidence and the likelihood of exaggeration, both intentional and unintentional, makes it difficult to go beyond saying that these were almost certainly people who lived to over 100 years old. In truth, chapters such as this one are as much of interest for the marvellous asides and ancillary information Gould provides as for the specific cases he cites.

'Old Parr', from Alderbury in Shropshire, was said to have been 152 years old at his death in 1635. Such was his celebrity as an old man in his final year that he was brought to London and introduced at court. Consequently more details of his life and supposed age have survived for scrutiny. He died in London and was buried in Westminster Abbey after a post-mortem by the famous William Harvey, discoverer of the circulation of the blood.

Katherine, Countess of Desmond, is said to have died in 1604 at the age of 140. She is known to have survived her husband by 70 years, but even this, along with much circumstantial evidence, and the fact that the couple had a daughter, cannot positively confirm her great age, nor even necessarily push it as far as 100.

With the example of Christian J. Drakenberg, who reputedly died at 146, Gould has rather more documented evidence. He was born in Blomsholm in Norway in 1626 and was a sailor by profession. There is a period in the chronology of his life when it has been suggested he died and another assumed his identity, but Gould considers this hypothesis and discounts it for a number of reasons. However, as in all these cases, he can only provide the evidence and leave us to make up our own minds.

The Landfall of Columbus

Then we move to material which is wholly Gould's forte, and on which we can relax in the knowledge that the evidence he provides has been sifted with truly professional expertise. This chapter, which had in fact been substantially published already in the *Geographical Journal* in May the previous year (1928), contains a thorough discussion of the much debated question of where in the Americas, or more precisely which island in the Bahamas, Columbus first sighted when he made landfall sailing across the Atlantic.

The several well documented and debated possibilities, 'the Big Six' as Gould dubs them, including Watling Island, in modern times the most favoured site, are all carefully considered. Columbus's own reported log of progress, and his detailed descriptions of the features of this first island and others he could see, and subsequently visited in turn, are all looked at. The probability of changes to the shorelines and geographical features of the existing islands is also factored into the equation. Gould concludes by proposing an entirely new choice, that of Conception Island, which he regarded, on the balance of probability, the true landfall of Columbus, and modern opinion now supports this view. In research carried out in the late 1980s by Dr Steven Mitchell, a professor of geology at California State College, he and three teams of volunteers visited the Bahamas and studied all the islands in question, both on land and on approaching by sea, also revisiting the historical evidence of Columbus's stated course and stops. Although not absolutely conclusive, what they found supported the Conception Island theory more strongly than the other possibilities. Nevertheless, in the second edition of *Enigmas*, Gould notes that he 'extensively recast' the writing of the chapter.

Bealings Bells

In contrast to the previous earthly and maritime mystery is this chapter concerning the very well- and carefully-documented nineteenth-century cases of poltergeist-type hauntings at a house in Bealings in Suffolk, one at Greenwich in London and one in Stapleton, near Bristol, where, in each case, the servants bells were constantly being rung. The source for the mystery is a book published by the unfortunate owner of the Bealings house, Major Edward Moor, F.R.S. Moor, needless to say, had all the obvious solutions to the strange events pointed out to him by his friends and neighbours but which he steadfastly assured them had been disproved as the explanation.

In his book, Moor cited other cases he was able to find, including one which occurred at Greenwich Hospital in 1834, and rumours of which were still current, Gould tells us, when he was studying there (then the Royal Naval College) in 1911. Lt William Rivers, in whose apartments the servants bells were ringing, recorded the whole strange story of how they rang, how the Clerk of Works and the bell-hanger, with all servants and others dismissed from the building, had both witnessed the ringing, including the violent operation of the bell handles (untouched by human hand) in all the rooms, and were completely unable to explain it. The 'haunting' ceased, as suddenly as it had begun, four days later.

The Stapletone case of July 1836 was very similar. One afternoon the bells in all rooms in the house began to ring. All the servants were gathered together in one room and the house was thoroughly searched to no avail. The bell-hanger was called, but could give no explanation, so the bells were tied up that night and while the wires continued to move for much of the night, things had ceased by the morning. Although the whole system had been newly installed just a year before, the whole thing was redone and the problem disappeared. Gould has no explanation but remarks that if these were hoaxes, which is the immediate assumption, then they were all carried out with very great skill.

The Strait of Anian

Returning to Gould's professional métier, Hydrography, this is the story of how the sixteenth-century navigator Gaspar de Corte Real in 1500 described the St Lawrence River as the Strait of Anian, believing it to connect to the Pacific Ocean. Gould narrates how this non-existent strait continued to be shown on charts and was believed in by some navigators and cartographers for more than two centuries. The question is related to the story of The Ships Seen on the Ice, (in *Oddities*), inasmuch as it concerned the issue of a northern route through to the Pacific, where a northwest passage over North America was being sought. So much did navigators wish to believe in such a strait that Gould is able to cite no less than ten accounts by deluded (or downright deceitful) navigators who claimed to have sailed down it and reached the Pacific.

The *Victoria* Tragedy

The second edition of *Enigmas* (1946) replaces the chapter on the Strait of Anian, which Gould fears 'may be of insufficient general interest', with an account of the sinking of the battleship *Victoria*, while on manoeuvres in the Mediterranean on 22 June 1893. The tragedy, which Gould describes as the Royal Navy counterpart to the *Charge of the Light Brigade*, resulted in the total loss of the Flagship of 'the world's crack fleet', her Commander-in-Chief, Sir George Tryon, and half the ship's company, some 365 men. *Victoria* sank because the ship was rammed by the second flagship *Camperdown* when following Tryon's specific, but disastrous instructions on the manoeuvres that ship was to make. The 'enigma' in the story is how on earth could Tryon make such a mistake in his calculations, and why on earth did the commander of the *Camperdown* obey Tryon's orders to the letter, knowing his ship must surely cause the disaster it did. Gould notes that the story 'has often been told, sometimes, temperately and accurately'. But he believes that the drawing he adds, showing the actual course of the fleet along with Tryon's intended manoeuvres, is a fresh contribution to the subject.

   In essence, Tryon's manoeuevres for the fleet that day had arranged the ships in two lines sailing together, parallel with each other, with *Victoria* and

Camperdown each leading one of the lines. The intention was that at the appropriate moment both lines should turn, through 180°, towards the other line, and sail back the way they had come, with both lines now naturally much closer to one another. A very simple manoeuvre, but the fatal mistake Tryon had made was to miscalculate the initial distance between the lines. They were too close together and the first two ships to turn, *Victoria* and *Camperdown*, simply ran into each other, *Camperdown* ramming *Victoria* just behind the bow and opening a huge hole in her side. Gould summarises the considerable quantity of evidence given at the Court-martial which formed the enquiry into the disaster, and dismisses the malicious rumour that Tryon had been drunk that afternoon. Nevertheless, he can only conclude that Tryon himself was correct in some of his last words, overheard just before the ship sank, when he remarked 'Its all my doing—all my fault'.

The Last of the Alchemists

In what is by far the largest chapter in the book, spanning 75 pages (over a quarter of all the chapters) Gould considers the evidence that in the past there have been chemists who, in spite of the received wisdom that it is impossible, have turned base metal into gold. He points out that no less a figure than Robert Boyle (1627–1691) 'the founder of modern chemistry' believed such a process might be possible. He reminds us that Boyle, 'By his rejection of all fanciful theories, and his patient accumulation of facts and observations . . . showed himself possessed of a scientific mind in the true and only sense of the term'.

Only a few examples of alleged transmutations are given, while Gould states that he could give many more. The Belgian, Jean Baptiste van Helmont (1577–1644)—'generally regarded as the greatest chemist of the 17ᵗʰ century'— (and who first coined the term 'gas') recorded having derived pure gold by mixing mercury with a small piece of yellow stone ('the philosopher's stone'), of a size considerably smaller than the resulting gold. Johann Frederic Helvetius, Physician to the Prince of Orange, described in great detail how he too had been given a small piece of 'the philosopher's stone' and how, still very sceptical, he had mixed it with half an ounce of lead and had produced the same quantity of pure gold.

Another case is that of the young chemist James Price (1752–1783) sometimes dubbed 'The Last of the Alchemists' (quite incorrectly, as Gould points out there were many 'pretenders'after him). Elected a Fellow of the Royal Society at the tender age of 29, he committed suicide just 2 years later, and his claim to have carried out the transmutation of gold rests entirely on one account he wrote, published in 1782. Many of his experiments to this end, carried out at his house in Guildford, were witnessed at every stage by distinguished 'natural philosophers' and scientists, and none could discover a deception.

Gould tells us the experiments were of three kinds: transforming mercury, with a certain powder, into silver or gold, transforming silver, with a powder, into an alloy of silver and gold, and making an amalgam of mercury with one of these powders (all of which he prepared himself) resulting in the formation of pure silver or gold. News of Price's achievements gained him great notoriety, but he refused either to explain the process or to divulge how he prepared the powders. A rumour began that the mercury he used already had some gold dissolved in it, and sceptical scientific opinion gradually began stacking against him.

It appears that the President of the Royal Society, Sir Joseph Banks, now put pressure on Price to either prove and explain his experiments to the Society in the proper way or resign, on the basis that he was bringing the Royal Society into disrepute. Price returned to Guildford, and within a few months he wrote his will, and took his own life. Gould then considers the various possibilities for what the real circumstances were surrounding this desperate solution. A deliberate deception about to be discovered? Lunacy, brought on by monomania? Did he, in fact, believe he had really managed the transmutations, but now began to fear he had himself been deceived by the suppliers of his materials?

Also included in the chapter are tales of deluded would-be alchemists from the mid-nineteenth century, including the French alchemist Theodore Tiffereau. His claims, Gould summarises, 'can be dismissed without much difficulty as the product of defective knowledge crossed with imperfect technique'. A similar case was that of Dr. S.H. Emmens of New York, who not only claimed to be making gold from another alloy he had discovered ('argentaurum'), but was actually selling it to the United States Mint! The *New York Herald* tried to expose what they were certain was a fraud by challenging Emmens to a public demonstration of his process, which he accepted, but no expert was prepared to be associated with such an event. Emmens published the story of his work in this field, without giving any clue as to how the process was supposed to have worked, except that it involved a 'force engine' and the physical hammering of silver, which already contained traces of gold.

As recently as 1924 there appeared in the press an account of a German professor, Dr A. Miethe, who claimed he had accidentally created small quantities of gold from mercury by the prolonged action of a high-tension electric current on it. This discovery seems to have been verified by experiments then carried out by Professor H. Nagaoka of Tokyo who believed that he had eliminated the possibility of accidental errors. Gould states however, that it was observed by F.W. Aston in 1925, that from the known isotopes of mercury, any gold obtained from it must have an atomic weight of 198 or over, but that Miethe's gold had an atomic weight of 197.2, that of ordinary gold, concluding it could not have been produced by transmutation.

However, the chapter ends by pointing out that at that date, in the 1920s, atomic theory was so rapidly changing, with new discoveries constantly being made, that 'one day perhaps we shall scrap the theory of the immutability of the elements and come, like the old alchemists, to regard them as varying forms of the same essential substance'.

New South Greenland

Back with Gould's professional expertise, this is the story of the American navigator Benjamin Morrell who claimed to have discovered land in the Antarctic regions which he named 'New South Greenland' during what was apparently a remarkable voyage made in 1823. His credibility as a navigator and reporter, however is very much in question; Gould tells us right at the outset that Morrell was popularly known as 'the biggest liar in the Pacific', and his accounts were often either inaccurate, or made up from the accounts of places he had from other navigators. Gould thus rather spikes his own guns, as the mystery, if it be one, seems likely to be simply an exaggeration or a poor record of the voyage. Nevertheless, the chapter proceeds to look at what Morrell said, in an attempt to sift out what may indeed be truth from fiction in the account. In fact, some parts appear to Gould not to be as incorrect or anomalous as other commentators have suggested, and he points out that at the time of writing (1929) modern knowledge of the Antarctic was 'still largely defective'.

The subject of the title appears in Morrel's account on 19 March 1823 when he states that they were close in with the north cape of an island he names New South Greenland, though no land has ever been recorded in the position he gives. Gould notes three distinct types of solution already expressed by various authors on this claim. First that he was simply lying, second that he did indeed sight land which has not yet been recorded on charts, and third that his estimated longitude was too far to the east and that in fact he sighted the island known as Graham Land, shown in both the maps he illustrates this story with. On the first, Gould comments that as Morrell had nothing whatever to gain by this small claim in an otherwise very large account, it seems highly improbable. On the second, Gould considers this most unlikely too, though comments that no later explorer had subsequently come within 60 miles of the position, so one cannot be certain. The third point does seem an attractive answer, given the similarities of the 'coastline positions' shown in Gould's map, which seem to echo Graham land's coastline, 14 degrees to the West. But Gould points out that Morrel's account is, at times, so very specific about longitudes that it seems likely he had a chronometer for determining them, and if so, an error of 14 degrees would be very difficult to explain. Nevertheless, of the three theories, the third is the only one that makes any kind of sense to Gould and it is this he suspects as the answer, ending the chapter with the observation that, at some future date, better knowledge of that part of the Antarctic may help us come to a more certain conclusion about what it was Morrel really saw.

Abraham Thornton Offers Battle

In the second edition of *Enigmas*, Gould considered the chapter New South Greenland had also been of 'insufficient general interest', and replaced it with the amazing story of the trial, in 1818, of a young Warwickshire bricklayer, Abraham Thornton. The case itself, which was really more of an *oddity* than an *enigma*, was simple enough: Thornton was accused of having raped and murdered a young woman with whom he had attended a dance one night; the evidence of many of the fellow dance-goers and neighbours producing a wealth of material for Gould to analyze.

The timings of the various pieces of evidence was naturally paramount, Gould reminding the reader of the rather uncertain time scales in use in the country at that date, making a clear sequence of events difficult to establish. Nevertheless, after much deliberation by the jury, in the absence of definite incriminating evidence, Thornton was found 'not-guilty', a verdict Gould considers fair, but one which the Magistrate and many of the local people felt was unjust. Thornton should have hanged for what was obviously murder.

The curious part in the story now appears, as the brother of the dead girl was persuaded by the Magistrate to invoke an ancient legal device, 'The Appeal of Murder', a system of appeal (usually by the next of kin of a murder victim) against a not-guilty verdict. Though ancient and very rarely invoked, The Appeal of Murder had never been formally removed from the statute books and was thus still legally valid. The traditional response to this ancient appeal was a 'Wager of Battle' in which the relative and the accused would literally fight it out to the death, or until one or other admitted defeat. If it were the accused they would then effectively be guilty and would be hanged. Appeals of Murder were occasionally heard at this time, but it had become customary for the accused to forego his right to battle and face a second Jury.

In this case however, Thornton refused and offered battle: the relative was a meek young labourer, no match for the well-built Thornton, and the relative was obliged to decline. Though officially 'off the hook' Thornton was then obliged to emigrate, the community at large still convinced of his guilt. The ancient system of Appeal of Murder, and its counterpart Wager of Battle were, as a result of this case, abolished the following year by Act of Parliament.

The Canals of Mars

In the final chapter of *Enigmas* Gould treads on controversial ground. At the outset he grabs the attention by engaging in an amusing discussion of the likelihood of there being life on other worlds. More seriously he then looks at the belief, in some serious astronomical circles, that linear marks, first observed in 1877 by G.V. Schiaparelli of Milan, on the surface of the planet Mars may in fact be artificial canals. This was compounded in 1881 by the same astronomer's claim that these canals now appeared to be *double* parallel canals. At first these claims were met with disbelief, but were then increasingly taken seriously, especially after the American astronomer Dr Percival Lowell supported the findings.

The key to understanding these observations is of course to consider just how accurately we see with our eyes, and this point is made clearly by Gould. He cites his own drawing of the various versions of the canals, interpreted from a number of sources, which he did to illustrate this very chapter in the book. The very close work entailed two weeks' intense drawing and he reported at the end of the period seeing 'canals' in all sorts of unlikely places, even on the billiard table! The illusion, undoubtedly caused by eyestrain he says, disappeared after a few days rest. Again the conclusion is that with increasingly good photography, this enigma should soon be laid to rest once and for all.

When sending copies of the book to the Astronomer Royal, Sir Frank Dyson, and his Assistant, William Bowyer, at the Royal Observatory, Gould remarked[405] 'you will notice that, with the courage of ignorance, I have tackled the problem of the Martian canals, and to Bowyer,[406] 'I expect the essay on the canals of Mars will get me into a good deal of hot water—however, I rather enjoy controversy' to which Bowyer replied: 'Canals on Mars is a dangerous subject if one is seeking peace and quietness, and I should not be surprised if you found yourself up against some rather strong minded observers both for and against'.

However, a postscript in the second edition makes no mention of hot water, Gould only notes that one Dr G.S. Brock FRSE (Fellow of the Royal Society of Edinburgh) suggested to him the possibility that the phenomena observed by Schiaparelli might in fact have been as a result of incipient disease in his eyes and did not in fact exist at all. Dr Robin Catchpole has commented again: 'Gould... rightly draws attention to the extent to which the human brain as much as the human eye, seeks out pattern and order where there may be none!' He notes: 'The best telescope images of Mars are obtained with the Hubble Space telescope and it is very difficult to see any similarity between the Hubble images and those labouriously reproduced by Gould... The bottom line: the "canals" are a figment of the imagination and Gould would not have been at all surprised to see the hubble and even the images obtained by satellites in orbit around Mars. He might even have been delighted to see that there are indeed large channels almost certainly carved out by the flow of water, but these are not seen in the drawings nor are they man made'.

APPENDIX FIVE

The Affair of the Queen's Watch

In the late 1920s, Gould became embroiled in something of a horological scandal, entirely hushed up at the time, and a sad tale of hubris, self-interest and deceit.[407] The affair concerned one of horological history's great icons, the watch made in 1769 by Thomas Mudge for King George III; presented by George to Queen Charlotte, the watch has however generally always been known as the 'Queen's watch'. The timepiece was of such importance because it was the first example of a pocket watch with Mudge's monumentally important invention, the *lever escapement*, the type which would be fitted to just about every ordinary mechanical wrist and pocket watch in the nineteenth and twentieth centuries. Along with Harrison's timekeepers, the 'Mostyn clock' by Thomas Tompion (the astonishing year-going, striking spring clock made for William III c.1690) and a few others, this watch is one of the most important horological objects in existence.

The story begins in April 1920, with a study of the watch undertaken by the expatriate Austrian, Heinrich Otto (1875–1966), a very talented watchmaker, and erstwhile instructor with the British Horological Institute in Clerkenwell, London. Otto had the opportunity of examining and photographing the Queen's watch while it was in pieces in the workshop of the Clerkenwell (London) watchmaker W. Alfred Curzon. Otto had also taken dimensions and other data with a view to drawing up the escapement and, one day, publishing his findings. This would be a useful and interesting service to horologists, and there is no doubt Otto also relished the idea of his association with such an important and prestigious piece. There was however a problem for him.

A.H. DYSON

In fact the watch had been sent by the Royal Household to A.H. Dyson, the watchmaker at Windsor (no relation to the Astronomer Royal at the time, Frank Dyson) who the Royal family used to look after the watches and clocks in the Castle. Unknown to the Royal household, Dyson had sub-contracted the repair work outside his shop, quite normal practice in fact, for such businesses. The watch had been sent for repair to Curzon (under whom Dyson had studied) and his colleague William Beckman was photographing the watch. Otto had been allowed in, confidentially, to study it, purely out of interest. Naturally, without the permission of the Royal Household, Otto could not publish any description, and seeking permission would mean disclosing that Dyson had allowed the watch off his premises, which had been expressly forbidden.

Nevertheless Otto, who had visited Gould in July 1921 (to see H4 at the flat in Campden Hill Court, see p. 111), told him about his study of the Queen's watch a couple of months before, and two years later, in his book, *The Marine Chronometer*, Gould made reference to the watch, thanking Otto for the information.

PAUL CHAMBERLAIN

At the time *The Marine Chronometer* was published, Gould's American friend Paul Chamberlain was working on a book about the lever escapement—it was this research which had brought him to England in the Spring of 1924 when he met Gould in his sick bed in Southsea (see p. 138). Knowing Otto had inspected the Queen's watch, Chamberlain visited him and asked for information. Otto was as courteous as he could be, but refused to give him the data, explaining that he intended to publish his own description of the watch, with a drawing of the all-important escapement and suggesting that Chamberlain wait until that time to get the data he needed. What he did not tell Chamberlain was that he could not publish, owing to the difficulty with Dyson, and the months, and the years, passed with no article appearing in the *Horological Journal*.

In August 1929 Chamberlain finally decided he would have to find another way of getting the data on the watch and wrote to General Charles G. Dawes, the US Ambassador to the Court of St. James, asking if he would request permission for a personal inspection to enable him to do a description of the watch. To his frustration, the following month Chamberlain heard, via the US Embassy, from the Lord Chamberlain's Office, that neither the King nor Queen (George V and Queen Mary) knew anything about such a watch. Unable to pursue this from across the Atlantic, but determined to get a result, Chamberlain cabled (sent a telegram to) Gould: 'Please find exact location of Queen Charlotte's watch', and wrote to the Embassy that he would soon be able to tell them where the watch in question was. Gould got straight down to it, cabling Chamberlain on 1 October 1929: 'Enquiries hitherto fruitless, proceeding, writing', but nine days later, after a personal visit to Windsor, he was able to report: 'Watch found, Windsor, writing details, anticipating no difficulties your examining it, regret delay', whereupon Chamberlain prepared to sail for England.

But Gould had traced the watch through Otto, who was now alerted to the fact that Chamberlain was about to steal his thunder and publish on the watch. So, on 9 October, Otto risked exposing Dyson, and sent a letter to the Royal Household, asking for permission to publish a paper about the watch, presumably not being specific about how much information he had, or from where he got it. Luckily, no difficult questions were asked and Otto received permission by letter of 15 October from Colonel Sir Clive Wigram, Assistant Private Secretary to the King.

Once in England, as part of his fact-finding mission on the lever escapement, Chamberlain visited Otto at his home in Streatham. Otto now tried to dissuade him from taking the watch to pieces at Windsor, for fear of an accident. Chamberlain politely refused to accept this, as Otto was still unwilling to part with his own data. The following day Otto tracked him down by telephone in London and again tried to dissuade him from going to Windsor, again expressing concern for the safety of the watch. Chamberlain later recorded that Otto went so far as to hint that taking Gould with him might be unwise, the 'innuendos regarding the danger of taking him with me could not be mistaken', Chamberlain recalled.

Apart from apparently wishing to discredit Gould, no doubt Otto wanted to prevent Chamberlain's parallel research and retain the main credit for this important and prestigious project. So now he offered, 'under certain restrictions', to share his information, if Chamberlain and Gould agreed not to dismantle the watch. But Chamberlain was having none of it. 'I had travelled too far and secured too valuable gracious assistance to turn in my path'.[408]

THE ARTICLE BY OTTO

Otto, now desperate to forestall Chamberlain's publication, rushed into print with a series of six articles in the *Horological Journal*, beginning in November 1929. The second of these, in December 1929, proudly presented Otto's illustration of the escapement, elaborately titled and giving credit for 'Kind Permission from His Majesty the King', and inscribed: "Drawn, for the First Time, From the Original, by H. Otto, FBHI. Lond.," and dated above "*1921*", the year he claimed to have drawn it.

When Gould had enquired of Otto about the watch he had been referred to Curzon, who duly introduced him to Dyson, who then took Gould to Windsor Castle. With staff from the Royal Household, Gould and Dyson searched for an hour before finding the watch. Gould reported: 'incidentally, the Castle is a perfect rabbit-warren. If I hadn't had a guide I should be there now'. Gould sent Chamberlain all the relevant correspondence on the matter, and now Chamberlain passed all this, with sketches of aspects he needed to research in the watch, to the Embassy, requesting they seek permission for him to closely study the watch. It seems he did not however directly state that he wished to dismantle the watch. The general request for study was gladly given by the King, and 21 November 1929 saw Chamberlain arrive at Windsor, with Gould whom he had invited to be present. Captain Marsh, the Inspector (the equivalent of Curator) greeted them cordially, but was adamant they could not take the watch to pieces, so a further delay of several days ensued while another letter, to Major General Sir John Hanbury-Williams, His Majesty's Marshal of the Diplomatic Corps, sought this permission.

STUDY AT WINDSOR

With permission granted from Windsor, Chamberlain and Gould returned to the Castle and Chamberlain dismantled the watch, studying, measuring parts and arranging them, particularly the escapement parts, for photography. This presented a problem as the official photographers decreed it had to be done using only daylight and the winter days in Berkshire were not obliging enough. Several days passed before the Sun came out, during which time the watch was all in pieces, a worrying time for the horologists, but the watch was soon back together safely, was duly adjusted to keep good time, and all was well—or so it seemed.

Chamberlain now returned to the United States, taking with him his notes and some useful references from the correspondence between Thomas Mudge and his patron, Count von Bruhl, which Gould had extracted for him, annotating them with many useful observations, all published, a few years later, in the mid-1930s.[409]

May 1930 saw the last part of Otto's series of articles on the Queen's watch published in the *Horological Journal*. Otto was evidently relieved that he had got his account of the watch published first and was very pleased with the result. As soon as he could, he had a number of off-prints bound at his own expense, sending one to King George himself, Colonel Wigram conveying to Otto on 30 July 'His Majesty's sincere thanks for this book, which both the King and Queen will peruse with interest . . .'.

All should have been well, but unfortunately for Otto it was here that things started to go wrong. There was in fact a rather serious error in his beautifully produced and published technical drawing of the Queen's watch escapement. The number of teeth embraced by the pallets, the most basic of escapement criteria, was shown as 5½ instead of the correct 4½. This was a fundamental and misleading mistake to have made and published in the BHI's own *Horological Journal*, especially by one of its own former Drawing Instructors. Otto probably suspected his drawing was wrong soon afterwards, but there was nothing he could do, unless he had a chance to confirm it by looking at the watch in pieces again, an opportunity which fate coincidentally delivered to him soon afterwards.

PROBLEMS WITH THE WATCH

In March 1931 the Queen's watch suddenly began losing time dramatically, and on being asked to look into it, Dyson found that extensive rusting had appeared on the lower balance spring of the watch (unusually, the watch has two balance springs, one above the other) and had caused the spring to break, a replacement spring now being necessary. Dyson was asked to carry out the repair and this time, on Curzon's recommendation, Dyson's son took the watch (the movement only in fact) to Otto to do the work, turning up at Otto's premises with it in his pocket.

From the outset though Dyson was worried about the situation and soon after Otto had the watch, Dyson wrote to him, on 4 March 1931, asking for it to be done quickly: 'you see I have undertaken not to let it out of my keeping—so I do hope you will keep it secret & not let it be shown about . . . Since you made the report about it they are so fussy about it' and adding suspiciously 'it is strange how that rust could have got on [the] hairspring'. The next day Dyson wrote to Otto again, pointing out that when last running (after Chamberlain and Gould had completed their study) the watch was running perfectly, gaining just 1 s in 24 h '. . . which my *patrons* like . . .', and asking him only to replace the balance spring and ensure the watch comes back performing the same as before.

On 8 March Otto replied with details of how the work was progressing, and commenting on the rust, adding darkly: 'I sincerely hope that onus will not be cast on you, for it would be a serious thing indeed . . .' and then strongly recommending Dyson go to the Inspector at Windsor and explain the whole story: 'be quite candid about it and tell him that you have handed the watch to me and refer him to my interest I always had for this watch . . . I have overhauled it thoroughly and I shall report about this later'.

Otto was in fact also having the watch photographed again by William Beckman and had taken it completely to pieces so he could study the escapement and have it carefully photographed. He must now have been in no doubt that his drawing was incorrect. Dyson replied immediately: 'I am sorry you have taken watch down, you see I only had the order to supply new hairspring [balance spring] in place of broken one. . . . I just hear H.M. the Q is expected here on Friday 13th so I will call on you on Thursday & hope it will be ready. So please don't disappoint me'.

Of course, the watch was nowhere near completion and Friday 13th must have seemed as unlucky as ever. However, somehow Dyson managed to keep the watch's absence from the Queen and the Inspector at Windsor a secret (did he show them the case and dial only? Otto only had the movement). His next letter stated: 'I cannot tell him I have handed it to you as I informed him & the Master of the Household I would keep it in my possession—& they would say it is no use Dyson's having the jobs, they have to send them to London. I will tell the Inspector . . . you have helped me get the hairspring. . .'.

Dyson wrote again on Monday 16th, hoping to collect it at the end of the week, but Otto replied that it would not be ready, in spite of his working on it until 11.30 p.m. each night. Dyson was now desperate to get the watch back, writing to Curzon (who had recommended Otto) that he had to have it by the following Monday morning, 'it is as much as my Royal Warrant is worth not to return it'. After fitting a succession of experimental new balance springs, supplied by the specialist maker Ganeval and Callard on City Road, Otto finally completed the restoration that weekend and the watch was duly returned to Dyson early the following week, as demanded.

HEINRICH OTTO

As Chamberlain and Gould probably realized by now, it seems Otto was one of those people with whom associations always seemed to get complicated. Evidently he had always regarded Gould as something of a hero, but reading between the lines it seems he also envied Gould his opportunity to restore the great Harrison timekeepers. He may also have rather resented the fact that a wholly unqualified man had been able to achieve such celebrity on a practical horological subject. The fact that this was partly thanks to Gould's social position, and certainly because he had the advantage of being British by birth, clearly rankled with the talented Austrian watchmaker.

But here was Otto's chance, in a small way, to emulate his hero whilst claiming a great horological achievement of his own. It might mean having to 'ride roughshod' over Dyson's 'delicate situation', but ambition urged that he somehow make sure his talents and hard work on the Queen's watch be properly known at the highest level. Otto decided therefore to write a full report on the whole saga of the Queen's watch, bringing in Curzon to add a ringing endorsement that 'this restoration was solely and willingly carried out by Mr H. Otto regardless of Time and Expense . . .' and that 'it required an enthusiast equipped with the theoretical knowledge and all the technical skill and patience . . .'.

The two-part report (first a description of the history of its needing a restoration, and second a technical report), addressed directly to Sir Clive Wigram, left nothing to the imagination, telling how the watch had strangely developed rust on the balance spring, and then just happening to mention that the last people who had examined it had been Chamberlain and Gould. Otto concluded that it must have resulted from their intervention (which actually may well have been the case) adding 'the watch would then be a 'restoration', whilst but a little while ago it was in its original state'.

Emphasizing the professionalism with which he was working (contrasting with both Chamberlain and Gould as amateurs), he described the processes he used in detail, even specifically requesting 'would you, Dear Sir Clive, acquaint His Majesty, at an opportune moment, of the restoration work . . .' going on to ask if the King would approve his mounting the broken balance springs and sending them to the Science Museum! He went one stage further, verging on insolence, by 'requiring' that the whole report be attached to the off-print he had sent to the King, 'and I should feel much obliged if the Royal Librarian would inform me in which way it could best be done'.

A copy of this report was first sent, on 23 March, to Dyson for comment, but that it would be sent to Sir Clive Wigram was not negotiable. Otto asked for the report and comments to be returned within two days 'This will give you time to find out the best means to inform the official with whom you are concerned'. Dyson remained friendly—for the time being—replying on the 29th that he was very grateful for all Otto's trouble and detail in the report, but that as the first part of the report was too long (for which read 'too revealing about Otto's involvement') it would be better if he, Dyson, wrote to Sir Clive instead.

Dyson had also had a chance to see the Inspector at Windsor and told him, without being too specific as to what had been done, how much time Otto had spent ('making and supplying the balance springs' was what had been agreed). The Inspector was impressed and agreed to advise him how much Otto should charge. Knowing that he had no choice, as Otto was virtually blackmailing him ('if you don't I will'), Dyson then handed the Inspector Otto's 'part two' (the technical report) and retired immediately, leaving the Inspector to peruse it at his leisure, when he would discover how much had in fact been done by Otto.

Otto now replied to Dyson in even more determined terms, saying he was unhappy with Dyson's alternative letter, remarking: 'So we will leave this report as it is', meaning he intends to send it unaltered to Sir Clive, apart from one or two slight changes he had agreed with Dyson beforehand. But this intention was overtaken by events, as Dyson sent a memorandum to Otto on 30 March 1931: 'I have today seen the Inspector he is *awfully annoyed* about watch & says I am to inform you not to *say* or *write* a word about it; the order to me was a hairspring only and if it could not have been done I ought to have informed them as I said I would keep it in my possesion. Should they [meaning the King and Queen] get to hear about it no doubt my warrant will be taken from me— and the Inspector would also get in trouble—I was afraid it would be so, but when it was returned in such good order I hoped he would see it in that light'.

Recognizing that Otto wanted something out of it, Dyson added 'Personally I am awfully sorry & feel I must make compensation for all your trouble—so please return the draft letter . . . and the old hairsprings . . . and your account'. Needless to say, Otto was not happy, replying the same day that for the time being he will 'refrain for the moment to send in my report'. It would seem that he then took the trouble to go and see Dyson to discuss the matter, and there was evidently a 'frank' exchange of views, the result being unsatisfactory for Otto.

On 4 April he wrote to Dyson asking for the name of the Inspector at Windsor and enclosing a draft of a letter he intended to send, explaining the whole matter to the Inspector including his intention never to charge for his work. He noted to Dyson 'You can charge if you care, your percentage of commission for handling the repair—but as far as I am concerned—I shall not accept anything as I told you at the end of our interview. This watch has cost me more than anybody else ever paid for since it was made'. Evidently the exchange of views at their meeting had been frank indeed! Otto continued: 'And you told me during our conversation "that this was done for my own glorification"—a remark I very much regretted and which made me assume that you can never have had any but material interests in Horology'.

Otto informed Dyson that he now intended to write formally 'a registered letter to the authorities' and enclosed a copy of the six-page 'statement' for Dyson to see. Dyson's heart must have sunk; the letter began with an introduction of his (Otto's) own horological virtues, going on to describe how Dyson and Curzon had had to pass on the work to him and how he 'very much regretted that such a restoration should have been needed'. He was also 'disconcerted that my technical reports should not be sent in and kept with my monograph at the Royal Library at Windsor. I should feel obliged to know the reason for this and I should be glad if you could help me'.

Dyson was horrified and replied at length on 6 April: 'Now I am asking you as a Gentleman and Fellow M.B.H.I. *not* to write to the Inspector . . .'. He states he will definitely lose his Royal Warrant if the matter is revealed, concluding: 'You cannot think what a wrong this has been to me—so for the present do please finish with it'. But Otto was not letting go, sending an eight-page reply on 11 April, in essence giving Dyson three months to prepare the Inspector and the Royal Household for the full story, which he intends then to send in. A point he makes clearly is that Dyson should have nothing to fear as the cause of the problem in the first place was evidently 'that this watch was handled by two gentlemen, in quite different stations, in December 1929'.

Dyson then allowed a pause to consider but, notwithstanding, his answer on 25 May gets rather carried away. Beginning that he was 'utterly sick of the matter . . . the way you keep writing quite alienates my sympathy with you', Dyson then notes that 'It seems you want to belittle Lt. Gould [sic] and aggrandise yourself—so please understand this is my last letter'. Of the Inspector he said 'I could see he did not wish you to *libel* Lt. Gould. It was His Majesty's orders the watch should be handed to him . . . And you impertinently think I should have objected to His Majesty's wishes—even if it had belonged to the Nation, & not personal property could I have dared to have done so . . .'. Dyson's tone then becomes increasingly shrill: 'Lt. Gould had to do the work in the Inspector's Office, a poor place for such a job. The Court photographer took the photographs & several callers—being interested had a look at the parts—Of course the Inspector might have sneezed—the photographer spilt some acid—or being in Decbr with fogs at night have got some damp on it—and yet although it went perfectly for several months after it was examined, you still want the World to know Lt. Gould was the cause of the breakage'.

He goes on, gratuitously, 'You think the watch is thought a lot of: It is never looked at; it is wound every morning—but having a centre seconds hand, it is much too bewildering for Her Majesty to tell the time by, so the clock is always used'. Then, rather unnecessarily pushing the knife in: 'You have remarked about your monograph being in the Royal Library; I asked the Librarian & he said it was with several other Horological works but had never been taken down since its arrival'. And finally on Gould: 'You have spoken of Lt. Gould as an amateur, no doubt His Majesty's Advisers found out all about him—Sir Clive has a brother, an Admiral in the Navy—& by the *Horological Journal* I see he has been overhauling some of the pieces at Greenwich—anyhow I don't see why you should be so bitter against him & should strongly advise you to let the matter drop . . .'.

Needless to say, Otto replied again, with a four-page letter on 2 June 1931. Asking 'Please do not make me the keeper of secrets . . .'. Naturally he challenged the remarks about Gould: 'If you know Lt. Commander Gould as well as I do and for whose technical interest I have a profound admiration—and whom I consider one of the intellectuals of the country—whose book on the Marine Chronometer I reviewed and where you can find my name, you would have been more guarded about that statement—for you do not know that he personally informed me in May 1930 . . . that he had nothing to do with the taking down and putting together, and who repeated this again to me on Friday March 20th 1931, when he chanced to call on me and when I showed him the rusted springs'. He noted that Gould had a copy of the article where he is mentioned and this 'should sufficiently put you at ease about Lt. Commander R.T. Gould, who is quite a power in stating a case if such should be needed'. Indeed.

The letter ends with complaints about the many discourteous remarks Dyson has made about him and states he intends to send in his report at the date he has warned Dyson of. Caution appears to have got the better of him however, and he does not seem to have submitted it. Meanwhile, Paul Chamberlain had discovered the error in Otto's drawing, something Otto had been hoping was quietly consigned to history. Chamberlain wrote to him on 27 April asking for a photograph of the escapement, duly sent by Otto on 29 May, Otto apparently expressing surprise and doubt to Chamberlain that his drawing might be incorrect, though the question was left open.

The story picks up again in 1935, directly after Gould's great Harrison lecture of 4 February that year. In early March Otto wrote to Paul Chamberlain enclosing a copy of the *Illustrated London News* coverage of the lecture, which Chamberlain had been regretfully unable to attend. Chamberlain replied to Otto on 19 May, and after telling him about a visit from the electric-clock pioneer Frank Hope-Jones (who had been awarded the medal of the Franklin Institute that month), Chamberlain gets down to the real business of the letter. He informs Otto about the imminent publication of his own article on the Queen's watch, which tells the story of Chamberlain's and Gould's study of the watch and points out the error in Otto's drawing, published five years before. Otto replied on 15 July asking him to defer publication until they were quite clear, asking for a copy of his drawing and saying that Gould had seen him last week and had asked for a copy of the escapement photograph. On 29 July Chamberlain answered that he was willing to defer and sent a sketch of his drawing. Otto replied on 10 August 1935, agreeing that Chamberlain's drawing was correct and enclosing a copy of a letter he had sent to the HJ as a correction.

This was published in the *Horological Journal* in August and, amongst a number of rather pointed statements, he noted the 'peculiar coincidence' that after Chamberlain's study, the watch should come to him (Otto) for respringing.[410] He then tried to explain away the error in his drawing by saying he had decided to do the drawing later and had difficulty counting the teeth from the photograph. He also stated, quite untruthfully, that the well-known historical print of the escapement by Pennington had also shown the escapement with this mistake 'and so did the others who used this as a source', a similar fiction. The notice ends with the legend 'To be continued', but no more was published in the HJ on the matter.

Continuing his letter to Chamberlain, Otto then, most unwisely, opens the whole Queen's watch 'can of worms' again by deciding to send Chamberlain copies of all the correspondence with Dyson, presumably thinking that Chamberlain should at least hear his side of the story. This prompted Chamberlain to reply with great frankness. 'I do not know the cause of your bitterness toward Commander Gould which I first discovered on the morning in December 1929 you called me by phone at Desoutters to dissuade me from going to Windsor and examining the Mudge watch. . . from the correspondence it is all too evident that you were trying to hang on him damage to the watch'. Chamberlain accepted that the rust may indeed have been caused by himself when taking the watch apart, but noted that it had never happened to hundreds of watches in his own collection. He then points out the rather devious way in which Otto has manipulated the facts to lighten his culpability in the drawing error, and goes on 'As to the publication of your having had the watch it is a matter for your own ethics to decide. The command from Windsor has evidently dissuaded you from trying to pin anything on Commander Gould but there still hangs the fate of Mr Dyson's Court Warrant. Would it not be a rather dearly bought revenge on Mr Dyson?'.

Before receiving this letter, Otto wrote again (29 August) asking Chamberlain to omit his name completely from the whole piece. This is soon followed by another letter after he had got Chamberlain's, complaining, 'I must confess that in my long association with Horology I never had a letter that contained so much hurt as yours does . . .'. He asks for all the correspondence to be returned to him so he could send the whole lot to Gould 'to settle this matter, for there is some misunderstanding'. Chamberlain now wrote (17 September) agreeing to Otto's requests, evidently hoping to have no more to do with the matter.

On 28 September 1935, Otto started the whole merry-go-round again, by sending everything to Gould, with a long note of explanation, but the parcel of letters was immediately returned by Gould, on 5 October, with the note 'I regret that I cannot see my way to commenting on them unless and until I receive Mr Dyson's personal assurance that he knows, and approves of your circulating his letters for other people to read. I have derived from them a strong impression that, when he wrote them, he did not for a moment contemplate their being thus treated. And I do not feel either inclined or entitled to comment upon information which has been brought to my notice by what I can only regard, at present, as a violation of confidence. I am forwarding a copy of this letter to Mr Dyson'.

Of course, Otto was duly 'surprised at the tenor of your letter . . .' (7 October), assuring Gould that he and Chamberlain, and 'one quite uninterested person', were the only ones who had seen the letters and assuring him he only wanted him to know that he bore no grievance against him. No further letters were sent to Gould, but one more very long one arrived with Chamberlain in October, answering his from 26 August, going over the same ground and protesting his innocence.

When Dyson saw Gould's copy of his letter to Otto, revealing that Otto had discussed the whole matter outside, this was the last straw and he called his solicitors. Otto duly received a letter from Lovegrove & Dunant of Windsor, dated 11 October, requiring him to desist from publishing Dyson's private correspondence. After apparently having a solicitor draft a reply, Otto sent a long answer, stating that he had not *published* such correspondence, adding that they had evidently had insufficient details laid before them, and then explaining again the whole saga, and his mistreatment in it, adding that he was still waiting for Dyson's written apology concerning the remarks about Cdr. Gould! A letter from Dyson's solicitors, dated 6 November 1935, accepts that this was an end to the matter.

One might have expected Otto to want to completely forget the whole saga after this but, as a little post script, in 1950 (by which time both Gould and Chamberlain were dead) he attempted a little re-writing of history. That year, the young horological historian Charles Allix was writing a piece about Thomas Mudge for *Apollo Miscellany*.[411] Consulting Otto for information, Otto proposed that Allix relate the story of the restoration of the Queen's watch, insisting that he dictate the exact lines to be used by Allix, and presenting a very biased account of the affair.

In the autumn of the same year, the antiquarian horologist and bookseller Malcolm Gardner included copies of the *Apollo Miscellany* in his catalogue and commented on this aspect of Allix's article; Gardner had heard the details from each of those involved and knew the truth behind this sad story. He remarked: 'These dissentions and acrimonies had almost been forgotten, and it is to be regretted that they should now be revived—involving as they did Mr Otto, Mr Dyson of Windsor, Commander Gould, Major Chamberlain and some officials of the King's Household. Were Commander Gould and Major Chamberlain still with us, doubtless we should be reading some spirited and diverting replies . . .'[412]

And so ends the saga of the Queen's watch, which has, for the last half a century, enjoyed the most scrupulous and attentive care at Windsor Castle.

APPENDIX SIX

Glossary of Technical Terms

(Note: Where terms in the glossary are used in the explanations of other terms, they are given once in *italics*)

Anchor Escapement. Invented in England in the late 1660s, this became the most common form of *escapement* for use in pendulum clocks. It is also known as the *recoil escapement* as the *escape wheel* is caused to move backwards, to recoil, after *impulse* has been given.

Anti-Friction (wheels / rolls). Smooth rimmed wheels, or segments of such *wheels*, upon which the *pivots* of train wheels (or other moving parts such as balances) roll, as opposed to sliding, when working, thus requiring no lubrication. A caged roller bearing is made up of a circle of such rolls, pivoted in a circular cage (see plate 8). Balance bearers, (sometimes called anti-friction segments), such as those used in H1, H2 and H3, are parts of such *wheels*, and which, usually in pairs, support the pivot of the *balance* (see plate 5).

Arbor. The term for an axle or spindle (e.g. of a *wheel* or *balance*) in a clock or watch mechanism.

'Artificial Gravity'. The name given by John Harrison to the effect of the *restoring force* of the *balance springs* in his timekeepers.

Axial wires (of H1-H3). The wires attached to the ends of the *balance arbors* and connected, in tension, to the frame of the *movement*, which prevent the balances from having any significant endways movement.

Balance. The oscillating wheel which, by its swings, counts out the time for the movement. In the case of H3 there are two balances in the form of wheels and in H1 & H2 the balances are in the form of bar balances, each shaped like a dumb-bell.

Balance Spring. The spring which is attached to the *balance* and which acts as its *restoring force*, making the balance oscillate to and fro when it is given motion.

Balance bearers. (see *anti-friction*)

Bimetal. Also known as a *bimetallic strip*, the bimetal is the temperature sensitive device invented by Harrison, which automatically compensates for temperature changes. Today the device is in common use in thermostats of all kinds. (see plate 8)

Caged roller bearing. (see *anti-friction*)

Chronometer. The term, first coined in its modern sense for a *timekeeper* by John Arnold in 1780 (though there were one or two isolated uses before this), has two main definitions. The English definition refers to a portable timepiece constructed specifically for scientific (usually navigational) use, with a detached, detent escapement as in the form designed by John Arnold or Thomas Earnshaw. A standard 'marine chronometer' (sometimes referred to as a 'box chronometer' or 'ship's chronometer') is of this kind, typically mounted in gimbals in a wooden box about 20cms cube. The Swiss definition of *chronometer* applies to any portable timepiece which has undergone official timekeeping trials and has met specific time-keeping limits. Thus, by the English definition, a *chronometer* which is a very poor timekeeper is still a *chronometer*, while according to the Swiss a 'Mickey Mouse' wrist watch which goes sufficiently well is also a *chronometer*.

Compensation (temperature compensation). A mechanism for ensuring that the effects of temperature on the timekeeping on a clock or watch are negated.

Constant Force Escapement. A type of *escapement* which delivers a perfectly uniform *impulse* to the *oscillator* (the balance or pendulum).

Cross-wires. The metal wires or ribbons which connect the two balances in H1, H2 and H3 (See plate 5).

Cylinder Escapement. Developed by George Graham from about 1719, this was the type of *frictional rest* escapement used in the better class of pocket watch throughout the 18th century. It was also referred to as the *horizontal escapement*.

Dead Beat Escapement. This *escapement*, which is of the *frictional rest* type, was developed for clocks by Richard Towneley and Thomas Tompion in the 1670s. The *escape wheel* advances in simple steps, without recoiling, and after *impulse* the wheel stops 'dead', hence the name dead beat.

Detached Escapement. A detached *escapement* is one where the *impulse* is delivered during a very short interval when the *oscillator* is passing the central point in its swing, where it has its maximum momentum and is thus least disturbed. At all other times the oscillator swings freely, without any inter-ference from the escapement, and in this sense it is the opposite of a *frictional rest escapement*.

Detent Escapement. Also known sometimes as a 'chronometer escapement', this is the form of *detached escapement* most often used in the *chronometer*. The detent itself is a little catch which literally detains the *escape wheel* until *impulse* is delivered. The early form, developed by Pierre Le Roy in France and John Arnold in England, was the 'pivoted detent escapement', with the detent running on pivots. The successful type, probably invented by Thomas Earnshaw in 1781, was the 'spring detent escapement', with the detent mounted on a single blade spring.

'Double-start'. Expression describing a spiral or a helix (e.g. like a screw thread) which has two 'threads' each following the other. A double 'spiral staircase' or the structure of DNA are examples of 'double start' helices. The 'spiral' grooves on the *fusee* of H1 is also of this type and thus has two chains running on it.

Escapement. The escapement is the part of a clock or watch *movement* which is the interface between the *train* of *wheels* and the *oscillator*. It feeds the energy from the wheels to the oscillator by giving it little pushes (*impulses*) to keep it swinging, while allowing the wheels of the movement to turn at a controlled rate, governed by the oscillator.

Escape Wheel. The escape wheel is the last in the *train* of *wheels* and also forms part of the *escapement*. In most escapements it is the escape wheel which delivers the *impulse* to the *oscillator*, either directly, or via *pallets*.

Frictional Rest Escapement. In this kind of escapement, the *escape wheel* is in virtually constant contact with the *oscillator*, even when *impulse* is not being delivered, tending to interfere with the timekeeping of the oscillator.

Fusee. A device designed to equalise the pull of the *mainspring* in a spring-driven clock or watch movement. The first wheel in the *train* (known as the 'great wheel') is not directly driven by the mainspring but is attached to a conical-shaped pulley (the *fusee* itself), which is cut with a helical groove around its body (reminiscent of a 'helter skelter' at the fairground). A chain (or cord) is attached to the fusee at its lower, wider end, and wraps around the groove in the fusee, coming off at the upper, narrower end. From here the chain is attached to a drum (known as the 'barrel') which contains the mainspring. When the mainspring is fully wound the chain is pulling on the upper, narrower end of the fusee, but as the clock movement runs, the mainspring in its barrel gradually turns, pulling the fusee and great wheel round, the chain coming off the fusee onto the barrel. As the spring unwinds, getting weaker all the time, so the chain pulls on a larger and larger diameter of the fusee and the increasing leverage counteracts the weakening spring force.

Gimbals. A suspension device mounted within the carrying box of an instrument (e.g. a compass or a marine chronometer movement) which acts like a universal joint, ensuring the instrument remains horizontal whatever angle the outer box may be at.

Grasshopper Escapement. A type of clock *escapement*, invented by John Harrison c.1722, which delivers the *impulse* from the *escape wheel* to the *pallets* with direct pushes, without sliding, thus requiring no lubrication. The escapement was first designed for use with a *pendulum*, but was adapted for use with *balances* in H1–H3.

Gridiron. A temperature-sensitive device, invented by John Harrison, which is used to compensate a clock for changes in temperature. A gridiron *pendulum* is one which does not change its effective length with temperature changes.

Horizontal Escapement. See *Cylinder Escapement*.

Impulse. The little 'pushes' given to an *oscillator* by the *escapement*, to ensure it keeps swinging. Without impulse given by the escapement, all oscillators would eventually stop swinging owing to friction and air resistance.

Isochronous. An *oscillator* is said to be isochronous when all swings, whether large or small, take the same time. In the developed *chronometer* the *balance* is made isochronous by adjustments made to the *balance spring*. Harrison attempted to force the oscillators in his timekeepers to be isochronous using mechanical means.

Mainspring. The primary power source for a spring-driven (as opposed to a weight-driven) *movement*.

Maintaining Power. The device fitted to some *movements* which have a *fusee* (sometimes also fitted to weight driven clocks) which ensures that the movement continues to work during winding, when the force of the mainspring (or weight) is temporarily removed. In its first, 17th century form, maintaining power had to be manually activated before winding, but in 1730 John Harrison invented the first fully automatic form of maintaining power.

Motion work. The motion work consists of the wheels under the dial which cause the hands to revolve at the correct rate for hours and minutes to be shown.

Movement. The movement is simply the professional term for the mechanism or the 'works' of a clock or watch.

Oscillator. A general term for the timekeeping part of the *movement* which swings to and fro and measures out the time. Almost all clocks, watches and chronometers have either a *pendulum* or a *balance* as an oscillator.

Pallets. The parts of an *escapement* which receive the *impulse* from the *escape wheel* and deliver it to the *oscillator*.

Pendulum. The most common form of *oscillator* in clockwork. Using gravity-which is highly constant-as its *restoring force*, the pendulum is a relatively good timekeeper.

Pinion. The pinions in the *wheel train* are the small toothed gears, formed as part of the arbor of a wheel, and usually driven by the previous wheel in the train. Usually made of steel, but in Harrison's large timekeepers formed with a circle of small wooden rollers.

Pivot. The pivots are the parts at each end of an *arbor*, upon which the arbor actually runs when working. In conventional *movements*, steel pivots of small diameter run in plain 'pivot holes' in the brass plates of the frame of the movement. Such pivots run with a sliding action and thus require lubrication. The pivots in Harrison's H1-H3 mostly roll on *anti-friction* wheels, and thus do not need oil.

Recoil Escapement (see *Anchor escapement*)

Regulator. A term used to describe a *timepiece* with a *pendulum* which is specifically designed to keep time as accurately as possible. To simplify the *motion work*, a 'regulator dial' has a single, central, minute hand with separate smaller dials to indicate hours and seconds.

Remontoir. A mechanism (usually only fitted to very sophisticated *movements*) which ensures as uniform a drive to the *escapement* as possible. It usually consists of a small, relatively delicate spring (or springs), fitted near the 'upper end' of the *wheel train*, which drives the escapement. Such a spring isolates the escapement from the variations in driving force coming from the train, but does require frequent winding up. The remainder of the wheel train, 'below' the remontoir, then merely has the task of keeping the remontoir wound up when it needs it. The remontoir 'fly', is a kind of air-brake, controlling the speed of this winding up.

Restoring Force. The force which causes the oscillator to keep trying to return to its position of rest. If a *pendulum* is pushed to one side and released, it is gravity, acting as its restoring force, which keeps bringing it back, and past, its position of rest. Similarly, a *balance* oscillates under the influence of the *balance spring*, acting as its restoring force.

Roller bearing (see anti-friction)

Saddle-piece. A unique device created by John Harrison for H3, intended to force the *balances* of that timekeeper to be *isochronous*.

Set-up ratchet. In a *movement* with a *fusee*, this is the part with which one sets up the correct tension of the *mainspring* in its barrel.

Stop-work. A device in the movement which prevents one from over-winding the mainspring or weight.

Timekeeper. A term which today is often used synonymously with *timepiece*, but which in the 18th century specifically referred to a portable timepiece of very high accuracy. After 1780, the term was gradually superseded by the expression *chronometer*, which had the same meaning.

Timepiece. A general term for a simple device which tells the time, but does not strike the hours or quarters.

Train. The train, or 'wheel train', is the series of *wheels* and *pinions* which feeds the driving force of the *mainspring* or weight from the 'lower end' of the *movement* up to the *escapement* and *oscillator*. The gearing of the train enables a large force to be divided up into a much smaller, but longer lasting, force to keep the oscillator going.

Verge Escapement. The first type of *escapement*, dating back to the late thirteenth century, but still the most commonly used in pendulum clocks and pocket watches until the beginning of the nineteenth century.

Wheel. In clock and watch work, the general term wheel means a toothed gearwheel. Usually the expression refers to one of the wheels in a wheel *train*, and actually consisting of the wheel itself mounted on its *arbor*, integral with a *pinion*.

Index

Note: page numbers in *italics* refer to illustrations, and those in **bold** refer to Glossary entries.

Kirkwall

Loch Ness (Inverness)

S C O T L A N D

Druimgigha
(Mull)

Jura Sound

Glasgow

Edinburgh

Newcastle

Carlisle

Belfast

Lancaster

Harrogate
Leeds

Bradford
Hull

Manchester

Wakefield
(Foulby)

Barrow

I R E L A N D

Dublin

Anglesea

Bangor

Liverpool

Derby

Nottingham

Sheringham

Norwich

Leicester

W A L E S

Birmingham

Coventry

Cambridge

Malvern

Northampton

Hereford

Ipswich

Cheltenham

Colchester

Swansea

Southend

Cardiff

Windsor

London

Bradford on Avon

Ashtead/Epsom/

Canterbury

Bath

Upper Hurdcott

Leatherhead

Dover

Shaftesbury

Winchester

Rye

Herstmonceux

Exeter

Southsea

Eastbourne

Tavistock

Portsmouth

Devonport

Weymouth

Bournemouth

Isle of Wight

Penzance

Dartmouth (Britannia)

Land's
End

Falmouth

Family Tree of Rupert Thomas Gould

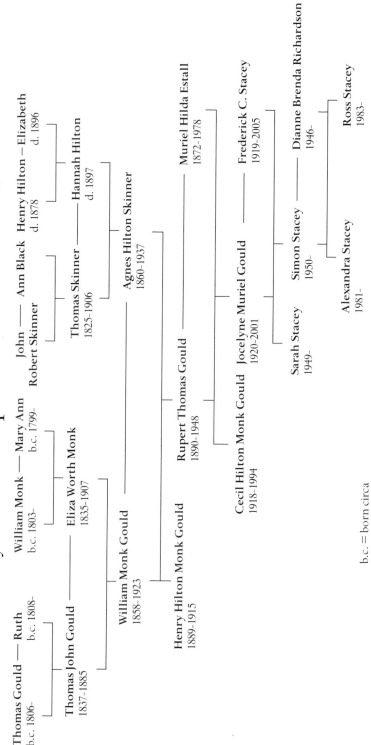

Thomas Gould — Ruth
b.c. 1806-   b.c. 1808-

William Monk — Mary Ann
b.c. 1803-   b.c. 1799-

John — Ann Black   Henry Hilton — Elizabeth
Robert Skinner   d. 1878   d. 1896

Thomas John Gould
1837-1885

Eliza Worth Monk
1835-1907

Thomas Skinner
1825-1906

Hannah Hilton
d. 1897

William Monk Gould
1858-1923

Agnes Hilton Skinner
1860-1937

Henry Hilton Monk Gould
1889-1915

Rupert Thomas Gould
1890-1948

Muriel Hilda Estall
1872-1978

Cecil Hilton Monk Gould
1918-1994

Jocelyne Muriel Gould
1920-2001

Frederick C. Stacey
1919-2005

Sarah Stacey
1949-

Simon Stacey
1950-

Dianne Brenda Richardson
1946-

Alexandra Stacey
1981-

Ross Stacey
1983-

b.c. = born circa